Dynamic Modeling and Econometrics in Economics and Finance

Volume 15

Editors
Stefan Mittnik
University of Munich
Munich, Germany

Willi Semmler
Bielefeld University
Bielefeld, Germany
and
New School for Social Research
New York, USA

For further volumes:
www.springer.com/series/5859

Elke Moser · Willi Semmler · Gernot Tragler ·
Vladimir M. Veliov
Editors

Dynamic Optimization in Environmental Economics

Springer

Editors
Elke Moser
Institute of Mathematical Methods
in Economics
Vienna University of Technology
Vienna, Austria

Gernot Tragler
Institute of Mathematical Methods
in Economics
Vienna University of Technology
Vienna, Austria

Willi Semmler
Department of Economics
The New School for Social Research
New York, NY, USA

Vladimir M. Veliov
Institute of Mathematical Methods
in Economics
Vienna University of Technology
Vienna, Austria

ISSN 1566-0419 Dynamic Modeling and Econometrics in Economics and Finance
ISBN 978-3-662-51192-3 ISBN 978-3-642-54086-8 (eBook)
DOI 10.1007/978-3-642-54086-8
Springer Heidelberg New York Dordrecht London

Softcover reprint of the hardcover 1st edition 2014

Printed on acid-free paper

Springer is part of Springer Science+Business Media (www.springer.com)

Preface

This book presents applications of optimal control theory and dynamic game theory in a broad range of problems associated with environmental economics. The book consists of 15 chapters, roughly half of which are based on research presented at the "12th Viennese Workshop on Optimal Control, Dynamic Games and Nonlinear Dynamics", which was held at the Vienna University of Technology (TU Wien) from May 30th to June 2nd, 2012. The workshop, which hosted more than 200 participants, was organized by Gustav Feichtinger, Josef L. Haunschmied, and Alexander Mehlmann, and two editors of this book, Gernot Tragler and Vladimir M. Veliov (all from TU Wien).

While that workshop provided the motivation to produce this book, the book cannot be considered as the proceedings thereof. Rather, for the purpose of providing a broader view of late-breaking applications of dynamic optimization in environmental economics, the chapters that stem from selected presentations at the workshop have been complemented by chapters from distinguished invited scientists in this field. The chapters are collected in two parts of the book and are ordered alphabetically according to the name of the first author within each part.

The first part, "Interactions between economy and climate", addresses the "economy $\mapsto$ pollution $\mapsto$ climate change $\mapsto$ economy" circle. The eight chapters in this part cover a variety of different approaches to modeling the feedbacks between the environment and the economy. For instance, some contributions describe the environment by its quality, concentration of pollutants, temperature, or a renewable resource stock, while others involve the environment only implicitly, represented by tax levy on emission or emission caps. Environmental policy instruments that are considered for the purpose of diminishing the climate change include (public) abatement, cap-and-trade, taxes, R&D, or technological change in several variants (e.g., exogenous versus endogenous, directed versus undirected).

The second part of the book, "Optimal extraction of resources", deals with optimal or rational utilization of renewable and non-renewable resources. The problems described in the seven chapters in this part include commercial fishery, forest management and biodiversity under climate change, the effects of resource exploitation and landowning on growth, export and import of fossil fuels, and harvesting of size-structured biological populations.

From a methodological perspective, the authors use various types of models and, therefore, various tools to analyze them appropriately. For instance, we find optimal control models in cases of a central planner, and dynamic games in cases of competing decision makers. While most of the models are deterministic, some also include stochastic uncertainties. In addition to standard problem formulations that rely on ordinary differential equations, there are also size-structured and spatially distributed systems. The tools used to analyze the problems include, but are not limited to, Pontryagin's maximum principle, nonlinear model predictive control techniques, nonlinear programming and the Karush-Kuhn-Tucker (KKT) theorem, the computation of Nash and Stackelberg equilibria, solution of Hamilton-Jacobi-Bellman equations, and numerical solution techniques such as the "Escalator Boxcar Train". Not only are some of the solution procedures innovative and sophisticated, but we also find complex solutions involving multiple equilibria and indeterminacy.

This book will be particularly interesting for economists, engineers, environmental managers, and applied mathematicians working on all kinds of dynamic optimization problems related to the interaction between environment, resources, and economic growth.

Finally, we wish to express our sincere gratitude to all of the authors of this book for their contributions, and the referees for their constructive suggestions on how to improve the individual chapters.

Vienna, Austria
November, 28, 2013

Elke Moser
Willi Semmler
Gernot Tragler
Vladimir M. Veliov

Contents

Contributors

Anton Bondarev Department of Business Administration and Economics, Bielefeld University, Bielefeld, Germany

Kirill Borissov Saint-Petersburg Institute for Economics and Mathematics (RAS), Saint-Petersburg, Russia; European University at St. Petersburg, Saint-Petersburg, Russia

Lucas Bretschger Center of Economic Research, CER-ETH, Zürich, Switzerland

Thierry Bréchet Louvain School of Management, Université catholique de Louvain, CORE, Louvain-la-Neuve, Belgium

William A. Brock Department of Economics, University of Wisconsin, Madisson, Wisconsin, USA; Department of Economics, University of Missouri, Columbia, Columbia

Francisco Cabo Departamento de Economía Aplicada (Matemáticas), IMUVA, Universidad de Valladolid, Valladolid, Spain

Carme Cañizares University of Girona, Girona, Spain

Christiane Clemens Department of Business Administration and Economics, Bielefeld University, Bielefeld, Germany

Renan Goetz University of Girona, Girona, Spain

Alfred Greiner Department of Business Administration and Economics, Bielefeld University, Bielefeld, Germany

Ben J. Heijdra Faculty of Economics & Business, University of Groningen, Groningen, The Netherlands

Pim Heijnen Faculty of Economics & Business, University of Groningen, Groningen, The Netherlands

Seiichi Katayama Department of Economics, Aichi Gakuin University, Aichi, Japan

Mika Kato Department of Economics, Howard University, Washington, DC, USA

Stéphane Lambrecht Université de Valenciennes et du Hainaut, Cambrésis and IDP, Valenciennes, France; EQUIPPE, Villeneuve d'Ascq, France; CORE, Université catholique de Louvain, Louvain-la-Neuve, Belgium

Ulla Lehmijoki University of Helsinki and HECER, Helsinki, Finland

Guiomar Martín-Herrán Departamento de Economía Aplicada (Matemáticas), IMUVA, Universidad de Valladolid, Valladolid, Spain

María Pilar Martínez-García Departamento de Métodos Cuantitativos para la Economía, Universidad de Murcia, Murcia, Spain

Stefan Mittnik Department of Statistics, Ludwig-Maximilians-Universität München, Munich, Germany

Hiroshi Ohta GSICS, Kobe University, Kobe, Japan

Tapio Palokangas University of Helsinki, HECER and IIASA, Helsinki, Finland

Alexia Prskawetz Wittgenstein Centre for Demography and Global Human Capital (IIASA, VID/ÖAW, WU), Laxenburg, Austria; Institute of Mathematical Methods in Economics, Research Unit Economics, Vienna University of Technology (TU), Vienna, Austria

Joan Pujol University of Girona, Girona, Spain

Daniel Samaan Department of Economics, New School for Social Research, New York, USA

Andreas Schaefer Institute of Theoretical Economics/Macroeconomics, University of Leipzig, Leipzig, Germany

Willi Semmler New School for Social Research, New York, USA

Sjak Smulders Department of Economics and CentER, Tilburg University, Tilburg, The Netherlands

Olli Tahvonen Department of Forest Sciences, University of Helsinki, Helsinki, Finland

Yacov Tsur Department of Agricultural Economics and Management, The Hebrew University of Jerusalem, Rehovot, Israel

Ngo Van Long Department of Economics, McGill University, Montreal, Canada

Alexandra Vinogradova Center of Economic Research, CER-ETH, Zürich, Switzerland

Franz Wirl Department of Business Studies, University of Vienna, Vienna, Austria

Angels Xabadia University of Girona, Girona, Spain

Anastasios Xepapadeas Department of International and European Economic Studies, Athens University of Economics and Business, Athens, Greece

Athanasios N. Yannacopoulos Department of Statistics, Athens University of Economics and Business, Athens, Greece

Amos Zemel Department of Solar Energy and Environmental Physics, The Jacob Blaustein Institutes for Desert Research, Ben Gurion University of the Negev, Sede Boker Campus, Israel

Part I
Interactions Between Economy and Climate

Climate Change and Technical Progress: Impact of Informational Constraints

Anton Bondarev, Christiane Clemens, and Alfred Greiner

Abstract In this paper we analyse a growth model that includes environmental and economic variables as well as technological progress under different informational constraints on the behavior of economic agents. To simulate the informationally constrained economy, we make use of the non-linear model predictive control technique. We compare models with exogenous and endogenous technical change as well as directed and undirected endogenous technical change under different informational structures. We show that endogenous technical change yields lower environmental damages than exogenous technical change with a fully informed social planner. At the same time, welfare may rise or decline depending on the efficiency of the technology in use. In the case of directed technical change, a green growth scenario generates a smaller temperature increase that, however, goes along with less output and lower welfare. This holds both for the informationally constrained market economy and for the social optimum. We find that the effects of informational constraints, with respect to the climate system, increase with the degree of endogeneity of technology in the model.

1 Introduction

In this paper we develop the simple dynamic endogenous growth model of the world economy which takes into account environmental damages. There are a great many such models in the literature, starting with the seminal paper by Nordhaus (2007). Some of these models are of integrated assessment type (IAM) and employ the detailed description of the economy under consideration together with many sectors and parameters which are then estimated. Other types of models are of simpler

A. Bondarev · C. Clemens · A. Greiner (✉)
Department of Business Administration and Economics, Bielefeld University, Universitätstraße 25, 33615 Bielefeld, Germany
e-mail: agreiner@wiwi.uni-bielefeld.de

A. Bondarev
e-mail: abondarev@wiwi.uni-bielefeld.de

C. Clemens
e-mail: cclemens@wiwi.uni-bielefeld.de

E. Moser et al. (eds.), *Dynamic Optimization in Environmental Economics*, Dynamic Modeling and Econometrics in Economics and Finance 15,
DOI 10.1007/978-3-642-54086-8_1,

structure and are employed to study some new approaches to the modeling of the environment in endogenous growth theory. In this second strand of literature there are two different approaches to modeling environmental damages and environmental threat for the economy: through the inclusion of environmental quality into the utility function of the representative household, as in the paper by Ligthart and Van Der Ploeg (1994), or through the assumption of productivity decreases due to the environmental degradation, or both. An example of such an approach is the paper by Bovenberg and Smulders (1995), where the notion of pollution-augmenting technical change is adopted. According to this classification, our paper belongs to the second approach.

The main focus of this paper is the influence of different forms of technical change on the evolution of the economy and on the environment under different informational regimes for the economy. Hence, there are two main departures from the majority of the literature on endogenous growth taking into consideration the environment. The first concerns the way the technological change is modeled and, the second, the way the representative household takes into account the environmental change in its decision making.

As concerns the first aspect, the technology in environmental models was usually modeled as an exogenous process of accumulation of knowledge according to some given function, without any influence from the part of the optimizing agent. Later on, there appeared a number of papers where the environmental variables are subject to the control of the agent together with the technology. These papers build up upon two well-known models of endogenous growth, namely that of Romer (1990) and that of Aghion and Howitt (1992). As an example for an endogenous growth model with environmental damages, based on variety expansion, one may take the paper by Barbier (1999), while papers by Grimaud (1999) and Grimaud and Rougé (2003, 2005) are based on the model of vertical innovations by Aghion and Howitt (1992). These and similar papers do not take into account the environmental friendliness of technologies being developed and deal only with productivity. At the same time, there is a discussion in the literature on the possibility of "green growth", where the productivity increase of the economy does not lead to environmental damages. In recent years, endogenous growth models have appeared that distinguish between "clean" and "dirty" technologies. This type of modeling uses the notion of directed technical change and the most recent example of such a literature is the paper by Acemoglu et al. (2012). The natural question one may ask is: what additional insight and implications follow from the inclusion of directed technical change into such a model. To answer it, we employ the same strategy as in the early paper by Smulders and Gradus (1996) and compare three simplified models in their predictions. We compare the results of the model with exogenous technical change, similar to the one employed in the paper of Bréchet et al. (2011), with those of undirected but endogenous technical change in the spirit of papers by Barbier (1999), Grimaud (1999) and with the outcome of the model featuring directed endogenous technical change with similar ideas as in Acemoglu et al. (2012). We come to the conclusion that in the absence of external stimuli, the planner will choose the more productive technology with higher environmental damages, rather than the cleaner one under directed technical change. At the same time, with undirected exogenous technical change, environmental damages may be lower than under directed change, given the "dirty" scenario of the economy.

Another aspect of interest for our research is the comparison of performance of the model under different informational regimes being allowed for. To this end, we employ the non-linear model predictive control (NMPC) approach which has been proposed for the environmental growth model in the paper by Bréchet et al. (2011), while developed earlier on in the literature on the NMPC technique, see the collection of contributions in Allgöwer and Zheng (2006) for reference. We compare the results of the model with an "optimal" (Pareto-optimal) behavior of the social planner, who cares about the environment to a full extent acting as a perfect-foresighted individual, with the outcome of a representative household with limited rationality, modeled as a receding planning horizon of the household.

As concerns the household sector, we assume a homogeneous household sector of mass one with household production, where each individual household has measure zero. Thus, the representative household has a negligible effect on aggregate emissions so that it neglects its emissions of greenhouse gases, which result as an external effect of production. Therefore, it does not invest in abatement but only chooses the optimal consumption share and the optimal share of investment in the creation of new technologies, which gives the laissez-faire or market solution. However, the household knows that the environment changes over time and, therefore, updates its optimal controls at certain discrete points in time, taking into account the new state of the environment. But, due to informational constraints, it does not continuously observe the changes in the environment. This makes our approach different from the usual modeling of externalities, where the representative household does not take into account the external effect but continuously observes the state of the environment, as in Greiner (1996, 2003) or more recently in Antoci et al. (2011).

It turns out that under receding horizon decision rules, the difference in terms of social welfare and environmental degradation between smart management of endogenous directed and undirected technological change and exogenously given pattern of technology is higher, compared to full information regime rules. At the same time, the directed technical change differs to a lesser extent from the undirected endogenous one (again in terms of welfare and environment) under informational constraints than under the full information regime. These differences in ordering of social welfare under different decision rules may help us to clarify the role that the management of technological progress plays with respect to the urgently desired switch of the equilibrium dynamics towards cleaner growth policies.

The rest of the paper is organised as follows. In the next section the formal description of all three versions of the model is given together with some necessary comments on the model structure. The main part is taken by the simulation results and their analysis, where the comparison between different models of technical change as well as different decision rules is made. The concluding section contains some brief discussion of results.

2 Model

We introduce the model of endogenous technical change in this section. First, we model undirected technical change by allowing for the productivity parameter, $A(t)$, to be controlled by the social planner, while leaving the emissions reduction technology, $e(t)$, exogenous which, later on, is controlled by the planner, too. The model presented below may be viewed as a straightforward extension of the model with exogenous technical change by Bréchet et al. (2011). We take this model with exogenous technical change as the benchmark.

2.1 Undirected Endogenous Technical Change

Consider first the model with only productivity being controlled by the social planner. There is also a gradual process of reduction of emission intensity, which is assumed to be exogenous for the time being. The social planner in the model represents some central authority (government). This planner has full information about the influence of economic activities on the environment. The economic part of the model is rather stylized and represented by the capital accumulation process. The climate change is represented by a pair of equations for the dynamics of temperature and greenhouse gas (GHG) concentrations.

The social planner optimally chooses the rate of consumption per capita and the rate of abatement activities to maximize social welfare and keep environmental degradation limited. The planner can also increase the productivity of the economy through R&D investments. With these assumptions the control problem of the planner contains 4 state variables (capital, temperature, GHG concentration and the state of technology) and 3 control variables (consumption rate, abatement rate and R&D investments per capita):

$$J^E = \max_{u,a,g}\left\{\int_0^T \mathrm{e}^{-rt}\left[\frac{[u(t)Y(t)]^{1-\gamma}}{1-\gamma}\right]dt\right\} \tag{1}$$

s.t.

$$\dot{k}(t) = -\delta k(t) + \left[1 - u(t) - c_1\big(a(t)\big) - c_1\big(g(t)\big)\right]Y(t), \tag{2}$$

$$\dot{\tau}(t) = -\lambda\tau(t) + d\big(m(t)\big) = -\lambda\tau(t) + \eta \ln\frac{m}{m_0^*}, \tag{3}$$

$$\dot{m}(t) = -\nu m(t) + \big(1 - a(t)\big)e(t)Y(t), \tag{4}$$

$$\dot{x}(t) = \beta g(t) - \delta_2 x(t), \tag{5}$$

$$Y(t) = A^E(t)\phi\big(\tau(t)\big)k(t)^\alpha = A^E(t)\left(\frac{1}{1+\theta_1\tau^{\theta_2}}\right)k(t)^\alpha, \tag{6}$$

$$A^E(t) = 1 + \omega x(t), \tag{7}$$

where:

J^E is the objective functional;
r is the discount rate;

$u(t)$	is the consumption rate per capita;
$Y(t)$	is the total output;
$k(t)$	is the total capital;
δ	is the depreciation rate of capital;
δ_2	is the depreciation rate of technology;
$a(t)$	is the abatement rate;
$g(t)$	are R&D investments;
$\tau(t)$	is the temperature increase from the preindustrial level;
λ	is the rate of temperature decrease due to natural causes;
$m(t)$	is the GHG concentration in the world's atmosphere;
ν	is the rate of recovery of the atmosphere due to natural absorption;
$e(t)$	is the reduction of intensity of emissions from economic activities;
$x(t)$	is the state of technology;
$A^E(t)$	is the productivity of the economy;
$\phi(\tau(t))$	is the damage function depending from the temperature increase;
α	is the parameter of capital productivity;
ω	is the rate of transformation of the current state of technology into the productivity of the economy.

In the model the evolution of state variables is given in the following way:

- Capital increases due to investments into capital, (2);
- Temperature increases as a function of the GHG concentration in the atmosphere, (3);
- GHG concentration increases due to economic activity in the economy (it is assumed that natural causes may be neglected), while the impact of economic activity is weakened through abatement and exogenous improvement in cleaning technologies, (4);
- Technology improves in a linear way from R&D investments while decreasing in the absence of such investments, (5);
- Output is of Cobb-Douglas type with labor supply normalized to unity with no population growth, (6);
- At last, productivity grows due to the transmission of a (fixed) proportion of technology into the production technology, (7).

It has to be noted that the original model of Bréchet et al. (2011) is easily obtained from this model by assuming a constant and linear increase in productivity, i.e. by substituting (7) with the linear technology $A^B = \kappa_1 t + \kappa_2$ and by setting $e(t) = \mathrm{e}^{-\iota_1 t - \iota_2}$ as well as dropping the (5) and the term $c_1(g(t))$ from (2).

The form of dynamics of technical progress itself is rather simple: the technology improves via the investments into the technological progress, $g(t)$ and declines in the absence of investments with some rate δ_2. Such a form of dynamics is rather simple and yet allows for the existence of steady state and endogenous technology.

In the case of (undirected) endogenous technical change we assume the same cost function for technology investments and for abatement $a(t)$:

$$c_1\big(g(t)\big) = 0.01\frac{g(t)}{1-g(t)},$$
$$c_1\big(a(t)\big) = 0.01\frac{a(t)}{1-a(t)}. \tag{8}$$

This specification guarantees that there will be some resources left for consumption with any positive values of abatement and R&D investments. However, negative investments into the capital are possible: in this case capital decreases in time.

Making use of this specification of cost functions, we write down the analytic form of optimal controls for the problem given by (1) s.t. (2)–(7):

$$u^E_{opt} = \frac{\psi_k(t)^{-\frac{1}{\gamma}}k(t)^{-\alpha}}{\frac{1}{1+\theta_1\tau(t)^{\theta_2}}(1+\omega x(t))},$$
$$a^E_{opt}(t) = 1-0.1\frac{\sqrt{-\exp(\iota_1 t+\iota_2)\psi_m(t)\psi_k(t)}}{\psi_m(t)}, \tag{9}$$
$$g^E_{opt}(t) = 1-0.1\frac{\sqrt{\beta_1\frac{1}{1+\theta_1\tau(t)^{\theta_2}}(1+\omega x(t))\psi_x(t)\psi_k(t)k^{\alpha}(t)}}{\beta_1\psi_x(t)}.$$

It can be seen that the optimal abatement rate depends only on the ratio of shadow costs of capital and environmental degradation (which coincides with the benchmark model), but the consumption rate now negatively depends on technical progress that is endogenous. This means that technology boosting the total output, makes consumption higher even with the same share of output being devoted to consumption and, thus, the faster is the technological change, the lower this consumption share has to be. Investments into technology depend on the level of technology achieved, on the capital level and on the ratio of shadow costs of capital and technology. The resulting dynamical system for the state variables is 4-dimensional and explicitly includes technical progress:

$$\dot{k}(t) = -\delta k(t) + \left[1-u^E_{opt}(t)-0.01\frac{a^E_{opt}(t)}{1-a^E_{opt}(t)}-0.01\frac{x^E_{opt}(t)}{1-x^E_{opt}(t)}\right]$$
$$\cdot\big(1+\omega x(t)\big)\frac{1}{1+\theta_1\tau^{\theta_2}}k(t)^{\alpha},$$
$$\dot{\tau}(t) = -\lambda\tau(t)+\eta\ln\frac{m}{m_0^*}, \tag{10}$$
$$\dot{m}(t) = -\nu m(t)+\big(1-a^E_{opt}(t)\big)\exp(\iota_1 t+\iota_2)\big(1+\omega x(t)\big)\frac{1}{1+\theta_1\tau^{\theta_2}}k(t)^{\alpha},$$
$$\dot{x}(t) = \beta_1 g^E_{opt}(t)-\beta_2 x(t),$$

where $u^E_{opt}(t), a^E_{opt}(t), g^E_{opt}(t)$ are given by equations in (9).

The main difference in this system compared to that of the model with exogenous technical change comes from the endogenous technology which increases productivity by the factor ω and which is governed by technological investments, rather than by the exogenous growth rate. In such a system emissions of GHGs depend not only on the capital accumulation but also on technological advances in the economy which creates a link between the technological and the environmental sector. However, the abatement rate does not depend on technology and is exactly the same as in the benchmark model. Thus, the model becomes of an endogenous growth type, where the growth rates are defined through technological change. This technological change, however, is described in rather a stylized way and does not take into account the impact of technology on the reduction of emission intensity, $e(t)$. This is achieved by further extending the basic model to account for the direction of technological change.

2.2 Directed Endogenous Technical Change

To model directed technical change we relax the assumption of the exogenous rate of emission intensity decrease, $e(t)$. Now, we allow this to depend on the endogenous technological development, too. To this end, we assume that a certain fraction of technological progress is devoted to the reduction of emission intensity without increasing productivity, while the other fraction is devoted to the increase in productivity without reducing emissions. We respecify the functions $e(t)$, $A(t)$ as follows:

$$e^D(t) = \frac{e_0}{1+(1-\varepsilon)\omega x(t)},$$
$$A^D(t) = 1+\varepsilon\omega x(t),$$

and the dynamic problem is formulated with the same constraints as in Eqs. (1)–(7). The parameter e_0 is set to the initial level of the emission intensity from the benchmark model, $e_0 = e(0)$. In such a formulation, the (exogenous) parameter $\varepsilon \in [0,\dots,1]$ measures the direction of technical progress. With $\varepsilon = 0$ all of the technical progress is devoted to the reduction of emissions from production without increasing productivity at all, while with $\varepsilon = 1$ all of the technical progress is going to the increase in productivity. Parameter ω, as before, is measuring the efficiency of technical progress for productivity increase.

The optimal controls in this case are:

$$\begin{aligned}
u^D_{opt} &= \frac{\psi_k(t)^{-\frac{1}{\gamma}} k(t)^{-\alpha}}{\frac{1}{1+\theta_1\tau(t)^{\theta_2}}(1+\varepsilon\omega x(t))},\\
a^D_{opt}(t) &= 1-0.1\frac{\sqrt{-\frac{e_0}{1+(1-\varepsilon)\omega x(t)}\psi_m(t)\psi_k(t)}}{\psi_m(t)},\\
g^D_{opt}(t) &= 1-0.1\frac{\sqrt{\beta_1\frac{1}{1+\theta_1\tau(t)^{\theta_2}}(1+\varepsilon\omega x(t))\psi_x(t)\psi_k(t)k^{\alpha}(t)}}{\beta_1\psi_x(t)}.
\end{aligned} \tag{11}$$

It can be seen that for the case of directed technical change, the abatement rates are different from the benchmark model as well as from undirected endogenous progress and include the evolution of technology as an argument. This links abatement efforts to technology, whereas in the undirected version of the model such a link is absent and technology influences the environment only through productivity increases in a negative way. In this case, the dynamical system for state variables changes and emissions accumulation depends on technical progress in an ambiguous way: it may decrease due to the evolution of clean technology or increase because of higher productivity and more production. The exact direction depends on the parameter ε:

$$\begin{aligned}
\dot{k}(t) &= -\delta k(t) + \left[1 - u^D_{opt}(t) - 0.01\frac{a^D_{opt}(t)}{1-a^D_{opt}(t)} - 0.01\frac{x^E_{opt}(t)}{1-x^E_{opt}(t)}\right] \\
&\quad \cdot \big(1+\varepsilon\omega x(t)\big)\frac{1}{1+\theta_1\tau^{\theta_2}}k(t)^{\alpha}, \\
\dot{\tau}(t) &= -\lambda\tau(t) + \eta\ln\frac{m}{m_0^*}, \\
\dot{m}(t) &= -\nu m(t) + \big(1-a^D_{opt}(t)\big)\frac{e_0(1+\varepsilon\omega x(t))}{1+(1-\varepsilon)\omega x(t)}\frac{1}{1+\theta_1\tau^{\theta_2}}k(t)^{\alpha}, \\
\dot{x}(t) &= \beta_1 g^D_{opt}(t) - \beta_2 x(t),
\end{aligned} \tag{12}$$

where $u^D_{opt}(t), a^D_{opt}(t), g^D_{opt}(t)$ are given by the equations in (11). For the case $\varepsilon = 0.5$ technical progress is environmentally neutral as the term $\frac{(1+\varepsilon\omega x(t))}{1+(1-\varepsilon)\omega x(t)}$ cancels out from the emissions equation. In this case, technology influences only the productivity growth, but productivity growth does not influence the environment, as its negative externality is exactly counterbalanced by the reduction in emission intensity from the clean technology. In all other cases, the technology is not neutral and influences emissions negatively or positively.

2.3 *Informational Regimes of the Economy*

The main focus of this paper is in the comparison of the dynamics of the environment and the economy under different informational regimes. Both versions of the model presented above assume that the social planner possesses full information about the links between economic and technological activities and the environment. In contrast to that, the laissez-faire or market solution assumes an informationally constrained representative household that does not continuously observe the influence of its activities on the environment. Instead, the household maximizes welfare over a certain period of time neglecting the environment. After the period has elapsed, it observes the state of the environment and its effect on output and solves a new optimization problem, again over a certain period of time neglecting environmental concerns. To model such an informationally constrained economy, we make use of the non-linear model predictive control (NMPC) technique.

In the modern world, economic agents should be aware of the changing environment. However, real-time online measurement of the state of the environment is costly and hence such a measurement might be made at some regular periods of time. Another argument for economic intuition might be that economic processes have much higher speed than environmental ones and the representative household in the economy may assume the state of environment to be constant for some periods.

With the help of the NMPC technique one may model such infrequent observations of the state of the environment. To this end, assume that the household under consideration is measuring the state of environment every k periods of time and revises its optimal controls over consumption and technology investments. In such a case, one has to consider the full dynamical system given by (1)–(7) for true state dynamics and the reduced one for the determination of optimal controls of the household under informational constraints of this type.

Following the idea of Bréchet et al. (2011), we define strategies of the informationally constrained household as Business-as-Usual (BaU) scenarios in the following way:

1. At the initial point in time, t_0, the household is solving the reduced dynamical problem (defined in the Appendix) on some fixed time horizon Θ (with Θ being some long but finite time horizon being chosen in such a way, as this length allows the system to be marginally close to the steady state) and defines its optimal controls;
2. These controls are then used to determine the evolution of the full dynamical system which includes environmental variables as well as economic ones for the same time horizon;
3. After $h = t_h - t_0$ time (being the step of measurement of the environment) the household measures the state of the full system at the time $t = t_h$ and revises its optimal controls with this state of the system given as an initial state from $t = t_h$ onwards till $t = t_h + \Theta$ (thus obtaining the new optimal policy for the whole planning horizon and not just till the next measurement time);
4. When the next measurement time t_{h+1} is reached, the household again revises its policies in the view of new information obtained about the state of the environment;
5. The procedure repeats until the terminal time is reached.

Note that such a procedure essentially requires a limited time horizon for the household since, otherwise, it could not be completed in a finite number of steps. This difficulty is resolved by choosing rather a long terminal time θ, as the system may arrive to its steady state within this time length.[1]

[1] See the Appendix for an illustration of NMPC for the model with directed technical change.

3 Numerical Simulation

3.1 Computational Issues and Calibration

To obtain solutions for all of the versions of the model we make use of the gradient projection method of simulations. The basic idea is as follows: at each simulation step the vector of optimal controls is computed iteratively as the preceding value plus the gradient increase. As the control approaches the optimal value, the gradient decreases. At the optimal point the gradient of the system is zero. In practice, iterations are performed until the gradient reaches a sufficiently small value. The steps of the algorithm of iterations is described below:

1. Set initial values for controls u^0, a^0, g^0;
2. Solve the system for state variables, one of Eqs. (10), (12), depending on the version of the model;
3. Calculated state variables k^0, m^0, τ^0, x^0 are used for the solution of the co-state system;
4. Solutions of both systems are used to compute the next-step gradient of the system, $\nabla \mathbf{X}^1$;
5. Next step controls have the form $u^1 = u^0 + b \cdot \nabla \mathbf{X}^1$;
6. The procedure repeats until $\nabla \mathbf{X}^k \to 0$.

The gradient of the system is given by first-order conditions for controls, (13), (14) or (15), depending on the model under consideration. The b parameter is chosen arbitrarily and is the scale of one iteration step, remaining constant for all iterations, but varying from system to system. This is determined experimentally and depends on the numeric scale of the gradient being computed.

Note that the algorithm above is valid only for the computation of solutions under full information, while for implementing the NMPC technique for informationally constrained economies it is not sufficient. In the latter case, the procedure above has to be repeated at each time step, t_i, along the whole time path. Otherwise, the algorithm remains the same. The values of the parameters ω and ε depend on the respective scenario under consideration and are explained below. As concerns the other parameters, these are the same for all different scenarios and for the calibration we set the parameters to the values given in Table 1.

Concerning the evolution of technology, we consider different scenarios with respect to the choice of exogenously given transformation rate ω and the proportions of clean and dirty technologies ε. These scenarios are summarized in Table 2.

Here, the upper part gives values of the transformation rate ω being considered in simulations for endogenous undirected technical change. Since the technology is transformed only in productivity increase, this is equivalent to setting the ε parameter to 1, as the table shows. The lower part of the table shows how different proportions of technical change, going into cleaning or more productive technologies, affect the resulting effective transformation rate, $\varepsilon\omega$, and the fraction of technical change in cleaner technologies, $(1 - \varepsilon)\omega$. Setting $\varepsilon = 0.5$ implies environmentally neutral technical change. With a productivity impact of $\omega = 0.2$ this yields

Table 1 Parameter values

Economic parameters		
Depreciation rate	δ	0.075
Inverse of elasticity of substitution	γ	2
Interest rate	r	0.015
Capital elasticity	α	0.45
Climate parameters		
Temperature re-absorption	λ	0.11
Climate sensitivity	η	0.59
Pre-industrial carbon concentration	m_0^*	5.964
Damage function parameter 1	θ_1	0.0057
Damage function parameter 2	θ_2	2
GHG re-absorption rate	ν	0.0054
Technological parameters		
Emission intensity reduction parameter 1	ι_1	0.00384
Emission intensity reduction parameter 2	ι_2	3.1535
Initial emission intensity reduction for endogenous models	e_0	0.0427
Linear technology parameter 1	κ_1	1
Linear technology parameter 2	κ_2	0.0014
Efficiency of technological investments	β_1	0.7
Decay of technology in absence of investments	β_2	0.1

Table 2 Simulated technological parameters values

Scenario	ω	ε	$\omega\varepsilon$	$(1-\varepsilon)\omega$
Undirected change				
Slow growth	0.05	1	0.05	0
Normal growth	0.10	1	0.10	0
Directed change				
Clean growth	0.20	0.1	0.02	0.18
Neutral growth	0.20	0.5	0.1	0.1
Dirty growth	0.20	0.9	0.18	0.02

the same overall productivity growth as for the undirected change ($\varepsilon \cdot \omega = 0.1$). Next, we consider the "green" or "clean growth" scenario, where the technological progress is biased towards the reduction of emission intensity, with $\varepsilon = 0.1$. In such a case, overall productivity growth is much slower than for the undirected technical change, $\varepsilon \cdot \omega = 0.02$, while emission intensity reduces with the factor

Table 3 Damage function

Temperature increase	Decrease in productivity
+2 °C	−2.23 %
+4 °C	−8.36 %
+6 °C	−17.02 %
+8 °C	−26.73 %

$1/(1+(1-\varepsilon)\cdot\omega) = 0.84$ from the initial state of cleaning technology $e(0)$. Finally, we consider the "dirty growth" scenario with $\varepsilon = 0.9$ and the resulting productivity growth $\varepsilon \cdot \omega = 0.18$ higher than for undirected change and higher emission intensity with only slight reduction to 0.98 level in 100 years.

3.2 *Discussion of the Damage Function*

Here we discuss our choice of parameters of the damage function, θ_1, θ_2. With the chosen functional form of this function, given by (6), the parameter θ_1 measures the linear impact of the temperature on the productivity of capital, while θ_2 is chosen due to the functional form considerations to provide a hyperbolic type decay rate for productivity with temperature increases. This specification follows the one assumed in the paper by Bréchet et al. (2011): an increase in the mean temperature by 2 °C leads to a 2.23 % decrease in productivity. However this effect is not linear but rises: the higher increase in temperature leads to even stronger decreases in productivity, as Table 3 shows.

It can be seen that the worst case scenario leads to an extreme rise in temperature of 8 °C and implies a reduction of productivity by more than 25 %. At the same time, the chosen specification of exogenous productivity growth in the benchmark model implies an increase of productivity by the same 25 % in 100 years. Thus, in the exogenous growth scenario the technology growth always has a higher significance than environmental damage, which is one reason for taking such a high damage function compared to Nordhaus (2007), where the damage is almost twice as low for the same temperature increase: for a 2 °C increase in temperature only a 1 % decrease in productivity is assumed there. In the view of recent data, however, such an assumption appears too optimistic, since it does not account for the additional losses in GDP due to the impact of higher temperatures on the sea level increase, which has already started. With this in mind, it might be the case that more pessimistic estimates, as adopted here, might be useful. Our calibration is more in line with the calibration of damage functions for Europe as in the model by Hassler and Krusell (2012), where it is claimed that environmental damages differ from region to region and appear to be higher for Africa and for the EU than for the US or China. There, it is assumed that a 2 °C increase leads to a 2.83 % productivity damage for Europe and to 3.91 % damages for Africa. Hence, our calibration values are in between the values used in the two papers above mentioned.

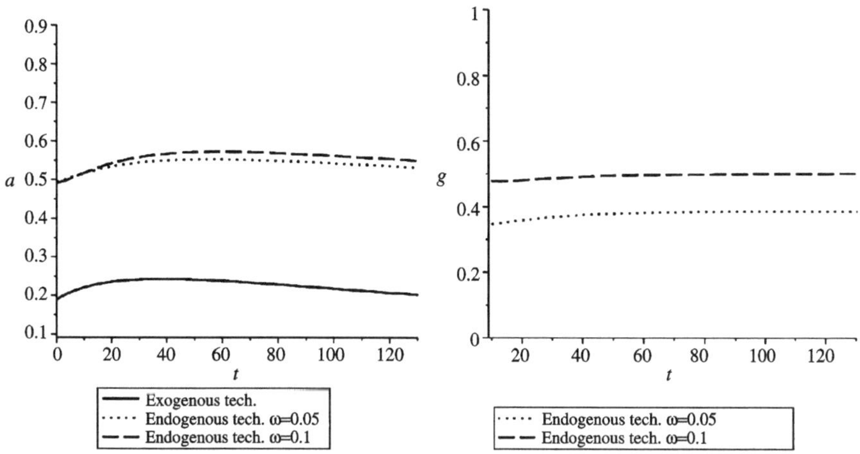

(a) Abatement rates

(b) R&D investments

Fig. 1 Optimal policies for the exogenous and endogenous technology

3.3 Economy with Full Information

3.3.1 Exogenous vs. Undirected Endogenous Technology

The introduction of the endogenous technology into the benchmark model, described by (1)–(7), allows a more efficient environmental policy of the social planner in the case of full information. The efficiency of technological progress in respect of increasing productivity plays a crucial role for social welfare in terms of consumption. We consider two values for this parameter, which give the productivity growth lower and higher than the exogenous linear growth in the benchmark model. Namely, we take $\omega = 0.05$ for low yield of technology for productivity growth and $\omega = 0.1$ for high yield.

First, consider the dynamics of optimal abatement and technology investments per capita in Fig. 1. One can see that abatement efforts are higher for both scenarios and the difference between low and high technology yields is rather small. Both are stabilized at the level between 0.5 and 0.6, while in the exogenous technology case it is much lower, at the level of 0.2. This differs from the dynamics of the benchmark model in Bréchet et al. (2011) substantially due to different values of parameters. One would expect higher abatement efforts for a higher technology impact. It is indeed so, since abatement efforts in Fig. 1a are given in per capita terms. In terms of final output these investments are higher for the high technology yield scenario, since the output itself is higher.

Technology investments in both endogenous technology scenarios are also almost constant in time with more investments being made for the higher omega parameter. There are more incentives to invest into technology if it has more impact on productivity, thus a higher output share is invested.

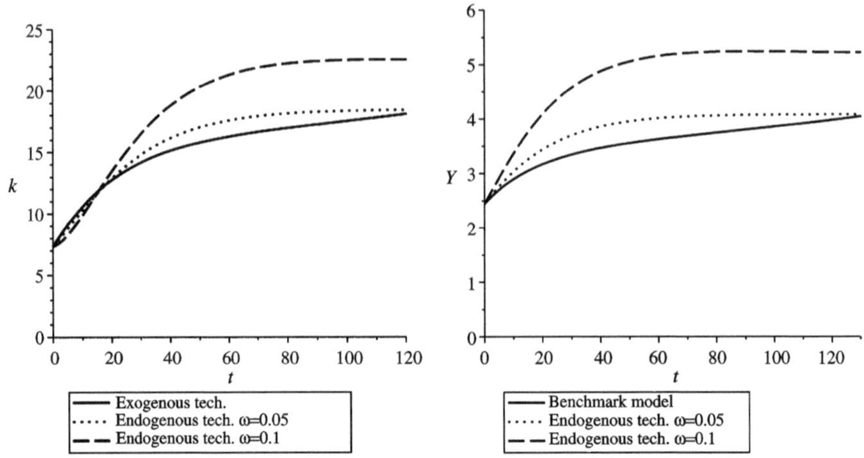

(a) Capital dynamics (b) Output dynamics

Fig. 2 Economic dynamics for the basic model and with endogenous technology

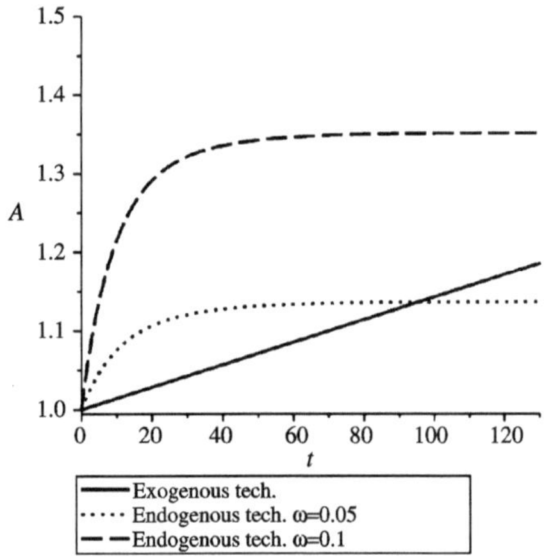

Fig. 3 Productivity for the basic model and with endogenous technology

Because of the reduced consumption share due to endogenous technology (see Fig. 5a), capital accumulation is boosted in comparison to the benchmark model as well as the total output of the economy. Furthermore, in the scenario with low impact of technology on productivity capital accumulation rates at later stages of development decline and are outperformed by the linearly rising exogenous technology of the basic model. The same is true for the output. This can be seen in Fig. 2.

This figure demonstrates the importance of parameter ω for the economy. With high efficiency of the transformation of technical change into the productivity growth, the growth of capital and output is stimulated by higher technological advances and by lower consumption shares, while for lower values of ω the reduction in consumption per se is not sufficient to outperform the exogenous technology. To see this, consider the relative productivity growth for all three scenarios in Fig. 3.

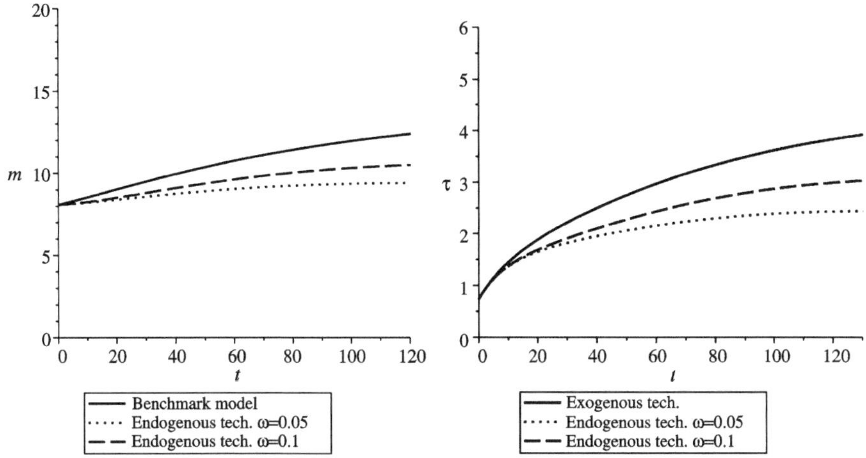

(a) GHG concentration

(b) Temperature increase

Fig. 4 Climate dynamics for the basic model and with endogenous technology

As a result of lower consumption shares and higher abatement investments per capita, the climate in the endogenous technology version of the model demonstrates much less drastic temperature increases and GHG concentrations than the benchmark model as Fig. 4 shows.

It should be noted that after 120 years of simulation, the model with endogenous technology tends to the stabilization of temperature and emissions at some lower level compared to the model with exogenous technical change. In particular, with slower advances in productivity ($\omega = 0.05$) the temperature increase amounts to not more than 2 °C, while for the benchmark model with exogenous technology this value is higher than 3.5 °C and approaches 4 °C. Slower environmental degradation together with higher economic performance of the endogenous technology model are the consequences of different dynamics of technical change in comparison with the linear one in the benchmark model. The highest increase of productivity happens in the first 20 years of simulation, while later on R&D investments are being made on the level just to support the achieved productivity level. In such a way the impact of technology on the environment is minimized and the environmental degradation slows down. Additional resources which are gained through this rapid technological advance are then devoted mainly to abatement activities further reducing the impact of the output on the environment. Thus, under endogenous technological change it appears to be optimal for the fully informed planner to "grow up first and clean up later", rather then gradually increase the productivity and invest into the abatements simultaneously. As a result, the environment suffers less, since abatement activities are initially and all over the simulation period higher than in the scenario with exogenous technology.

In terms of consumption and welfare, the scenario with a high impact of technology on productivity delivers greater consumption to the representative consumer

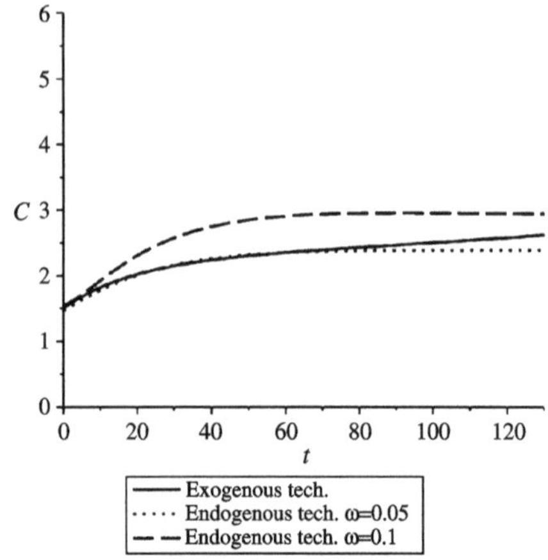

Fig. 5 Consumption dynamics for the basic model and with endogenous technology

Table 4 Present value consumption changes with endogenous technology (relative to exogenous)

ω	$t = 50$	$t = 100$
0.05	−0.96 %	−0.80 %
0.1	13.78 %	17.27 %

than the benchmark scenario, while this is not true for the low impact scenario. This is demonstrated in Fig. 5.

At last, we compute welfare gains or losses expressed as relative changes of present value consumption. To be precise, we compute the present value of the necessary change in the consumption stream (in percent) that makes welfare in the scenario with exogenous technology equal to welfare in the scenario with endogenous technology. The results over 50 and 100 years of simulations are shown in Table 4. The planning horizon for the planner is the same for all the scenarios and equals 180 years. Thus, we compare present value consumption changes along the same optimal trajectory in two different time points.

For longer time horizons the model with exogenous technology outperforms the model with endogenous technology with low impact ($\omega = 0.05$) in terms of consumption. The scenario with higher technology impact yields a much higher consumption path, increasing to more than 17 % in 100 years above the benchmark model. For shorter time horizons, this difference is less drastic with an almost 14 % increase in consumption in the case of endogenous technology with a high transformation rate. It should be noted, that these differences may be explained by the different assumptions on the form of technical change. In our endogenous technology model, technical change has an exponential form, while in the benchmark model it is linear. However, these simulations demonstrate, that exponential-type technical change is better for the environment and for consumption if the planner controls technical change.

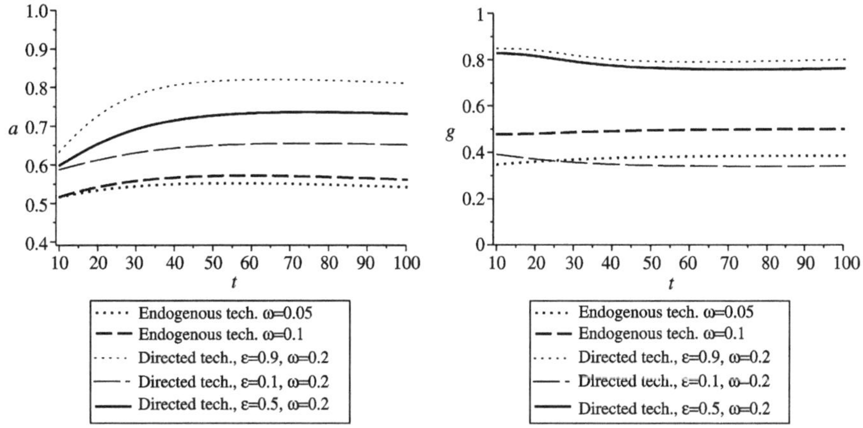

(a) Abatement rates

(b) R&D investments rates

Fig. 6 Optimal controls with directed and undirected technical change

3.3.2 Directed vs. Undirected Endogenous Technological Change

Next we consider the extension of the model to the case of directed endogenous technical change. To this end we introduce the parameter of the direction of technical change through ε. We consider 3 different scenarios and compare them with the endogenous undirected change, discussed above, with a productivity impact of $\omega = 0.1$, since this is the value which allows for productivity growth comparable to the exogenous one and is the "medium" scenario with the respect to economic and environmental dynamics. All the configurations of technological parameters are given in the Table 2.

Consider abatement and investments into technology in per capita terms displayed in Fig. 6. The first thing to note is that abatement activities are increased for all scenarios of directed technical change in comparison with the undirected one. The highest abatement rate is obtained for the dirty growth scenario, while the lowest (among directed growth scenarios) for the clean growth. This seems rather intuitive: the higher productivity growth with dirty technology frees more resources for abatement activities, while with clean technology productivity grows much slower but, at the same time, the environmental damage is also lower such that abatement activities are not that necessary. However, in the case of undirected growth with comparable productivity, abatement rates are lower than for the neutral technical change which gives the same rate of productivity increase as the undirected change with $\omega = 0.1$. This points to the difference between undirected and directed technical change models: with an exogenously given reduction of emissions, which is not part of the technical change managed by the planner, the planner has lower concerns for abatement activities even with a comparable productivity growth. The abatement rate dynamic is displayed in Fig. 6a.

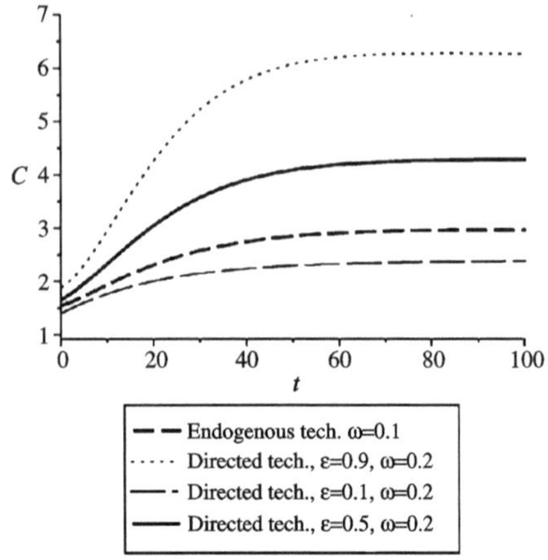

Fig. 7 Consumption with directed and undirected technical change

The interesting difference can be observed for technology investments. These are also constant for all scenarios after 100 years. In the case of clean growth, technological investments are lower than for both scenarios of undirected growth, while they are higher for neutral and dirty growth. In these last two scenarios, technological investments are almost the same, although the return for such investments in terms of productivity is twice as high as for dirty growth. The economic intuition for this result may be as following. After achieving some sufficiently high productivity level, new additional resources are rather spent for abatement activities and consumption. It becomes more profitable, in terms of consumption gains, for the planner to invest additional resources into abatement to decrease damages, rather than to boost productivity further, since at the high level of productivity achieved, additional R&D investments would increase productivity to rather a small extent. At the same time, additional abatements will significantly slow down the environmental degradation, thus, decreasing threats to the output and consumption coming from the $\phi(\tau)$ damage function. This also means that the threat of dirty technology is at least partially counterbalanced by reduced productivity growth (and the associated environmental threat) at later stages of development in the dirty growth scenario. R&D investments are displayed in Fig. 6b.

The level of consumption is the lowest one for the clean growth scenario and the highest one with dirty growth. Undirected technical change, with technology impact $\omega = 0.1$, yields lower consumption than the neutral directed growth scenario, which has the same overall impact of technology on productivity, $\varepsilon\omega = 0.1$. The difference in steady state consumption levels between clean and dirty growth scenarios is almost 300 %. This is the direct consequence of lower capital accumulation and output for the clean growth scenario, since the productivity growth is much slower there. The dynamics of consumption is displayed in Fig. 7.

The dynamics of capital and output is displayed in Fig. 8. One can observe that in the case of clean growth, capital accumulation and output of the economy are lower compared to undirected technical change, while these two are higher both for neutral and dirty growth scenarios. In all scenarios, the steady state levels of both variables

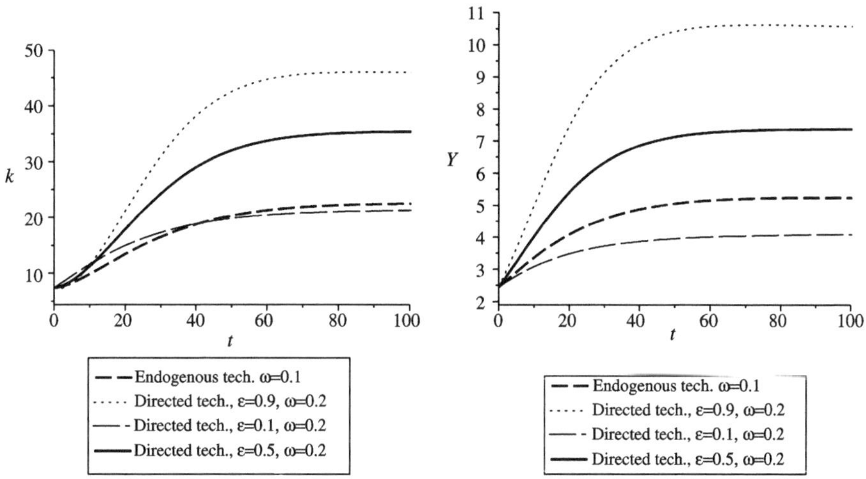

(a) Capital dynamics

(b) Output dynamics

Fig. 8 Economic dynamics with directed and undirected technical change

are achieved after 100 years of simulations and remain constant afterwards. This is different from the benchmark model with exogenous technology because technology growth is not linear but rather of exponential type. Despite of almost equal technology investments in per capita terms for neutral and dirty scenarios, the capital and output dynamics in the latter case are higher by roughly 25 %. This is the effect of higher productivity. Technology for the growth model with directed technical change is described by two variables rather than by one: emission intensity reduction due to cleaner technology and productivity growth, displayed in Fig. 9.

As it can be seen, productivity growth is higher in the case with directed technical change only for the dirty growth scenario, while directed technical change in the clean scenario generates smaller productivity growth than directed neutral and undirected technical progress. In the case of dirty growth, productivity grows twice in 100 years while in all other cases the growth is below 40 %. This is the explanation why in the dirty growth case technology investments are the same as for the neutral case: higher capital accumulation gives more investments in absolute value with the same share and still the productivity grows much faster. Emissions reduction for all directed growth scenarios is less intensive than for the exogenous function, even for the case of clean growth where 90 % of technological progress is going into the emissions reduction. It is important to note that for the neutral case, the emissions reduction is not constant, as it is displayed on the graph, but the total influence of productivity growth plus emissions reduction technology is constant. One can conclude that the bias towards clean technology is not sufficient to achieve the same emissions reduction ratio as for the model with undirected technical change while losses in economic variables are substantial in comparison with dirty growth, as it is displayed in Fig. 8.

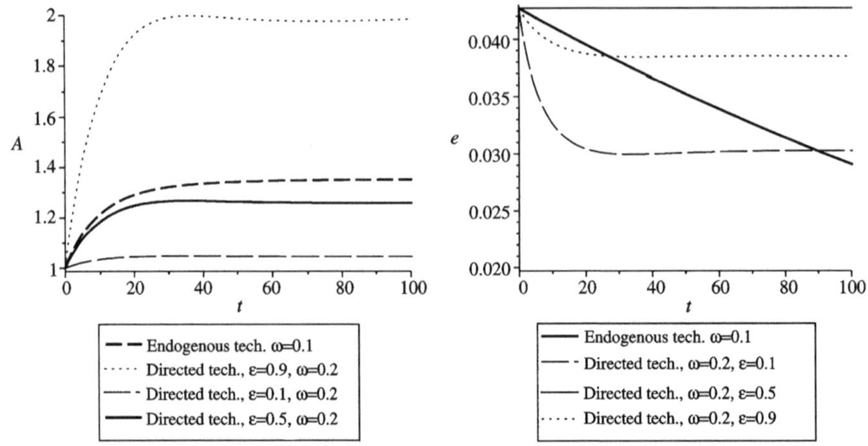

(a) Productivity growth (b) Reduction in intensity of emissions

Fig. 9 Technology dynamics with directed and undirected technical change

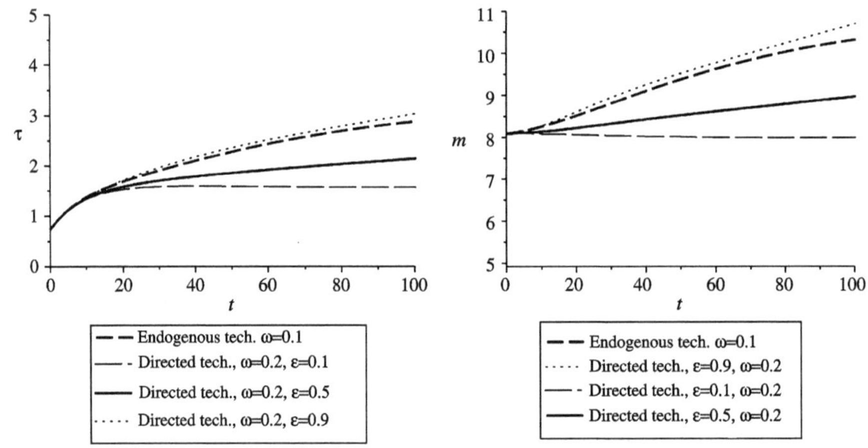

(a) Temperature increase (b) GHG concentration

Fig. 10 Environmental dynamics with directed and undirected technical change

At last, consider the dynamics of the environmental part of the model in Fig. 10. One realizes that, in terms of environmental damages, the dirty growth scenario is very close to the undirected change, while the productivity is almost twice as high. At the same time neutral and clean growth scenarios provide a better environment but at the cost of economic losses. As a result, one may conclude that the dirty growth scenario is the most beneficial for the economy by the total of economic and climate characteristics.

Table 5 Present value consumption changes with directed technical change (relative to undirected)

ε	$t = 50$	$t = 100$
0.1 (clean)	−14.42 %	−17.16 %
0.5 (neutral)	28.63 %	29.98 %
0.9 (dirty)	79.17 %	84.35 %

To obtain this aggregate measure, we compare welfare of the economy in the case of directed technical change with the undirected one, with welfare again expressed as percentage change of present value consumption. The change in present value consumption is calculated for a time period of 50 and 100 years as above. Changes are computed for directed technical change scenarios with $\omega = 0.2$ relative to the undirected endogenous technical change model with $\omega = 0.1$. Consumption changes are displayed in the Table 5.

One can see that the dirty growth scenario is by far the most beneficial one with an 80 % rise in consumption in 50 years in comparison to the undirected change. The case of neutral technical progress also gives some improvement of roughly 30 %. This happens due to lower environmental damages and higher economic dynamics in this scenario than for undirected change. The main drawback of the neutral technical progress is that it may happen only for exactly one value of the direction parameter, ε, and this is not easy to achieve in practical implementations of environmental policy. However, one may conclude, that if to choose between clean and dirty technological scenarios, the dirtier is better, $\varepsilon \geq 0.5$, since the increase in productivity sets free resources for partial compensation of environmental damages through increases in abatement, rather than reduction in emission intensity.

3.4 *Informationally Constrained Economy*

3.4.1 Full Information vs. Informationally-Constrained Scenarios

Here, we compare the simulation results for the fully informed social planner with the informationally constrained (BaU) behaviour of the representative household. It turns out that in all cases, the environmental damages for BaU scenarios are higher than for the economy with full information. This is rather intuitive, since the main feature of the informationally constrained economy is the neglect of the influence of economic variables on the environment. The dynamics of the state variables in the benchmark model with exogenous technology under full information against the BBaU (basic BaU) scenario are illustrated in Fig. 11.

From this figure, it may be clearly seen that in the case of the BBaU scenario, GHG accumulation and the temperature increase are higher than for the benchmark model with full information whereas the capital stock is higher too. This is the typical feature for the majority of informationally constrained scenarios: higher environmental damages and higher capital growth. For the benchmark model with our set of parameters the general claim of the paper Bréchet et al. (2011) holds: the in-

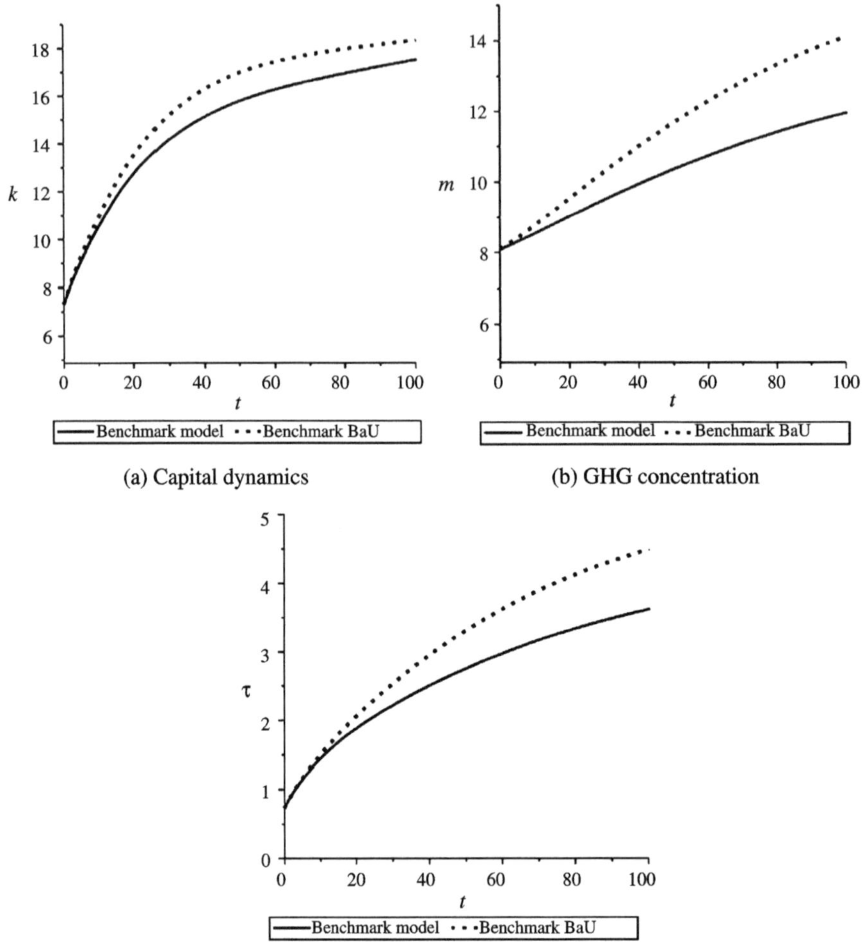

(a) Capital dynamics (b) GHG concentration

(c) Temperature increase

Fig. 11 BaU dynamics compared to full information (exogenous technology)

formationally constrained economy yields lower consumption paths. However, the difference is very small. It amounts to a 0.68 % decline in consumption in 50 years and to 1.12 % decline in 100 years, in comparison with the full information scenario.

In the case of endogenous technology (EBaU), the difference is more drastic, since not only the influence of economic variables is neglected, but also the effect of technological variables on the environment. Technology influences the environment through productivity growth which boosts the emissions accumulation, while the household is unaware of this influence when determining its policy. As a result, the difference in dynamics between full information and informationally constrained scenarios is larger than for the exogenous model. This can be seen in Fig. 12.

Even more differences between the two solutions are revealed for the case of directed technical change (EDBaU), since now there is another additional influence

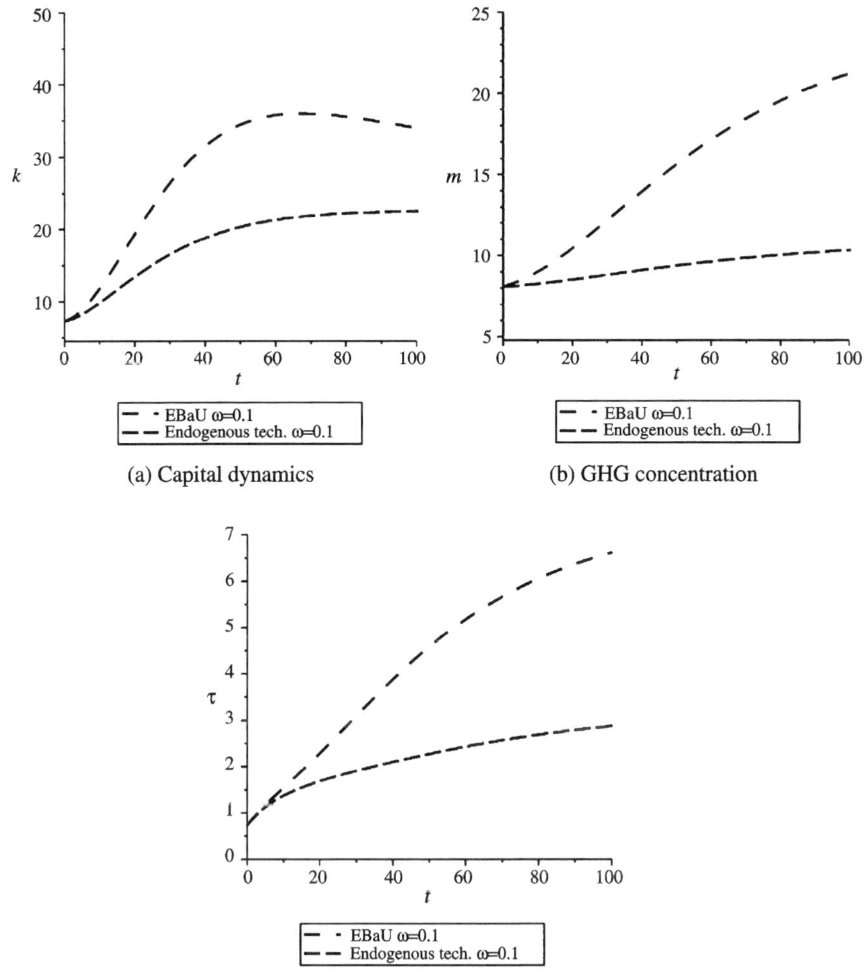

(a) Capital dynamics

(b) GHG concentration

(c) Temperature increase

Fig. 12 BaU dynamics compared to full information (endogenous technology)

of technology on the environment the household in the BaU scenario is unaware of: the emissions reduction intensity, $e(t)$, which is also endogenous in this version of the model. As a result, not only environmental damages are higher, as for undirected technical change, but the capital accumulation is lower for the EDBaU scenario than under the full information. This is seen in Fig. 13 for the dirty growth scenario.

3.4.2 Comparison of Different Technological Change Scenarios for the Informationally Constrained Economy

Finally, we compare all of the computed scenarios with informational constraints with each other to find out possible gains and losses in social welfare as well as the

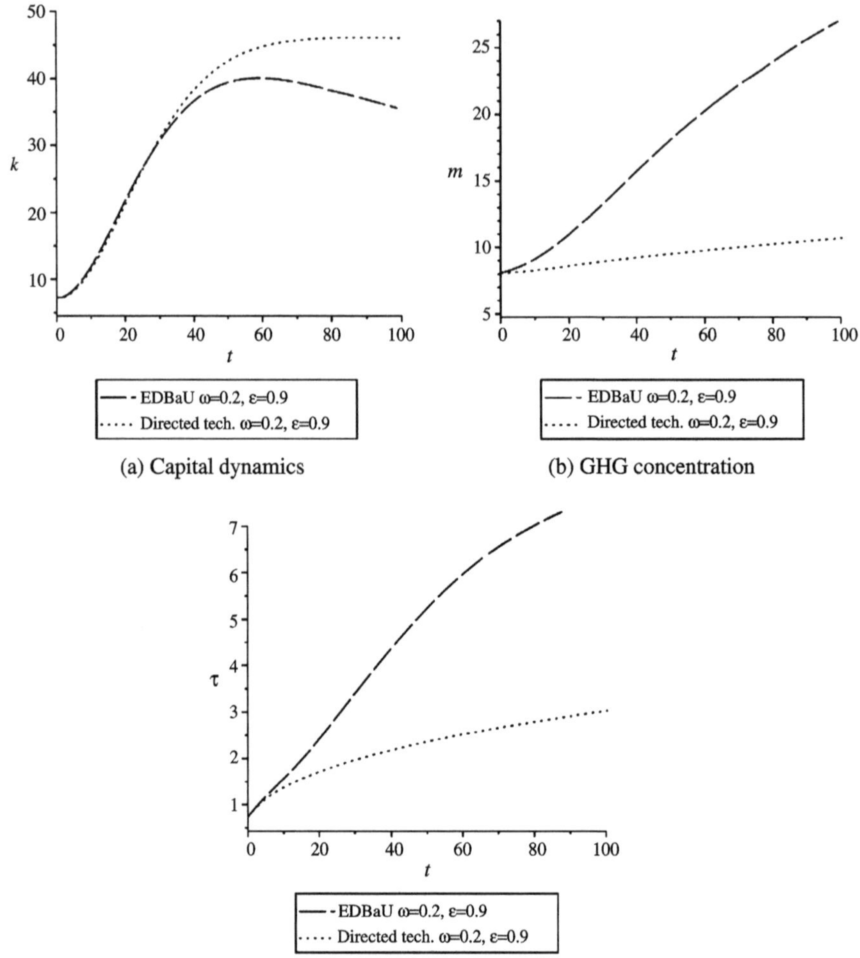

(a) Capital dynamics

(b) GHG concentration

(c) Temperature increase

Fig. 13 BaU dynamics compared to full information (directed technical change)

dynamics of the economic-environmental system. First, consider the dynamics of technology investments for BaU systems in Fig. 14.

Abatement rates are not controlled for in BaU scenarios since the household is unaware of the dynamic link of the economy and the environment and, thus, cannot influence the degree of environmental damages. Technology investments are the highest for the BaU scenario with undirected endogenous technology (EBaU) (with $\omega = 0.1$), while minimal for the clean growth BaU scenario with directed endogenous technology (EDBaU) (with $\varepsilon\omega = 0.02$). Thus, one may conclude that the level of technology investments depends not only on the total productivity parameter (ω and $\varepsilon \cdot \omega$ for EBaU and EDBaU scenarios, respectively), since this one is higher for the dirty growth scenario than for the EBaU scenario, but also on the achieved

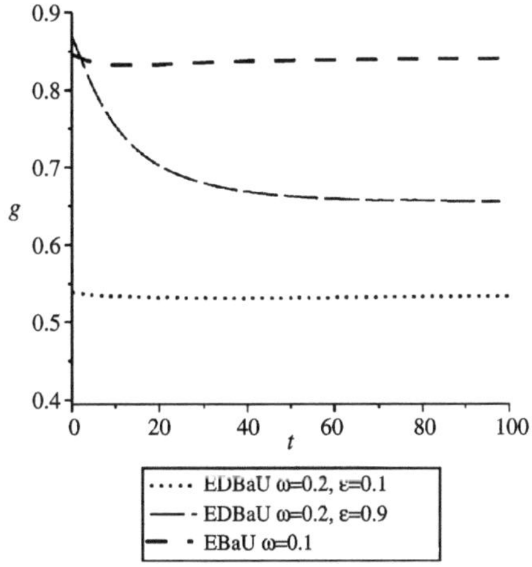

Fig. 14 Technology investments for BaU scenarios

level of technology. There is a maximal level of technology which is sufficient for the household and it does not continuously increase productivity, in the same way as for the full information scenarios above.

The economic dynamics of BaU scenarios is displayed in Fig. 15. One may see that the dynamics of capital accumulation and output for the case of the clean growth EDBaU system is very close to the one of the BBaU scenario, while the consumption dynamics in general follows the same pattern as that of output. The ordering of consumption, output and capital accumulation paths is the same as for full information systems. Further, the capital accumulation for the EDBaU scenario is lower than for the full information directed growth case. In the case of undirected technical change, capital accumulation is higher for the EBaU scenario than for full information undirected growth. Thus, the difference between the dirty growth EDBaU scenario and EBaU is smaller than the difference between these scenarios under full information. To see that, just compare capital accumulation for EBaU and dirty EDBaU in Fig. 15a and for their full information counterparts in Fig. 8a.

Technology for BaU models is endogenous only for EBaU and EDBaU scenarios and its evolution is displayed in Fig. 16. As it can be seen, productivity growth for EDBaU models may be lower or higher than for the EBaU model in the same way as for their optimal counterparts. In the clean growth scenario, only 2 % of the total technical progress are going to the increase in productivity and the latter grows less than 10 % in 100 years. With dirty growth, productivity grows almost twice which, however, is lower than for the optimal case since in the EDBaU scenario the effect of technological investments for productivity is overestimated. As a result, both capital accumulation as well as productivity growth are lower than in the full information case. On the other hand, productivity growth in the EBaU scenario is higher than for the full information strategy of undirected growth (roughly 50 % against 30 %) which leads to higher capital accumulation.

Such a difference appears because in the undirected growth case, the emissions reduction technology is exogenous and partially dissipates the effect of the productivity growth. As a result, the underestimation of the effect of technological invest-

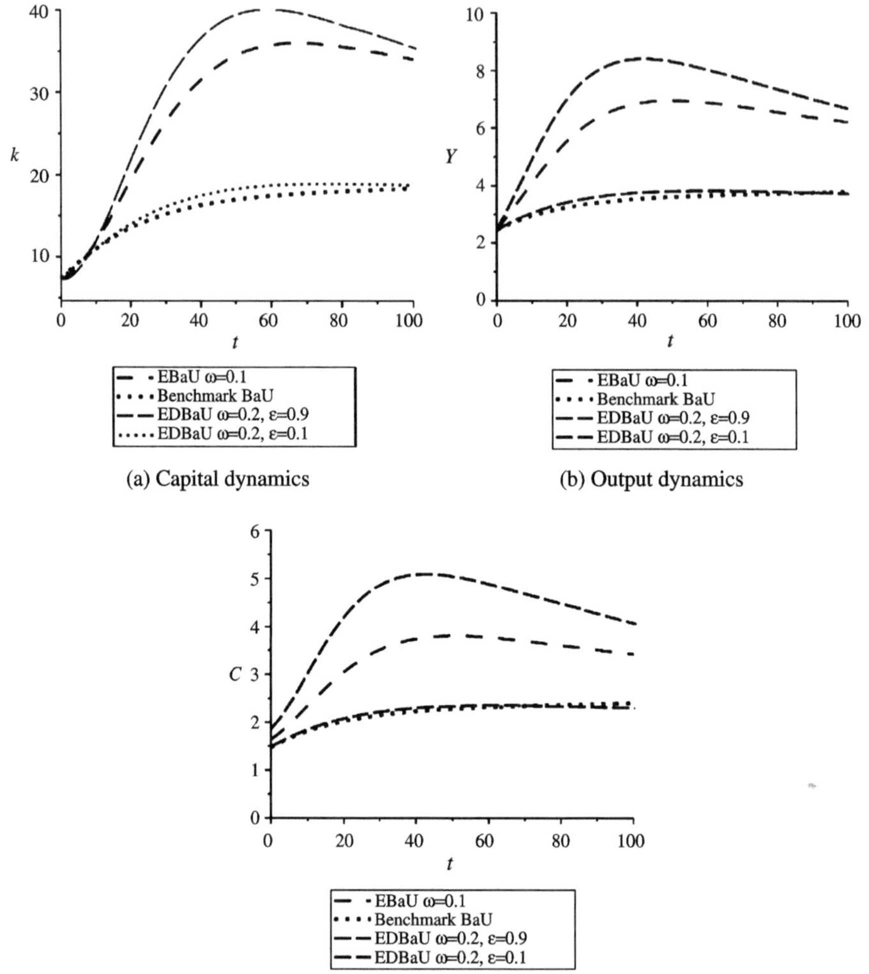

(a) Capital dynamics

(b) Output dynamics

(c) Consumption dynamics

Fig. 15 Economic dynamics for BaU scenarios

ments by the household in the EBaU scenario leads to lower capital accumulation and, consequently, to lower environmental damages compared to the EDBaU scenario. In the latter case, emissions reduction intensity is rather low, especially for the dirty growth scenario in comparison to the exogenous reduction technology.

Finally, consider the climate dynamics for BaU scenarios in Fig. 17. Again, one can see that for the case of directed technical change the outcome ranges from catastrophic, in the case of dirty growth with a temperature increase up to 8 °C, to moderate for the optimistic one of the clean growth scenario with an increase of only 3 °C. BaU model and EBaU model dynamics lie within this range in the same way as for full information models. At the same time, the increase in emissions and tempera-

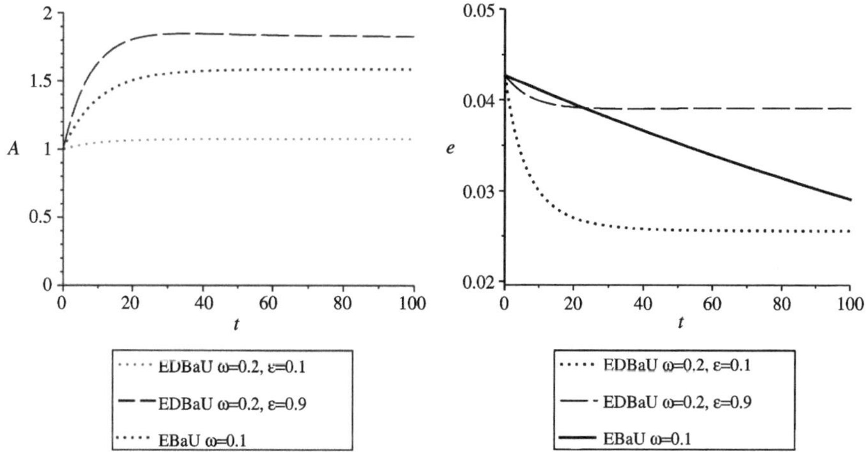

(a) Productivity growth (b) Reduction of emission intensity

Fig. 16 Technology for BaU models

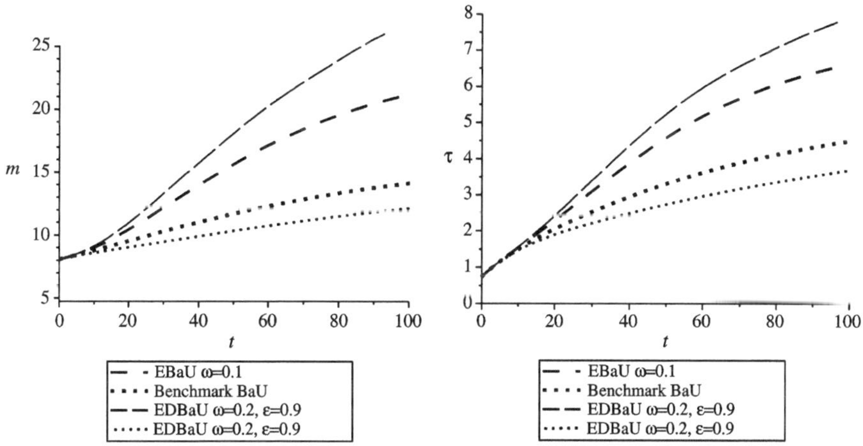

(a) GHG concentration (b) Temperature increase

Fig. 17 Climate module for BaU scenarios

ture for BaU scenarios is much higher for all three models being considered and is not stabilizing in the long-run. In the full information case, there exists at least one scenario (clean growth) with a stabilizing temperature, while this is not the case for BaU simulations.

Table 6 gives the relative welfare losses and gains, expressed in present value consumption, for the EBaU and EDBaU scenarios in comparison with the BBaU and EBaU scenarios. From this table, one can see that both the model with undirected and directed endogenous technical change yield higher social welfare than the basic

Table 6 Present value consumption changes between BaU scenarios

Scenario	$t = 50$	$t = 100$
EBaU/BBaU	48.89 %	51.64 %
EDBaU(dirty)/BBaU	98.15 %	98.23 %
EDBaU(neutral)/BBaU	41.58 %	43.56 %
EDBaU(clean)/BBaU	2.43 %	1.48 %
EDBaU(dirty)/EBaU	33.09 %	30.72 %
EDBaU(neutral)/EBaU	−4.90 %	−5.32 %
EDBaU(clean)/EBaU	−31.20 %	−33.08 %

model with exogenous technology, even if environmental damages are higher in almost all cases (except the clean growth scenario). The clean growth scenario yields almost the same welfare as the BaU scenario with exogenous technical change, but with stabilized emissions and temperature. In addition, the dirty growth scenario with directed technical change yields higher social welfare than the scenario with undirected technical change in BaU scenarios in the same way as in the case of optimal strategies. For clean and neutral technical change, one sees that directed technical change leads to a loss of social welfare compared to the scenario with undirected technical progress for BaU strategies, whereas with optimal strategies the neutral growth scenario with directed technical change exhibits higher welfare than the model with undirected technical change.

One can conclude that even in BaU scenarios there is a way to improve the environment without incurring social welfare losses. This is the case in the clean growth scenario where capital accumulation is reduced in order to invest in clean technologies that generate less GHG emissions. On the other hand, accelerated productivity growth in the scenario with directed technical change can lead to social welfare gains, however, at the expense of higher environmental losses for BaU strategies. The best performance among BaU scenarios is obtained for the undirected endogenous technical change, while models with directed technical progress display the highest diversity of possible outcomes and, thus, a high potential for policy.

4 Conclusion

In this paper we have analysed how technical change affects climate dynamics and economic variables in a basic growth model. We found that endogenous undirected technical change yields less greenhouse gas emissions and a lower temperature increase than the model with exogenous technical progress. This holds for the version of the model with full information but not for the informationally constrained version, where the optimizing representative household neglects the influence of economic and of R&D activities on the environment. Concerning welfare, a better outcome in the case of endogenous technical change can be only guaranteed for a sufficiently high efficiency of the technology in use in the social optimum.

In the case of directed technical change, where a certain fraction of the technical progress raises efficiency of production while the rest is devoted to the emission intensity reduction, results are more complicated. In the green growth scenario with a large fraction of technical progress devoted to the emission intensity reduction, the rise in temperature is clearly smaller compared to the model of undirected endogenous technical change. However, that goes at the cost of output and consumption such that the green growth scenario implies lower welfare than the model with undirected technical progress. That also holds for the informationally constrained version of the model.

The introduction of informational constraints decreases the consumption paths in all versions of the model and under all of the scenarios of technical change being considered. The higher is the degree to which the central authority may influence the technology, the more drastic are the differences between the outcome of the social optimum with full information and the laissez-faire or market economy with informational constraints of the type being considered here. However, in the class of informationally constrained economies it is possible to implement the clean growth scenario, since this one yields a higher present value consumption, i.e. higher welfare, than the scenario with exogenous technical progress. At the same time, the scenario of dirty growth is preferable under both full information and under informational constraints on the economy. If the informationally constrained household is allowed to choose between undirected and directed technical change, it will choose the dirty growth scenario. However, the simulations demonstrate that the fixed direction of technical change might be the key factor for the dirty growth alternative to be preferred by the household. The option of control over this direction of technical change may stimulate some dynamic adjustments in the R&D policy of the household after some initial period of accumulation of productivity.

Acknowledgements Authors thank A. Belyakov and V. Veliov for their advice with numeric algorithms and an anonymous referee for comments. Financial support from the Bundesministerium für Bildung und Forschung (BMBF) is gratefully acknowledged (grant 01LA1105C). This research is part of the project 'Climate Policy and the Growth Pattern of Nations (CliPoN)'.

Appendix

Optimality Conditions

The basic model with exogenous technical change contains 3 state variables and 2 controls. The (current-value) Hamiltonian associated with this problem is:

$$\begin{aligned}
&\mathscr{H}^B(k,\tau,m,\psi_k,\psi_\tau,\psi_m)\\
&\quad=\frac{[u(t)A(t)\phi(\tau(t))k(t)^\alpha]^{1-\gamma}}{1-\gamma}\\
&\qquad+\psi_k(t)\big[-\delta k(t)+\big[1-u(t)-c_1\big(a(t)\big)\big]A(t)\phi\big(\tau(t)\big)k(t)^\alpha\big]\\
&\qquad+\psi_\tau(t)\big[-\lambda\big(m(t)\big)\tau(t)+d\big(m(t)\big)\big]\\
&\qquad+\psi_m(t)\big[-\nu m(t)+\big(1-a(t)\big)e(t)A(t)\phi\big(\tau(t)\big)k(t)^\alpha+E\big(\tau(t)\big)\big].
\end{aligned}$$

Yielding first-order conditions on controls $u(t), a(t)$:

$$\begin{aligned}\frac{\partial \mathcal{H}^B}{\partial u} &= \frac{[A(t)\phi(\tau(t))k(t)^\alpha]^{1-\gamma}}{u} - \psi_k(t)A(t)\phi\big(\tau(t)\big)k(t)^\alpha = 0,\\ \frac{\partial \mathcal{H}^B}{\partial a} &= -A(t)\phi\big(\tau(t)\big)k(t)^\alpha\left(\psi_k(t)\frac{\partial c_1(a)}{\partial a} + \psi_m e(t)\right) = 0,\end{aligned} \tag{13}$$

and co-state equations:

$$\begin{aligned}\dot{\psi}_k &= \big(-\phi(\tau)eA\alpha k^{\alpha-1} + a\phi(\tau)eA\alpha k^{\alpha-1}\big)\psi_m\\ &\quad + \big(r + \phi(\tau)A\alpha c_1(a)k^{\alpha-1} + \delta - \phi(\tau)A\alpha k^{\alpha-1} + u\phi(\tau)A\alpha k^{\alpha-1}\big)\psi_k\\ &\quad - u\phi(\tau)\big(uA\phi(\tau)k^\alpha\big)^{-\gamma}A\alpha k^{\alpha-1},\\ \dot{\psi}_\tau &= \left(-e\cdot A\frac{\partial\phi(\tau)}{\partial\tau}k^\alpha + eA\frac{\partial\phi(\tau)}{\partial\tau}k^\alpha a\right)\psi_m\\ &\quad + \left(-A\frac{\partial\phi(\tau)}{\partial\tau}k^\alpha + uA\frac{\partial\phi(\tau)}{\partial\tau}k^\alpha + A\frac{\partial\phi(\tau)}{\partial\tau}k^\alpha c_1(a)\right)\psi_k\\ &\quad + \big(r + \lambda(m)\big)\psi_\tau - uAk^\alpha\big(uA\phi(\tau)k^\alpha\big)^{-\gamma}\frac{\partial\phi(\tau)}{\partial\tau},\\ \dot{\psi}_m &= (r+\nu)\psi_m - \psi_\tau\frac{\partial d(m)}{\partial m} + \psi_\tau\frac{\partial\lambda(m)}{\partial m}\tau.\end{aligned}$$

These equations are non-linear and do not separate from the state equations, which makes analytic closed-form solution difficult to achieve. Therefore, we have used numerical simulations to approximate the dynamics.

In the case of undirected endogenous technical change the (current-value) Hamiltonian is given by:

$$\begin{aligned}&\mathcal{H}^E(k,\tau,m,x,\psi_k,\psi_\tau,\psi_m,\psi_x)\\ &\quad= \frac{[u(t)(1+\omega x(t))\phi(\tau(t))k(t)^\alpha]^{1-\gamma}}{1-\gamma}\\ &\qquad + \psi_k(t)\big(-\delta k(t) + \big[1 - u(t) - c_1\big(a(t)\big) - c_1\big(g(t)\big)\big]\\ &\qquad \cdot \big(1+\omega x(t)\big)\phi\big(\tau(t)\big)k(t)^\alpha\big)\\ &\qquad + \psi_\tau(t)\big(-\lambda\big(m(t)\big)\tau(t) + d\big(m(t)\big)\big)\\ &\qquad + \psi_m(t)\big(-\nu m(t) + \big(1-a(t)\big)e(t)\big(1+\omega x(t)\big)\phi\big(\tau(t)\big)k(t)^\alpha\big)\\ &\qquad + \psi_x(t)\big(\beta_1 g(t) - \beta_2 x(t)\big).\end{aligned}$$

First-order conditions on controls $u(t), a(t), g(t)$:

$$\begin{aligned}\frac{\partial \mathcal{H}^E}{\partial u} &= \frac{[(1+\omega x(t))\phi(\tau(t))k(t)^\alpha]^{1-\gamma}}{u} - \psi_k(t)\big(1+\omega x(t)\big)\phi\big(\tau(t)\big)k(t)^\alpha = 0,\\ \frac{\partial \mathcal{H}^E}{\partial a} &= -\big(1+\omega x(t)\big)\phi\big(\tau(t)\big)k(t)^\alpha\left(\psi_k(t)\frac{\partial c_1(a)}{\partial a} + \psi_m(t)e(t)\right) = 0,\\ \frac{\partial \mathcal{H}^E}{\partial g} &= -\big(1+\omega x(t)\big)\phi\big(\tau(t)\big)k(t)^\alpha\psi_k(t)\frac{\partial c_1(g)}{\partial g} + \beta_1\psi_x(t) = 0.\end{aligned} \tag{14}$$

With directed endogenous technical change, the first-order conditions from the Hamiltonian (of the same type as for undirected change) are:

$$
\begin{aligned}
\frac{\partial \mathscr{H}^D}{\partial u} &= \frac{[(1+\varepsilon\omega x(t))\phi(\tau(t))k(t)^\alpha]^{1-\gamma}}{u} - \psi_k(t)\big(1+\omega x(t)\big)\phi\big(\tau(t)\big)k(t)^\alpha = 0,\\
\frac{\partial \mathscr{H}^D}{\partial a} &= -\big(1+\varepsilon\omega x(t)\big)\phi\big(\tau(t)\big)k(t)^\alpha\\
&\quad\cdot\left(\psi_k(t)\frac{\partial c_1(a)}{\partial a} + \psi_m(t)\frac{e_0}{1+(1-\varepsilon)\omega x(t)}\right) = 0,\\
\frac{\partial \mathscr{H}^D}{\partial g} &= -\big(1+\varepsilon\omega x(t)\big)\phi\big(\tau(t)\big)k(t)^\alpha \psi_k(t)\frac{\partial c_1(g)}{\partial g} + \beta_1\psi_x(t) = 0.
\end{aligned}
\tag{15}
$$

The NMPC Technique

The full dynamical system, which describes the evolution of economic and environmental variables, consists of 3 (in the case of the basic model) or 4 (for endogenous technology) dynamical equations. Consider, for example, the dynamical system for directed endogenous technology:

$$
\begin{aligned}
\dot{k}(t) &= -\delta k(t) + \big[1-u(t)-c_1\big(a(t)\big)-c_1\big(g(t)\big)\big]Y(t),\\
\dot{\tau}(t) &= -\lambda\tau(t) + \ln\frac{m(t)}{m_0^*},\\
\dot{m}(t) &= -\nu m(t) + \big(1-a(t)\big)e^D(t)\big(1+\varepsilon\omega x(t)\big)k^\alpha(t),\\
\dot{x}(t) &= \beta_1 g(t) - \beta_2 x(t).
\end{aligned}
\tag{16}
$$

At the same time the household solves the optimization problem that depends only on economic and technology variables for each period $[t_h,\ldots,t_h+\Theta]$, assuming environmental variables being constant on the level of the last measurement:

$$
J^{EDBaU} = \max_{u,g}\left\{\int_{t_h}^{t_h+\Theta} \mathrm{e}^{-rt}\left(\frac{[u(t)Y(t)]^{1-\gamma}}{1-\gamma}\right)dt\right\}
$$

s.t.

$$
\begin{aligned}
\dot{k}_i(t) &= -\delta k_i(t) + \big[1-u(t)-c_1\big(g(t)\big)\big]Y(t),\\
\dot{x}(t) &= \beta_1 g(t) - \beta_2 x(t),\\
Y_i(t) &= A^D(t)\phi(\tau_i)k_i(t)^\alpha,\\
A^D(t) &= 1+\varepsilon\omega x(t),\\
\tau_i &= \tau(t_i),
\end{aligned}
$$

where $Y_i(t), k_i(t)$ are different from the true evolution of capital, $k(t)$, and output, $Y(t)$, and are defined from the reduced problem without environmental constraints. This "capital" defines the optimal consumption share of the household, while the consumption is defined from the true capital and output, given by the evolution of the system (16). With such a problem the Hamiltonian of the household contains only two constraints (on "capital" and technology):

$$\begin{aligned}
&\mathscr{H}^{EDBaU}(k_i, x, \psi_{k_i}, \psi_x)\\
&\quad = \frac{[u(t)(1+\varepsilon\omega x(t))\phi(\tau_i)k_i(t)^{\alpha}]^{1-\gamma}}{1-\gamma}\\
&\qquad + \psi_{k_i}(t)\big(-\delta k_i(t) + \big[1-u(t)-c_1\big(g(t)\big)\big]\big(1+\varepsilon\omega x(t)\big)\phi(\tau_i)k_i(t)^{\alpha}\big)\\
&\qquad + \psi_x(t)\big(\beta_1 g(t) - \beta_2 x(t)\big),
\end{aligned}$$

and one may define only consumption share and technology investments, but not abatement rates from such a problem. Abatement rates are equal to zero for all BaU problems considered under this scheme.

The same type of logic of construction is applied for all three versions of the model: basic one, with undirected endogenous technical change and with the directed one. To obtain solutions in the BaU case we make use of the numerical methods, since no analytic solution may be derived for this NMPC technique. We also obtain numeric solutions for full problems of the type (1) s.t. (2)–(7).

References

Acemoglu, D., Aghion, P., Bursztyn, L., & Hemous, D. (2012). The environment and directed technical change. *The American Economic Review*, *102*(1), 131–166.

Aghion, P., & Howitt, P. (1992). A model of growth through creative destruction. *Econometrica*, *60*(2), 323–351.

Allgöwer, F., & Zheng, A. (2006). *Nonlinear model predictive control*. Berlin: Springer.

Antoci, A., Galeotti, M., & Russu, P. (2011). Poverty trap and global indeterminacy in a growth model with open-access natural resources. *Journal of Economic Theory*, *146*(2), 569–591.

Barbier, E. (1999). Endogenous growth and natural resource scarcity. *Environmental & Resource Economics*, *14*(1), 51–74.

Bovenberg, L., & Smulders, S. (1995). Environmental quality and pollution-augmenting technological change in a two-sector endogenous growth model. *Journal of Public Economics*, *57*(3), 369–391.

Bréchet, T., Camacho, C., & Veliov, V. (2011). Model predictive control, the economy, and the issue of global warming. *Annals of Operations Research*. doi:10.1007/s10479-011-0881-8.

Greiner, A. (1996). Endogenous growth cycles—Arrow's learning by doing reconsidered. *Journal of Macroeconomics*, *18*(4), 587–604.

Greiner, A. (2003). On the dynamics of an endogenous growth model with learning by doing. *Economic Theory*, *21*(1), 205–214.

Grimaud, A. (1999). Pollution permits and sustainable growth in a Schumpeterian model. *Journal of Environmental Economics and Management*, *38*(3), 249–266.

Grimaud, A., & Rougé, L. (2003). Non-renewable resources and growth with vertical innovations: optimum, equilibrium and economic policies. *Journal of Environmental Economics and Management*, *45*(2), 433–453.

Grimaud, A., & Rougé, L. (2005). Polluting non-renewable resources, innovation and growth: welfare and environmental policy. *Resource and Energy Economics*, *27*(2), 109–129.

Hassler, J., & Krusell, P. (2012). Economics and climate change: integrated assessment in a multiregion world. *Journal of the European Economic Association*, *10*(5), 974–1000.

Ligthart, J. E., & Van Der Ploeg, F. (1994). Pollution, the cost of public funds and endogenous growth. *Economics Letters*, *46*(4), 339–349.
Nordhaus, W. D. (2007). *The challenge of global warming: economic models and environmental policy*. Yale University.
Romer, P. (1990). Endogenous technological change. *Journal of Political Economy*, *98*(5), 71–102.
Smulders, S., & Gradus, R. (1996). Pollution abatement and long-term growth. *European Journal of Political Economy*, *12*(3), 505–532.

Environmental Policy in a Dynamic Model with Heterogeneous Agents and Voting

Kirill Borissov, Thierry Bréchet, and Stéphane Lambrecht

Abstract We consider a population of infinitely-lived houscholds split into two: some agents have a high discount factor (the patients), and some others have a low one (the impatients). Polluting emissions due to economic activity harm environmental quality. The governmental policy consists in proposing households to vote for a tax to maintain environmental quality. By studying the voting equilibrium at steady states we show that the equilibrium maintenance level is the one of the median voter. We also show that (i) an increase in total factor productivity may produce effects described by the Environmental Kuznets Curve, (ii) an increase in the patience of impatient households may foster environmental quality if the median voter is impatient and maintenance positive, finally (iii) a decrease in inequality among the patient households leads to an increase in environmental quality if the median voter is patient and maintenance is positive. We show that, when the median income of the median voter is lower than the mean (which is empirically founded),

K. Borissov
Saint-Petersburg Institute for Economics and Mathematics (RAS), Saint-Petersburg, Russia
e-mail: kirill@eu.spb.ru

K. Borissov
European University at St. Petersburg, 3 Gagarinskaya Str., Saint-Petersburg 191187, Russia

T. Bréchet (✉)
Louvain School of Management, Université catholique de Louvain, CORE, 34 voie du Roman Pays, Louvain-la-Neuve 1348, Belgium
e-mail: thierry.brechet@uclouvain.be

S. Lambrecht
Université de Valenciennes et du Hainaut, Cambrésis and IDP, Les Tertiales, Rue des Cents Têtes, Valenciennes 59313, France
e-mail: stephane.lambrecht@univ-valenciennes.fr

S. Lambrecht
EQUIPPE, U. Lille 1, Cité scientifique, Bât SH2, 59655 Villeneuve d'Ascq, France

S. Lambrecht
CORE, Université catholique de Louvain, 34 Voie du Roman Pays, Louvain-la-Neuve 1348, Belgium

E. Moser et al. (eds.), *Dynamic Optimization in Environmental Economics*, Dynamic Modeling and Econometrics in Economics and Finance 15,
DOI 10.1007/978-3-642-54086-8_2,

our model with heterogeneous agents predicts a lower level of environmental quality than what the representative agent model would predict, and that increasing the public debt decreases the level of environmental quality.

1 Introduction

With the growing importance of global environmental issues, such as global warming, and the emphasis put on the general question of sustainable growth and development, environmental policies and their financing have become a major subject of concern in many developing or developed countries. As a response, economic theory, and especially in macro-economics, elaborated dynamic models based on the representative agent assumption to disentangle the nexus between economic growth and pollution, or more generally environmental quality (see among many others, Gradus and Smulders 1993; Stokey 1998, or Xepapadeas 2005). Though, it is striking to notice that the public debate about environmental policies and their financing very often focus on the distributive aspects of the policies, and more precisely on the distribution of their burden among heterogenous agents. To capture that dimension, economists must get rid of the representative agent and must start considering heterogeneous agents in their macrodynamic models. There exist several ways of introducing heterogeneity, e.g. in wealth (Kempf and Rossignol 2007), in individual labor productivity (Jouvet et al. 2008), or in age with overlapping generations (John and Pecchenino 1994; Jouvet et al. 2008).

In this paper we consider heterogeneity in agents' discount factors.[1] We assume that the population is exogenously divided into two groups, one with patient households and the other with impatient households. Each individual votes in favor, or against a public policy for environmental maintenance. Maintenance is a public policy financed by a tax on households, and pollution flows from firm's activity. We define a voting equilibrium and the related general equilibrium of the economy at the steady state.

Our main results can be summarized as follows. First, if the policy choice were one-dimensional (i.e. static with one homogeneous agent) then the median-voter theorem would straightforwardly apply. Unfortunately, it cannot be applied in our dynamic multidimensional. We show that, at the steady state, a voting equilibrium coincides with the solution the one that would result from the median voter theorem. We thus provide a logically consistent definition of the median voter theorem in a dynamic setting. This establishes the applicability of the median voter theorem on steady state equilibria. This is an important theoretical result because the current literature always *assumes* that the median voter theorem can be applied after the steady state is defined, though the steady state equilibrium should itself depend on the voting outcome (see e.g. Kempf and Rossignol 2007; Corbae et al. 2009).

[1]For a general survey of the literature on models of economic growth with consumers having different discount factors, see Becker (2006).

Our theoretical contribution is to prove that a dynamic voting equilibrium coincides with the application of the median voter theorem. Furthermore, to highlight the advantages of considering heterogenous agents, we compare our results with what the representative agent framework would provide. The results differ in many respects.

Beyond the theoretical aspects, we also contribute to the literature on political economy and environmental policy. With some comparative statics, we show many novel results. We first show that, if the median voter is impatient, she consumes all her revenue, and the maintenance level is zero. But if the median voter is patient, then maintenance is positive but not uniquely determined. Then we go further and stress that there exist two channels through which discount factors shape agents' choices on maintenance, a direct one and an indirect one. In our model, the higher the agents' discount factor, the larger is her desired level of maintenance (it is the direct channel). But in the same time, the richer the agent, the stronger her desired level for maintenance. It is well-established in the literature that only agents with a high discount factor have positive savings in the long run. Those with a low discount factor save nothing. Thus, the former become wealthy in the long run and are prone to ask for high levels of environmental maintenance. In the meantime the latter become poorer and ask for lower levels of maintenance (it is the indirect equilibrium channel). Combining these two channels provides us with new results about the relationship between economic development and environmental quality through the voting process, i.e. a new rationale for the so-called Environmental Kuznets Curve (see e.g. Dasgupta et al. 2002; Prieur 2009). As far as inequality among agents is concerned, we also show that when the median voter is patient, then reducing inequality has a positive effect on environmental quality.

Actually, this discussion also relates to the broad debate about how discounting impacts the choice of environmental policies.[2] Although discounting is generally considered as a normative issue, it also has a positive content, as stressed by Dasgupta: "discount rates on consumption changes combine *values* with *facts*. Dasgupta (2008, p. 144) or by Arrow et al. (1995) who distinguishes *prescriptive* and *descriptive* positions. In environmental economics, a high discount factor leads to modest and slow environmental maintenance levels, while a low discount rate leads to immediate and strong action. The common characteristics in all this literature is to rely on the assumption that there exists a *representative agent* in the economy whose preferences are considered as given by a benevolent social planner. This agent further acts as a benevolent social planner.[3] We depart from the representative agent hypothesis by considering an economy populated with heterogeneous agents. Then, we are able to provide a microeconomic rationale to determine the implicit global discount rate in this economy. This departs from the normative discussion on what the discount rate *should* be. In our analysis we take heterogeneous agents' preferences beforehand and we scrutinize how heterogeneity shapes the policy in the

[2]Recently this debate has experienced a strong revival after the publication of the Stern Review (Stern 2006, 2008). Prominent economists have contributed to the debate, like Dasgupta (2008), Nordhaus (2008) or Weitzman (2007).

[3]Or the *social evaluator*, to take Dasgupta's words.

global economy. This is a novel contribution to the debate on discounting based on a positive approach.

Applying the median voter theorem to dynamic models requires a suitable analytical redesign of the political settings in this model. Models of such a kind are much harder to analyze than their static counterparts or than the usual intertemporal models without political ingredients. There is a growing interest in the recent literature on the analysis of the performance of majoritarian settings in dynamic frameworks, see e.g. Baron (1996), Krusell et al. (1997), Cooley and Soares (1999), Rangel (2003) and Bernheim and Slavov (2009). The stage of development of the theory is still in its infancy. In particular, there is no consensus about how to model dynamic majoritarian voting. Without going into detail in this introduction, it should be stressed that our approach to voting is different from the approaches used in the above-mentioned papers. We propose a novel definition of voting equilibrium, which is related to Kramer-Shepsle equilibrium concept (Kramer 1972; Shepsle 1979). This definition will allow us to provide new theoretical results about voting equilibrium in a dynamical setting.

These results also yield a discussion about alternative financing schemes of the environmental maintenance policy. We look at the different impacts on heterogenous households and especially on the median voter, of financing maintenance both with taxes and with issuance of public bonds. We show that, under common assumption about income distribution, an increase in the public debt leads to a lower environmental quality.

The paper is organized as follows. In Sect. 2 we present the model, define the competitive equilibria and describe steady-state equilibria for a given policy. In Sect. 3 we endogenize the voting procedure on environmental maintenance, define the intertemporal and steady state voting equilibria, and show the logical consistency between the median voter theorem and the voting equilibrium in dynamic general equilibrium. In Sect. 5 the comparison with the representative agent framework is proposed. In Sect. 4 we perform comparative statics exercises to analyze how environmental quality is impacted by an increase in total factor productivity, an increase in patience, and a decrease in inequality. The discussion about the impact of public debt on the environmental quality is carried out in Sect. 6. Section 7 is the conclusion.

2 The Model

Our objective is to define and to study the intertemporal competitive equilibria with voting on maintenance. We define voting equilibria in two steps. In this section, we do the first step as we determine the competitive equilibrium production and consumption paths for a given maintenance policy. The second step will be presented in the next section where, among these competitive equilibria, the ones for which a voting equilibrium exists will be selected. We use a discrete-time framework of infinitely-lived consumers who inelastically supply one unit of labor at each time period, with a representative polluting firm and a global public bad, a stock pollution.

2.1 Production and Pollution

Output is determined by means of a production function $aF(K_t, L_t) = Laf(k_t)$, where a is total factor productivity, K_t and L_t are capital and labor at time t, $k_t = K_t/L$ is capital intensity, $f(k) = aF(k, 1)$ is the production function in intensive form. Capital is assumed to fully depreciate each period. Output can be used for consumption, investment or environmental maintenance. For the sake of simplicity we will forget about the total factor productivity TFP a until Sect. 4 where it really becomes useful. The dynamics of capital is given by

$$K_{t+1} = F(K_t, L_t) - C_t - M_t,$$

where C_t is aggregate consumption and M_t is aggregate maintenance. The pollution flow, P_t, is proportional to output:

$$P_t = \lambda F(K_t, L) = \lambda L f(k_t), \quad \lambda > 0. \tag{1}$$

Let Q_t be an index of environmental quality defined as $\bar{Q} - S_t$, where $\bar{Q}$ is some pre-industrial (prior to global warming) quality level and S_t is the cumulative pollution stock at time t. The dynamics of Q_t is given by the following function:

$$Q_{t+1} = \Psi\left(Q_t - \kappa P_t + \frac{M_t}{\mu}\right), \tag{2}$$

where $\Psi : \mathrm{R}_+ \to \mathrm{R}_+$ is a concave increasing function, $\kappa > 0$ and $\mu > 0$ are two exogenously given coefficients (for dimensional issues). Because the "marginal environmental productivity" of maintenance ($\partial Q_{t+1}/\partial M_t = \Psi'(\cdot)/\mu$) depends negatively on μ, one may interpret $1/\mu$ as the environmental efficiency of maintenance. Let $\bar{Q}$ be a unique positive solution to the following equation: $\Psi(Q) = Q$, i.e. the stationary value of environmental quality with no pollution and no maintenance. For example, the following specifications of $\Psi(X)$ can be used: $\Psi(X) = X^{\nu}\bar{Q}^{1-\nu}$, with $0 < \nu < 1$, or $\Psi(X) = \nu X + (1-\nu)\bar{Q}$, with $0 < \nu < 1$. Let $\Phi(\cdot) = \Psi^{-1}(\cdot)$. We can rewrite (2) as:

$$\mu\Phi(Q_{t+1}) = \mu(Q_t - P_t) + M_t.$$

It should be noticed that $\mu\Phi'(Q)$ is the marginal cost of quality improvement.

The representative firm maximizes its profit π_t under the constraint of the technology $F(K_t, L_t)$ by choosing K_t and L_t and by taking real wage (w_t) and interest rates (r_t) as given. The firm's problem reads:

$$\max_{K_t, L_t} \pi_t = F(K_t, L_t) - (1 + r_t)K_t - w_t L_t. \tag{3}$$

The first-order conditions are $F'_K(K_t, L_t) = 1 + r_t$ and $F'_L(K_t, L_t) = w_t$, or in intensive terms: $f'(k_t) = 1 + r_t$ and $f(k_t) - f'(k_t)k_t = w_t$.

2.2 The Consumers

Population consists of L consumers, with L an integer and odd. Each consumer is endowed with one unit of labor force. The objective function of consumer i is:

$$\sum_{t=0}^{\infty} \beta_i^t \left[u(c_t) + v(Q_t)\right],$$

where c_t is consumption at time t and β_i is a discount factor. Let us assume that $u(c)$ and $v(Q)$ satisfy the following conditions:

$$\begin{aligned} u'(c) > 0, \qquad u''(c) < 0, \qquad u'(0) = \infty, \\ v'(Q) > 0, \qquad v''(Q) < 0, \qquad v'(0) = \infty. \end{aligned}$$

Each consumer i is either patient ($\beta_i = \beta^h$) or impatient ($\beta_i = \beta^l$), with $0 < \beta^l < \beta^h < 1$. The set of patient consumers (with a discount factor equal to β^h) is H_h, and the set of impatient consumers (those with β^l) is H_l. Consumers pay a tax $m_t = M_t/L$ to finance the public provision of environmental maintenance. The budget constraint of a consumer at time t is thus:

$$\begin{aligned} c_t + s_t + m_t \le w_t + (1 + r_t)s_{t-1}, \\ c_t \ge 0, \; s_t \ge 0, \end{aligned} \tag{4}$$

where w_t is the wage rate, r_t is the interest rate, and s_t are her savings at time t.[4]

Consumers' utility depends on variables on which she has full control (c_t and s_t) but also on maintenance m_t, which will be determined by voting (yet to be introduced). At this stage, the result of voting is taken as given by the agents. Hence, we need to solve the consumer's program (to choose the optimal values for c_t and s_t, $\forall t$), considering m_t as given.

Suppose that at time τ consumer i is given her predetermined level of savings $\hat{s}^i_{\tau-1}$, the predetermined level of environmental quality $\hat{Q}_\tau$, the stream of pollution $(P_t)_{t=\tau}^{\infty}$ and some maintenance policy which is represented by a sequence $\mathbf{m} = (m_t)_{t=0}^{\infty}$ of non-negative numbers. The problem $\mathcal{P}_1(\tau)$ of this consumer reads as follows:

$$\max_{(c_t)_{t=\tau}^{+\infty}, (s_t)_{t=\tau}^{+\infty}} \sum_{t=\tau}^{\infty} \beta_i^t \left[u(c_t) + v(Q_t)\right]$$

subject to:

$$\begin{aligned} &\mu\Phi(Q_{t+1}) = \mu(Q_t - \kappa P_t) + Lm_t, \quad t = \tau, \tau+1, \\ &c_t + s_t + m_t \le w_t + (1 + r_t)s_{t-1}, \quad t = \tau, \tau+1, \\ &s_{\tau-1} = \hat{s}^i_{\tau-1}, \; Q_\tau = \hat{Q}_\tau, \\ &c_t \ge 0, s_t \ge 0, \; t = \tau, \tau+1. \end{aligned}$$

[4]Consumers are forbidden to borrow against their future labor income. Hence, their savings must be non-negative.

It must be noticed that, since $\mathbf{m} = (m_t)_{t=0}^{\infty}$ is given, the sequence $(Q_t)_{t=\tau}^{+\infty}$ is in fact predetermined by $\hat{Q}_\tau$ and $\mathbf{m}$. Hence, the utility consumer i derives from environmental quality, $\sum_{t=\tau}^{\infty} \beta_i^t v(Q_t)$, does not depend on her choice. In what follows, when we will define voting equilibrium, it will be key to keep in mind that, if $(s_{t-1}^{i*}, c_t^{i*}, Q_t^*)_{t=0}^{\infty}$ is a solution to problem $\mathcal{P}_1(0)$, then $(s_{t-1}^{i*}, c_t^{i*}, Q_t^*)_{t=\tau}^{\infty}$ is also a solution to problem $\mathcal{P}_1(\tau)$ at $\hat{s}_{t-1}^i = s_{t-1}^{i*}$ and $\hat{Q}_t = Q_t^*$.

2.3 Competitive Equilibrium Paths and Steady-State Equilibria

Let at time 0 the environmental policy be represented by some given sequence $\mathbf{m} = (m_t)_{t=0}^{\infty}$ of non-negative numbers. Let an initial state $\{(\hat{s}_{-1}^i)_{i=1}^L, \hat{k}_0, \hat{Q}_0\}$ also be given. Here, $\hat{s}_{-1}^i \geq 0$ stand for the initial savings of consumers $i = 1, \ldots, L$, $\hat{k}_0 > 0$ is the initial per capita stock of capital ($\sum_{i=1}^L \hat{s}_{-1}^i = L\hat{k}_0$), and $\hat{Q}_0 > 0$ is the initial value of environmental quality.

Definition 1 (Competitive equilibrium path) Given $\mathbf{m}$, the sequence $\mathcal{E}^{\mathbf{m}} = \{k_t^*, 1 + r_t^*, w_t^*, (s_{t-1}^{i*}, c_t^{i*})_{i=1}^L, P_t^*, Q_t^*\}_{t=0}^{\infty}$ is called a *competitive equilibrium path* starting from $\{(\hat{s}_{-1}^i)_{i=1}^L, \hat{k}_0, \hat{Q}_0\}$ if:

1. capital and labor markets clear at the following prices: $1 + r_t = 1 + r_t^* = f'(k_t^*)$, $w_t = w_t^* = f(k_t^*) - f'(k_t^*)k_t^*, t = 0, 1, \ldots;$
2. for each household $i = 1, \ldots, L$ the sequence $(s_{t-1}^{i*}, c_t^{i*}, Q_t^*)_{t=0}^{\infty}$ is a solution to problem $\mathcal{P}_1(0)$ at $1 + r_t = 1 + r_t^*, w_t = w_t^*, t = 0, 1, \ldots;$
3. $\sum_{i=1}^L s_{t-1}^{i*} = Lk_t^*, t = 0, 1, \ldots;$
4. $P_t^* = \lambda L f(k_t^*), t = 0, 1, \ldots;$
5. $\mu \Phi(Q_{t+1}^*) = \mu(Q_t^* - \kappa P_t^*) + Lm_t, t = 0, 1, \ldots.$

Notice that, at each time t, maintenance m_t is given and smaller than the wage rate w_t. We will not discuss the existence of equilibrium paths. Our main emphasis is on steady-state equilibria.

Definition 2 (Competitive steady state equilibrium) Let an $m \geq 0$ be given and let $\mathbf{m} = (m_t)_{t=0}^{\infty}$, with $m_t = m, t = 0, 1, \ldots$. We call a tuple $E^m = \{k^*, 1 + r^*, w^*, (s^{i*}, c^{i*})_{i=1}^L, P^*, Q^*\}$ *a competitive steady-state equilibrium* if the sequence $\{k_t^*, 1 + r_t^*, w_t^*, (s_{t-1}^{i*}, c_t^{i*})_{i=1}^L, P_t^*, Q_t^*\}_{t=0}^{\infty}$ given for all $t = 0, 1, \ldots$ by

$$k_t^* = k^*, \qquad 1 + r_t^* = 1 + r^*, \qquad w_t^* = w^*, \tag{5}$$

$$\left(s_{t-1}^{i*}, c_t^{i*}\right)_{i=1}^L = \left(s^{i*}, c^{i*}\right)_{i=1}^L, \tag{6}$$

$$P_t^* = P^*, \qquad Q_t^* = Q^*, \tag{7}$$

is an equilibrium path starting from the initial state $\{(\hat{s}_{-1}^i)_{i=1}^L, \hat{k}_0, \hat{Q}_0\} = \{(s^{i*})_{i=1}^L, k^*, Q^*\}$.

Provided the above definition, the following proposition describes the structure of steady-state equilibria. It is an adaptation of the well-established results of Becker (1980, 2006) to our model.

Proposition 1 (Structure of steady state equilibrium) *A tuple* $E^m = \{k^*, 1+r^*, w^*, (s^{i*}, c^{i*})_{i=1}^L, P^*, Q^*\}$ *satisfying* $m < w^*$ *is a steady-state equilibrium if and only if*

$$\beta^h = \frac{1}{1+r^*}, \qquad 1+r^* = f'(k^*), \qquad w^* = f(k^*) - f'(k^*)k^*, \tag{8}$$

$$P^* = \lambda L f(k^*), \tag{9}$$

$$\mu\Phi(Q^*) = \mu(Q^* - \kappa P^*) + Lm, \tag{10}$$

$$s^{i*} = 0, \quad i \in H_l, \tag{11}$$

$$s^{i*} \geq 0, \quad i \in H_h, \tag{12}$$

$$\sum_{i=1}^{L} s^{i*} = \sum_{i \in H_h} s^{i*} = Lk^*, \tag{13}$$

$$c^* + s^* + m = w^* + (1+r^*)s^*. \tag{14}$$

Proof See Sect. 8.1. □

In this proposition, (8) shows that the steady-state capital intensity, interest rate, and the wage rate are determined by the discount factor of the patient consumer. Equations (11)–(12) tell us that impatient consumers have zero savings. It means that all the capital is owned by the patient consumers. As a consequence, in a steady-state equilibrium all impatient consumers have the same income and savings levels. In contrast, the distribution of savings among the patient consumers is indeterminate in a steady state. As shown by (13), only aggregate savings is determined.

3 Voting Equilibria

There is no reason for heterogenous agents to agree on the desired level of environmental maintenance. One way to solve this problem is to choose maintenance by majority voting. If policy choices were one-dimensional, one could refer to the median voter theorem. But this theorem does not apply here. In this section we propose a definition of voting equilibrium and we prove that the level of maintenance that comes out at the voting steady-state equilibrium is the one that would have been chosen by the median voter.

Let $\mathbf{m} = (m_t)_{t=0}^{\infty}$ be an environmental policy. The optimal value of problem $\mathcal{P}_1(\tau)$ for consumer i is a function of $\hat{s}_{\tau-1}$, $\hat{Q}_\tau$ and $\mathbf{m}$. We will denote this optimal value by $V_{i,\tau}(\hat{s}_{\tau-1}, \hat{Q}_\tau, \mathbf{m})$.

Definition 3 (Preferred change in environmental maintenance) Suppose that the environmental policy is represented by some sequence $\bar{\mathbf{m}} = (\bar{m}_t)_{t=0}^{\infty}$ of non-negative numbers and that at $\mathbf{m} = \bar{\mathbf{m}}$ the function $V_{i,\tau}(\hat{s}_{\tau-1}, \hat{Q}_\tau, \mathbf{m})$ is differentiable in m_τ. We say that consumer i is *in favor of increasing* m_τ if $\frac{\partial V_{i,\tau}(\hat{s}_{\tau-1}, \hat{Q}_\tau, \mathbf{m})}{\partial m_\tau} > 0$ and is *in favor of decreasing* m_τ if $\frac{\partial V_{i,\tau}(\hat{s}_{\tau-1}, \hat{Q}_\tau, \mathbf{m})}{\partial m_\tau} < 0$ and $\bar{m}_\tau > 0$.

Let us assume that, for an equilibrium path

$$\mathcal{E}^{\bar{\mathbf{m}}} = \left\{k_t^*, 1 + r_t^*, w_t^*, \left(s_{t-1}^{i*}, c_t^{i*}\right)_{i=1}^{L}, P_t^*, Q_t^*\right\}_{t=0}^{\infty}$$

the function $V_{i,\tau}(s_{\tau-1}^*, Q_\tau^*, \mathbf{m})$ is differentiable in m_τ at $\mathbf{m} = \bar{\mathbf{m}}$. We denote by $N_\tau^+(\mathcal{E}^{\bar{\mathbf{m}}})$ the number of consumers who are in favor of *increasing* $\bar{m}_\tau$, and by $N_\tau^-(\mathcal{E}^{\bar{\mathbf{m}}})$ the number of consumers who are in favor of *decreasing* $\bar{m}_\tau$. We are now equipped to define intertemporal voting equilibria.

Definition 4 (Intertemporal voting equilibrium) Let $\mathbf{m}^* = (m_t^*)_{t=0}^{\infty}$ be a maintenance policy and $\mathcal{E}^{\mathbf{m}^*}$ be an equilibrium path constructed at this policy. We call the couple $(\mathbf{m}^*, \mathcal{E}^{\mathbf{m}^*})$ *an intertemporal voting equilibrium path* if at $\mathbf{m} = \mathbf{m}^*$ $\forall \tau = 0, 1, \ldots$ the function $V_{i,\tau}(s_{\tau-1}^*, Q_\tau^*, \mathbf{m})$ is differentiable in m_τ, and

$$N_\tau^+\left(\mathcal{E}^{\mathbf{m}^*}\right) < \frac{L}{2}, \qquad N_\tau^-\left(\mathcal{E}^{\mathbf{m}^*}\right) < \frac{L}{2}, \quad \forall \tau = 0, 1, \ldots.$$

According to this definition, an intertemporal voting equilibrium is reached if, at each time period there exists neither a majority of agents who are in favor of increasing maintenance, nor a majority of agents who are in favor of decreasing maintenance. And because we take the number of agents as odd by assumption, then there exists an agent for whom the maintenance level is optimal in equilibrium.

This definition is in line with the usual way of defining intertemporal equilibrium, as articulated by Hicks (1936) and, more recently, by Grandmont (1983). In our model, any intertemporal voting equilibrium can be seen as a sequence of temporary voting equilibria in which agents perfectly anticipate the whole future, including voting results. Indeed, let $(\mathbf{m}^*, \mathcal{E}^{\mathbf{m}^*})$ be an intertemporal voting equilibrium. Suppose that at time τ the agents are asked to vote on m_τ and that they correctly anticipate m_t^* for all $t = \tau + 1, \tau + 2, \ldots$. Then it is clear that all the conditions for the median voter theorem hold in this one-dimensional voting, and that the preferred value of m_τ for the median voter coincides with m_τ^*. A key implication is that intertemporal voting equilibria are time consistent.

In the rest of the paper we shall focus on steady state voting equilibria. Consider a couple (m^*, E^{m^*}), where $m^* \geq 0$ and $E^{m^*} = \{k^*, 1 + r^*, w^*, (s^{i*}, c^{i*})_{i=1}^{L}, P^*, Q^*\}$ is a steady-state equilibrium constructed at the maintenance policy $\mathbf{m}^* = (m_0^*, m_1^*, \ldots)$, $m_t^* = m^*, t = 0, 1, \ldots$. Let $\mathcal{E}^{m^*}$ be an equilibrium path corresponding to E^{m^*}.

Definition 5 (Steady state voting equilibrium) We call the couple (m^*, E^{m^*}) *a steady state voting equilibrium* if the couple $(m^*, \mathcal{E}^{m^*})$ is an intertemporal voting equilibrium path.

To answer the question of whether a couple (m^*, E^{m^*}) is a steady state voting equilibrium or not, it is sufficient to know which consumers are in favor of an increase in $m_0^* = m^*$ at time 0, and which ones are in favor of a decrease. We know that, for each consumer i, the sequence $(\tilde{s}_{t-1}^i, \tilde{c}_t^i, \tilde{Q}_t)_{t=0}^{\infty}$ given by

$$\tilde{s}_{t-1}^i = s^{i*}, \qquad \tilde{c}_t^i = c^{i*}, \qquad \tilde{Q}_t = Q^*, \tag{15}$$

is a solution to

$$\max_{(c_t)_{t=0}^{+\infty}, (Q_t)_{t=0}^{+\infty}} \sum_{t=0}^{\infty} \beta_i^t \big[u(c_t) + v(Q_t)\big], \tag{16}$$

$$\mu\Phi(Q_{t+1}) = \mu\big(Q_t - \kappa P^*\big) + L m_t^*, \quad t = 0, 1, \ldots, \tag{17}$$

$$c_t + s_t + m_t^* \le w^* + \big(1 + r^*\big)s_{t-1}, \quad t = 0, 1, \ldots, \tag{18}$$

$$s_{-1}^i = \hat{s}_{-1}^i, \; Q_0 = \hat{Q}_0, \tag{19}$$

$$c_t \ge 0, \; s_t \ge 0, \; Q_t \ge 0, \; t = 0, 1, \ldots \tag{20}$$

at $\hat{s}_{-1}^i = s^{i*}$, $\hat{Q}_0 = Q^*$.

Lemma 1 (Differentiability of value function w.r.t. maintenance and sign of derivative) *Let for some i the sequence $(\tilde{s}_{t-1}^i, \tilde{c}_t^i, \tilde{Q}_t)_{t=0}^{\infty}$ given by* (15) *be a solution to problem* (16)–(20) *at given $m_t^* = m^* \in [0, w^*)$, $t = 0, 1, \ldots$ and at $\hat{s}_{-1}^i = s^{i*}$, $\hat{Q}_0 = Q^*$. Then $V_{i,0}(s^{i*}, Q^*, \mathbf{m}^*)$ is differentiable in m_0^* and*

$$\frac{\partial V_{i,0}(s^{i*}, Q^*, \mathbf{m}^*)}{\partial m_0^*} \gtreqless 0 \Leftrightarrow \beta_i L v'\big(Q^*\big) \gtreqless \mu u'\big(c^{i*}\big)\big(\Phi'\big(Q^*\big) - \beta_i\big). \tag{21}$$

Proof See Sect. 8.2. □

The interpretation of Lemma 1 runs as follows. Consider the first inequality in (21) at a given maintenance m_0^* and suppose that the left-hand side is higher than the right-hand side. In this case, out of a marginal change in maintenance, the induced marginal utility of environmental quality (i.e. the LHS of (21)), is larger than the induced marginal utility of consumption (i.e. the RHS of (21)). This is likely to happen when the given maintenance level m_0^* is low. In such a case the consumer will be in favor of an increase in maintenance. In the opposite case, the given maintenance m_0^* is likely to be large so that the induced marginal utility of consumption is higher than the induced marginal utility of quality and the consumer is in favor of decreasing maintenance.

To check whether a couple (m^*, E^{m^*}) is a voting steady-state equilibrium or not, let us consider the following problem $\mathcal{P}_2$ in which household i is free to determine her preferred level of maintenance m_t:

$$\max_{(c_t)_{t=0}^{+\infty},(s_t)_{t=0}^{+\infty},(m_t)_{t=0}^{+\infty},(Q_t)_{t=0}^{+\infty}} \sum_{t=0}^{\infty} \beta_i^t \left[u(c_t) + v(Q_t)\right],$$

subject to:

$$\begin{aligned}
&\mu\Phi(Q_{t+1}) \leq \mu\left(Q_t - \kappa P^*\right) + Lm_t, \quad t = 0, 1,\\
&c_t + s_t + m_t \leq w^* + \left(1 + r^*\right)s_{t-1}, \quad t = 0, 1,\\
&s_{-1} = \hat{s}_{-1}, Q_0 = \hat{Q}_0,\\
&c_t \geq 0, s_t \geq 0, m_t \geq 0, Q_t \geq 0, t = 0, 1.
\end{aligned}$$

Let $(\tilde{s}, \tilde{c}, \tilde{m}, \tilde{Q}) \in \mathrm{R}_+^4$ *determine a steady-state solution* to this problem if the sequence $(\tilde{s}_{t-1}, \tilde{c}_t, \tilde{m}_t, \tilde{Q}_t)_{t=0}^{\infty}$ given by

$$\tilde{s}_{t-1} = \tilde{s}, \qquad \tilde{c}_t = \tilde{c}, \qquad \tilde{m}_t = \tilde{m}, \qquad \tilde{Q}_t = \tilde{Q} \tag{22}$$

is its solution at $\hat{s}_{-1} = \tilde{s}$ and $\hat{Q}_0 = \tilde{Q}$.

Prior to formulating the following lemma, remind that $\beta^h(1+r^*) = 1$ and hence that $\beta_i(1+r^*) < 1, \forall i \in H_l$, and $\beta_i(1+r^*) = 1, \forall i \in H_h$.

Lemma 2 (Characterization of steady state solution to $\mathcal{P}_2$) *The tuple* $(\tilde{s}, \tilde{c}, \tilde{m}, \tilde{Q}) \in \mathrm{R}_+^4$ *determines a steady-state solution to* $\mathcal{P}_2$ *if and only if*

$$\beta_i\left(1 + r^*\right) < 1 \Rightarrow \tilde{s} = 0, \tag{23}$$

$$\beta_i L v'(\tilde{Q}) \leq \mu u'(\tilde{c})\left(\Phi'(\tilde{Q}) - \beta_i\right) \; (= \textit{if } \tilde{m} > 0), \tag{24}$$

$$\tilde{c} = w^* + r^*\tilde{s} - \tilde{m}, \tag{25}$$

$$\mu\left(\Phi(\tilde{Q}) - \tilde{Q} + \kappa P^*\right) = L\tilde{m}. \tag{26}$$

Proof See Sect. 8.3. □

For the sake of simplicity we can get rid of $\tilde{m}$ by noticing that $\tilde{m} > 0 \Leftrightarrow \tilde{c} < w^* + r^*\tilde{s}$. We can thus rewrite conditions (24)–(25) as follows:

$$\tilde{c} = \left(w^* + r^*\tilde{s} - \frac{\mu\kappa}{L}P^*\right) + \frac{\mu}{L}\left(\tilde{Q} - \Phi(\tilde{Q})\right), \tag{27}$$

$$\tilde{c} \leq w^* + r^*\tilde{s}, \tag{28}$$

$$\beta_i L v'(\tilde{Q}) \leq \mu u'(\tilde{c})\left(\Phi'(\tilde{Q}) - \beta_i\right) \left(= \text{if } \tilde{c} < w^* + r^*\tilde{s}\right). \tag{29}$$

Equation $\beta_i L v'(Q) = \mu u'(c)(\Phi'(Q) - \beta_i)$ shows that c in increasing in Q. As for equation $c = (w^* + r^*\tilde{s} - \frac{\mu\kappa}{L}P^*) + \frac{\mu}{L}(Q - \Phi(Q))$, it shows that, for any given $\tilde{s}$, the relationship between c and Q is either always decreasing, or first increasing ($\Phi'(Q) < 1$) and then decreasing ($\Phi'(Q) > 1$). Suppose we are given $m^* \in [0, w^*)$, where w^* is given by (8). Let $E^{m^*} = \{k^*, 1 + r^*, w^*, (s^{i*}, c^{i*})_{i=1}^{L}, P^*, Q^*\}$ be a

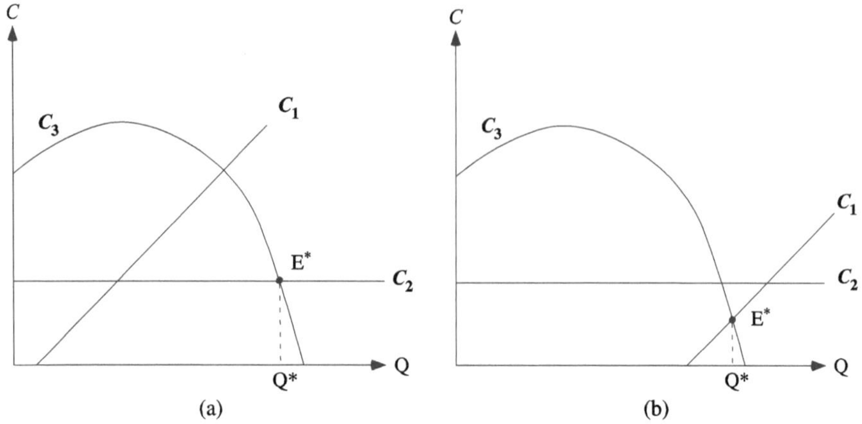

Fig. 1 *Left*: zero-maintenance equilibrium (Regime 1)—*right*: positive maintenance equilibrium (Regime 2)

steady-state equilibrium constructed at the maintenance policy $\mathbf{m}^* = (m^*, m^*, \ldots)$. Put all households in ascending order of their savings and take the median one, i_m.[5] Lemmas 1 and 2 lead to the following theorem.

Theorem 1 (Steady state voting equilibrium and median voter) *The couple* (m^*, E^{m^*}) *is a steady-state voting equilibrium if and only if for* $i = i_m$, *the tuple* $(s^{i*}, c^{i*}, m^*, Q^*)$ *is a steady-state solution to problem* $\mathcal{P}_2$.

The economic interpretation of the theorem runs as follows. We know from Proposition 1 that the *per capita* stock of capital in a steady-state voting equilibrium (k^*) and the wage and interest rates (w^* and r^*) are determined by the discount factor of patient households (β^h). In the meantime, Theorem 1 says that maintenance and environmental quality do depend on the median discount factor and the median savings. Combining the two yields the following outcome. If the median value of the discount factor is β^l, and so $\beta_{i_m} = \beta^l$, then maintenance and environmental quality are determined by β^l and w^*, because the savings of agent i_m are unambiguously zero. But if the median value of the discount factor is β^h and hence $\beta_{i_m} = \beta^h$, then they are determined by β^h, w^*, r^* and the savings of agent i_m, s^{i_m*}, which can be either zero or positive.

It follows from this theorem that there exist two possible equilibria depending on whether $c^{i_m*} = w^* + r^* s^{i_m*}$ ($\Leftrightarrow m^* = 0$) or $c^{i_m*} < w^* + r^* s^{i_m*}$ ($\Leftrightarrow m^* > 0$). They are illustrated by the left and right panel of Fig. 1, in which we take s^{i_m*} as given. On these graphs the three curves C_1, C_2 and C_3 are defined as follows:

---

[5] More formally, we can put the set of households in an order such that, if $\beta_i < \beta_j$ and if $s^{i*} < s^{j*}$, then i precedes j. Such an order exists because the impatient consumers do not save in a steady-state equilibrium. Now take the household median in the sense of the introduced order, i_m.

Curve $\mathcal{C}_1$: $\beta_{i_m} Lv'(Q) = \mu u'(c)(\Phi'(Q) - \beta_{i_m})$
Curve $\mathcal{C}_2$: $c = w^* + r^* s^{i_m *}$
Curve $\mathcal{C}_3$: $c = (w^* + r^* s^{i_m *} - \frac{\mu\kappa}{L} P^*) + \frac{\mu}{L}(Q - \Phi(Q))$.

Let us describe these two regimes.

Regime 1 Zero-maintenance. The equilibrium point $(Q^*, c^{i_m *})$ is at the intersection of the $\mathcal{C}_2$ curve and the $\mathcal{C}_3$ curve (see Fig. 1a) and, as far as curve $\mathcal{C}_1$ is concerned, we have $\beta_{i_m} Lv'(Q^*) < \mu u'(c^{i_m *})(\Phi'(Q^*) - \beta_{i_m})$.

Regime 2 Positive-maintenance. The equilibrium point $(Q^*, c^{i_m *})$ is at the intersection of the $\mathcal{C}_1$ curve with the $\mathcal{C}_3$ curve (see Fig. 1b) and, as far as curve $\mathcal{C}_2$ is concerned, we have $c^{i_m *} < w^* + r^* s^{i_m *}$.

In combination with the above-mentioned two regimes, two cases must be distinguished:

Case 1 Impatient median voter. $\beta_{i_m} = \beta^l$ and savings of the median voter are uniquely determined, $s^{i_m *} = 0$.

Case 2 Patient median voter. $\beta_{i_m} = \beta^h$ and the savings of the median voter, $s^{i_m *}$, are not unique: they can take any value in the interval $[0, \frac{2}{L+1} Lk^*]$.

In both cases, the regime of equilibrium maintenance can be nil or positive. In Case 1, the equilibrium level of maintenance and environmental quality are uniquely determined. As for Case 2, if there exists at least one equilibrium with positive maintenance, the equilibrium levels of maintenance and environmental quality are indeterminate since there is a continuum of these.

The very existence of steady-state voting equilibria deserves some comments. It is clear that if the majority of consumers is impatient, then steady-state voting equilibria exist for any distribution of savings among patient consumers because, in this case, the solution to problem $\mathcal{P}_2$ for the median voter, $(\tilde{s}, \tilde{c}, \tilde{m}, \tilde{Q})$, unconditionally satisfies $\tilde{m} < w^*$. But if the majority of consumers is patient, then steady-state voting equilibria do exist for any distribution of savings among patient consumers, where the savings of the median voter are nil or small enough.

4 Some Comparative Statics on Preferences, Income Inequality, and Technology

In our model, comparative statics requires some caution. As stressed above, if the median voter is patient, in a steady state the savings of the median voter are not determined uniquely. They can take any value in the interval $[0, \frac{2}{L+1} Lk^*]$. Therefore, when making a comparative statics exercise, a change in a parameter will have an indeterminate effect on the savings of the median voter. To circumvent this problem we will assume that k^* is does not change with $s^{i_m *}$. We will also assume that the ratio $s^{i_m *}/(\sum_{i=1}^{L} s^{i *})$, and hence the ratio $s^{i_m *}/k^*$, remain unchanged when a parameter changes (notice that since $k^* = (\sum_{i=1}^{L} s^{i *})/L$ shows the mean savings, $s^{i_m *}/k^*$ shows the proportion between the median and mean savings).

4.1 An Increase in $s^{i_m *}$ Other Things Equal

Let us first carry out a comparative statics exercise relevant in Case 2, when the median voter is patient and his savings can be positive. Assume that k^* is kept unchanged and $s^{i_m *}$ increases. The increase in $s^{i_m *}$ translates a change in the distribution of savings among the patient consumers only. Consequently, it leads to a different income distribution (an increase in the median income relative to the mean).

- Under *Zero-Maintenance Equilibrium* (Regime 1), a small increase in $s^{i_m *}$, other things equal, will shift $\mathcal{C}_2$ and $\mathcal{C}_3$ upwards by the same magnitude. Hence, consumption of the median voter $c^{i_m *}$ will increase, but environmental quality Q^* will remain unchanged. A larger increase in $s^{i_m *}$ may shift the economy to Regime 2.
- Under *Positive-Maintenance Equilibrium* (Regime 2), a small increase in $s^{i_m *}$, other things equal, will shift $\mathcal{C}_3$ upwards, while letting $\mathcal{C}_1$ untouched. Hence the environmental quality Q^* will increase.

Following the literature in political economy and income inequality (see e.g. Meltzer and Richard 1981), the "more equal" the income distribution, the higher the median income relative to the mean (this is only reasonable in the case where the median income does not exceed the mean, which is considered as a typical situation). In our model, it means that, in developed economies where maintenance is positive, lower inequality has a positive effect on environmental quality. Conversely, in developing economies where there is no maintenance, inequality itself does not effect the environmental quality.

4.2 An Increase in Total Factor Productivity

In the following sub-sections of this section, we consider a Cobb-Douglas production function, $f(k) = k^\alpha, 0 < \alpha < 1$. We also assume that the fraction of output necessary to remove all emissions is lower than the labor share in output, $1 - \alpha > \mu\lambda$. Geometrically, the latter assumption implies that the curve $\mathcal{C}_3$ shifts upwards after an increase in capital intensity.

Let us first consider an increase in the total factor productivity by introducing a scale parameter a in the production function, $aF(K, L) = Laf(k)$. The impact of an increase in total factor productivity will depend on the regime the economy follows in equilibrium.

Regime 1. Zero-Maintenance Equilibrium In this regime, a small increase in a leads to an increase in k^*, w^* and $w^* + r^* s^{i_m *}$. It will also increase the output $Lf(k^*)$ and pollution P^* levels, but it cannot make maintenance positive. As a consequence, the environmental quality Q^* decreases. Graphically (see Fig. 2,

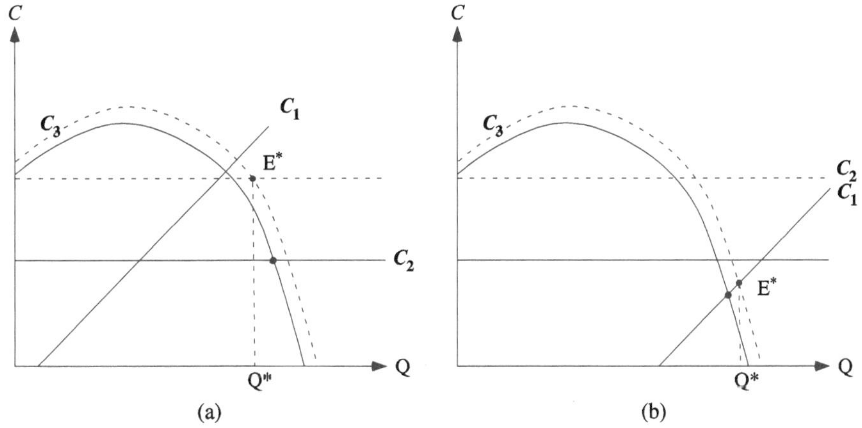

Fig. 2 An increase in total factor productivity in Regime 1 (*left*) and Regime 2 (*right*)

left panel), $\mathcal{C}_2$ shifts upwards due to the increase in $w^* + r^* s^{i_m *}$. $\mathcal{C}_3$ also shifts upwards, but to a smaller extent because both w^* and P^* increase. If the increase in a becomes large enough, then the economy switches to Regime 2, namely the Positive-maintenance Equilibrium.

Regime 2. Positive-Maintenance Equilibrium In this regime, an *increase* in a will shift $\mathcal{C}_3$ upwards, as shown in Fig. 2, right panel, and hence to an *increase* in Q^*.

To sum up, if the economy starts in Regime 1, then an increase in a from 0 to $+\infty$ first leads to a decrease in environmental quality Q^*, and then to an increase, as shown in Fig. 2. If one considers that developing countries most likely correspond to Regime 1 and rich countries to Regime 2, then this conclusion means that technological progress first goes with a decrease in environmental quality, and after some stage of development with an increase in environmental quality. This result provides a new rationale for an Environmental Kuznets Curve (see e.g. Stokey 1998; Dasgupta et al. 2002 or Prieur 2009) with heterogeneous agents and voting.

4.3 Patient Agents Become More Patient: An Increase in β^h

We first consider an increase in β^h, which means that patient agents become even more patient. The effects on the environmental quality will depend on which regime the economy experiences.

Under Zero-Maintenance Equilibrium (Regime 1), a small increase in β^h leads to an increase in capital intensity k^*, wage rate w^*, output $Lf(k^*)$ and pollution P^*, but it cannot make maintenance positive. Hence Q^* decreases as β^h increases under Regime 1. Graphically (see Fig. 2, left panel), $\mathcal{C}_2$ shifts upwards due to the increase in w^*; $\mathcal{C}_3$ also shifts upwards, but to a smaller extent (w^* will increase but P^* will

Fig. 3 Impatient agents become less impatient

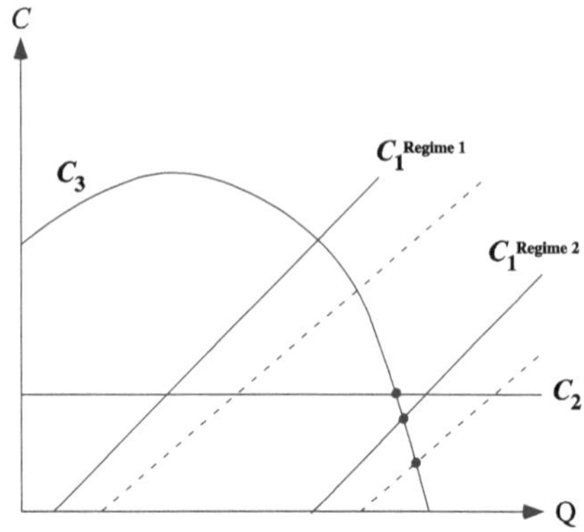

also increase). If the median voter is patient (Case 2) then, C_1 shifts to the right. As a consequence the economy may well switch to the Positive-maintenance regime (Regime 2).

Under Positive-Maintenance Equilibrium (Regime 2, see Fig. 2b) an increase in β^h will lead to an upward shift of C_3 and, in Case 2, to a shift of C_1 to the right. Hence Q^* will increase.

4.4 *Impatient Agents Become Less Impatient: An Increase in β^l*

Let us now consider the case where impatient agents become less impatient, i.e. an increase in β^l. The effect on Q^* will depend on whether the median consumer is impatient or patient, what we referred to as Case 1 and Case 2, respectively. In the case where the median voter is impatient (Case 1, $\beta_{i_m} = \beta^l$), then the two regimes have to be considered.

- Under Zero-Maintenance Equilibrium (Regime 1), a small increase in β^l does not change k^*, w^*, $Lf(k^*)$ or P^*. It neither changes Q^*. This case results in a shift of C_1 to the right, as illustrated in Fig. 3. Still, if the increase in β^l becomes large enough, then the economy switches to Regime 2.
- Under Positive-Maintenance Equilibrium (Regime 2), a small increase in β^l does not change k^*, w^*, $Lf(k^*)$ or P^*, but it does increase Q^*, as illustrated in Fig. 3.

In the case where the median voter is patient (Case 2, $\beta_{i_m} = \beta^h$), then it is clear that changing β^l has no effect on Q^*.

5 How Agents' Heterogeneity Shapes Environmental Maintenance

In this section we compare the level of environmental quality in voting steady-state equilibria with that in steady-state equilibria of a similar economy, but populated with symmetric agents. We constrain our consideration to the case where the equilibrium values of the capital stock, and hence the output level, are the same in both models. The question we address is the following: how does agents' heterogeneity in discount factor and wealth shape environmental maintenance when agents are asked to vote?[6]

How to reduce our heterogeneous agent model to an homogeneous agent one is not straightforward. We have one alternative: either to assume that all agents have the same high discount factor (β^h), or the same low discount factor (β^l). The important issue is that these options do not yield the same outcome. Actually, the latter option cannot be considered because it will be associated to different levels of macroeconomic variables in equilibrium. What we are interesting in is the analysis of the effect of heterogeneity on pollution, so we need to keep all other macroeconomic variables unchanged. Thus, the only solution is to assume that all agents have the same—high discount factor.

Moreover, the steady-state equilibria in the homogenous population model coincides with the symmetric voting steady-state equilibria in this particular case, i.e. equilibria where the level of savings for all consumers is the same and, consequently, the level of consumption is also the same. To be more precise, voting is irrelevant in symmetric equilibria, because it is unanimous.

Let $\{k_S^*, 1 + r_S^*, w_S^*, (s_S^{i*}, c_S^{i*})_{i=1}^L, P_S^*, Q_S^*\}$ be a symmetric steady-state voting equilibrium with $\beta_i = \beta^h,\ i = 1, \ldots, L$, and $\{k^*, 1 + r^*, w^*, (s^{i*}, c^{i*})_{i=1}^L, P^*, Q^*\}$ be a steady-state voting equilibrium with arbitrary discount factors. By "symmetric" we mean that $s_S^{1*} = \cdots = s_S^{L*}$. It must be noticed that

$$k_S^* = k^*, \qquad r_S^* = r^*, \qquad w_S^* = w^*,$$

and that, by assumption,

$$s_S^{i*} = k^*, \quad i = 1, \ldots, L.$$

The last equation says that the savings of agents in the symmetric steady-state voting equilibrium with $\beta_i = \beta^h, i = 1, \ldots, L$, are equal to the mean of the savings in the heterogeneous-agent economy. We assume that the discount factor shared by all consumers in the former model is β^h, and not β^l. This is simply because, otherwise, the equilibrium level of capital stock and output would differ between the two models.

Let

$$\begin{aligned} m^* &= w^* + r^* s^{i_m *} - c^{i_m *}, \\ m_S^* &= w_S^* + r_S^* k_S^* - c_S^* (= w^* + r^* k^* - c_S^*), \end{aligned}$$

[6]Note that this is different from the question raised by Caselli and Ventura (2000): under which condition does a model with heterogenous agents "admits" a representative agent model, namely a model with homogenous agents displaying the same aggregate and average behavior. Indeed, in our case, by assumption, we assume capital intensity to be the same in both models. On the other hand we do not fix the maintenance level, nor do we look at the representative agent version of the model which would yield the same maintenance level.

where $c_S^* = c_S^{1*}(= \cdots = c_S^{L*})$. The following proposition can be proved with the same argument as in the previous section.

Proposition 2 (Homogeneous *vs.* heterogeneous population equilibria) (1) *Suppose that* $\beta_{i_m} = \beta^l$ *and hence* $s^{i_m*} = 0$ *in the heterogenous-agent economy. Then*:

1. *if* $m_S^* = 0$, *then* $m^* = 0$ *and* $Q^* = Q_S^*$;
2. *if* $m_S^* > 0$, *then* $m^* < m_S^*$ *and* $Q^* < Q_S^*$.

(2) *Suppose that* $\beta_{i_m} = \beta^h$ *in the heterogenous-agent economy. Then*:

1. *if* $s^{i_m*} \leq s_S^{i*} = k^*$, *then*:
   (a) *if* $m_S^* = 0$, $m^* = 0$ *and* $Q^* = Q_S^*$;
   (b) *if* $m_S^* > 0$, $m^* < m_S^*$ *and* $Q^* < Q_S^*$;
2. *if* $s^{i_m*} \geq s_S^{i*} = k^*$, *then*:
   (a) *if* $m^{i*} = 0$, $m_S^* = 0$ *and* $Q^* = Q_S^*$;
   (b) *if* $m^* > 0$, *then* $m^* > m_S^*$ *and* $Q^* > Q_S^*$.

In the last section we have seen why developed countries are likely to vote in favor of environmental maintenance, and developing countries against. According to the proposition, the predictions of the two models coincide for developing countries: no maintenance in steady-state equilibria irrespective of whether the median voter is patient or impatient. However, the models' predictions differ for developed countries. If the majority of agents is impatient, then the maintenance equilibrium levels and environmental quality are lower than those predicted by the homogenous-agent model. If the majority of agents is patient, then the median saving or income must be compared with the mean ones. If the median savings are lower than the mean (or, equivalently, if the median income is lower then the mean income) then the maintenance equilibrium levels and environmental quality are lower in the heterogenous-agent model that in the homogenous-agent model. Otherwise, the opposite outcome holds.[7]

To sum up, our comparison shows that, because of heterogeneity, in most cases in the real world the observed levels of environmental maintenance and quality will be lower than what the homogenous-agent model would predict. Taking heterogeneity into account is key, even when interested in macroeconomic outcomes.

6 Debt-Financed Versus Tax-Financed Maintenance Policy

Until now we have assumed that the maintenance policy was financed by a pay-as-you-go tax (τ_t), a so-called *tax-financed* scheme. An alternative scheme could

[7]The case where the median income is lower than the mean is usually considered as typical on empirical grounds.

be a *debt-finance* one. It would consist for the government to issue public bonds to finance the environmental maintenance. Comparing the two schemes makes sense in our setting because heterogenous households are likely to be hit differently by the taxes needed to finance the public debt and by the interest earned on public bonds. The median voter may well depend on the funding scheme. On the government side, reducing the funding of environmental maintenance by taxes could improve its political acceptability. Finally, because introducing a public debt in our infinitely-lived agent model does not impact on the equilibrium steady state capital intensity, we can focus on its impact on environmental quality.

In this section we assume that the government is able to raise the voted tax τ_t or to issue one-period public bonds d_{t+1} to finance environmental maintenance m_t. The repayment of interests and principal of public bonds will appear in its budget constraint. It is assumed that public bonds and physical capital are perfect substitute and bear the same market interest rate r_t.

Let $d_t \geq 0$ be the per capita public debt and $\tau_t \geq 0$ be the lump-sum tax at time t. The government budget constraint reads:

$$\tau_t + d_{t+1} = m_t + (1 + r_t)d_t,$$

and the consumer's budget constraint (see (4)) becomes:

$$c_t + s_t + \tau_t \leq w_t + (1 + r_t)s_{t-1}, \quad s_t \geq 0.$$

One can easily update the definitions of competitive equilibrium path consequently. The only thing that deserves attention is that condition 3 (equilibrium in the capital market) now turns to:

$$\sum_{i=1}^{L} s_{t-1}^{i*} = L\big(k_t^* + d_t\big), \quad t = 0, 1, \ldots.$$

Suppose that the public debt is constant over time, $d_t = d, t = 0, 1, \ldots$. Then we can define the competitive steady-state equilibrium. Consider such an equilibrium, (m^*, E^{m^*}), where $E^{m^*} = \{k^*, 1 + r^*, w^*, (s^{i*}, c^{i*})_{i=1}^L, P^*, Q^*\}$. As in Mankiw (2000), government debt does not affect the steady-state capital stock and national income. So, as in the case with no governmental debt, we have:

$$\beta^h = \frac{1}{1 + r^*}, \quad 1 + r^* = f'\big(k^*\big), \qquad w^* = f\big(k^*\big) - f'\big(k^*\big)k^*.$$

In the meantime, the governmental debt does influence the distribution of income among households. The higher the debt, the higher the level of taxation to pay for the interest payments on that debt. The tax falls on both patient and impatient consumers, but the interest payments entirely go to the patient consumers, just because only patient consumers have positive savings in a steady-state equilibrium. In the steady-state equilibrium the budget constraint of the government becomes

$$\tau_t + d = m^* + \big(1 + r^*\big)d.$$

Hence, $\tau_t = \tau(d), t = 0, 1, \ldots,$ where

$$\tau(d) = m^* + r^*d.$$

Therefore, the budget constraint of a consumer in the steady-state equilibrium is as follows:

$$c_t + s_t \leq w^* - \tau(d) + \left(1 + r^*\right)s_{t-1}, \quad s_t \geq 0.$$

If the median voter is impatient, then we have $s^{i_m *} = 0$ in a steady-state equilibrium, and hence

$$c^{i_m *} + m^* = w^* - r^* d.$$

As a result, an increase in d is equivalent for the median voter to a decrease in the post-tax wage rate. It follows that (in the case where maintenance is positive) an increase in public debt unambiguously leads to a decrease in maintenance and environmental quality in the voting steady-state equilibrium if the majority of agents is impatient.

If the median voter is patient, in a steady state the savings of the median voter are not determined uniquely. Hence a change in d will have an indeterminate effect on the savings of the median voter. Let us assume that the ratio $s^{i_m *}/(\sum_{i=1}^{L} s^{i*})$ does not change. Since, in equilibrium, $(\sum_{i=1}^{L} s^{i*})/L = k^* + d$, it implies that the ratio $\gamma = s^{i_m *}/(k^* + d)$ (which is the proportion between the median and the mean savings) remains unchanged. Under this assumption, the key parameter becomes γ because we now have:

$$c^{i_m *} + m^* = w^* + r^* s^{i_m *} - r^* d = w^* + r^*\left(\gamma k^* + (\gamma - 1)d\right).$$

It is clear from the previous equation that an increase in d leads to a decrease in $c^{i_m *} + m^*$, if $\gamma < 1$, and to an increase in $c^{i_m *} + m^*$, if $\gamma > 1$.

Thus, in the case where maintenance is positive, $m^* > 0$, if the median savings and income are lower than the mean ($\gamma < 1$), then an increase in public debt leads to a decrease in maintenance and environmental quality. But if the median savings and income are higher than the mean ($\gamma > 1$), then an increase in public debt increases environmental maintenance quality. As noticed above, the case where the median savings and income are lower than the mean is usually considered as common.

7 Conclusion

In this paper we assumed that the population is exogenously divided into two groups: one with patient households and the other with impatient households. The environmental maintenance is voted by the households. We introduce the notion of voting equilibrium, look for steady state voting equilibria and find that the median voter theorem applies to them. If the majority of households is impatient, then the equilibrium level of maintenance and environmental quality is determined uniquely, but if the majority of households is patient, there can exist a continuum of these. We also fulfill comparative statics analysis and we show that (i) an increase in total factor productivity may produce a so-called Environmental Kuznets Curve, (ii) an increase in the patience of impatient households may improve the environmental quality if the median voter is impatient and maintenance positive, (iii) in the case where the median voter is patient and maintenance positive, and in the case where the median income is lower than the mean one (which is empirically grounded), then a shrink in inequality can lead to an increase in the environmental quality.

We also compare our model with a representative agent model, which is defined as a particular case of our model where all consumers are patient and savings are distributed evenly across them. We show that, in the case of impatient median voter, the level of environmental quality predicted by the heterogeneous-agent model is lower than the one predicted by the representative agent model. The same holds true if the median voter is patient but the median income lower that the mean, which is the common case.

Finally, some policy implications of our model are discussed. In this purpose we introduce public debt as an alternative source of financing environmental maintenance. We show that, if the median income is lower than the mean, then an increase in public debt leads to a lower environmental quality in the long run.

Acknowledgements We thank an anonymous referee for his careful reading and his suggestions. We also thank Raouf Boucekkine for discussions on a preliminary version. Part of this research was conducted during several short visits of K. Borissov at the Université Lille 1 Sciences et Technologies, laboratoire EQUIPPE—Universités de Lille, at CORE, Université catholique de Louvain and at IDP, Université de Valenciennes et du Hainaut-Cambrésis. He is grateful to the Russian Foundation for Basic Research (grant No. 11-06-00183) and Exxon Mobil for financial support. Preliminary versions of the paper circulated at the EAERE annual conference, at the CORE—EQUIPPE Workshop on "Political Economy and the Environment", Louvain-la-Neuve, and at the PET 2010 conference, Istanbul.

Appendix

8.1 Proof of Proposition 1

It is sufficient to notice that since in a steady-state equilibrium we have $\mu\Phi(Q^*) = \mu(Q^* - \kappa P^*) + L\bar{m}$ and for each i, the sequence $(\tilde{s}^i_{t-1}, \tilde{c}^i_t)_{t=0}^{\infty}$ given by $\tilde{s}^i_{t-1} = s^{i*}$, $\tilde{c}^i_t = c^{i*}$ is a solution to

$$\max \sum_{t=0}^{\infty} \beta_i^t u(c_t), \quad c_t + s_t \leq (w^* - \bar{m}) + (1 + r^*)s_{t-1},\ s^i_{-1} = s^{i*},$$
$$c_t \geq 0,\ s_t \geq 0$$

and to refer to Becker (1980, 2006).

8.2 Proof of Lemma 1

We have:

$$\frac{\partial V_{i,0}(s^{i*}, Q^*, \mathbf{m}^*)}{\partial m_0^*} = \frac{\partial \Lambda_{i,0}(Q^*, \mathbf{m}^*)}{\partial m_0^*} + \frac{\partial \Gamma_{i,t}(s^{i*}, \mathbf{m}^*)}{\partial m_0^*},$$

where the functions $\Lambda_{i,0}$ and $\Gamma_{i,0}$ are defined as follows:

$$\Lambda_{i,0}\big(Q_0, \mathbf{m}^*\big) = \max_{(Q_t)_{t=1}^{\infty}} \Bigg\{\sum_{t=0}^{\infty} \beta_i^t v(Q_t) \mid \mu\Phi(Q_{t+1}) \le \mu\big(Q_t - \kappa P^*\big) + Lm_t^*,$$
$$Q_{t+1} \ge 0,\ t = 0, 1, \dots\Bigg\},$$

$$\Gamma_{i,0}\big(s_{-1}, \mathbf{m}^*\big) = \max_{(c_t)_{t=0}^{\infty}, (s_t)_{t=0}^{\infty}} \Bigg\{\sum_{t=0}^{\infty} \beta_i^t u(c_t) \mid c_t + s_t + m_t^* \le w^* + \big(1 + r^*\big)s_{t-1},$$
$$c_t \ge 0,\ s_t \ge 0,\ t = 0, 1, \dots\Bigg\}.$$

It is not difficult to check that

$$\frac{\partial \Lambda_{i,0}(Q^*, \mathbf{m}_t^*)}{\partial m_t^*} = \beta_i \frac{Lv'(Q^*)}{\mu(\Phi'(Q^*) - \beta_i)},$$
$$\frac{\partial \Gamma_{i,0}(s^{i*}, \mathbf{m}_t^*)}{\partial m_t^*} = -u'\big(c^*\big).$$

Therefore,

$$\frac{\partial V_{i,0}(s^{i*}, Q^*, \mathbf{m}^*)}{\partial m_t^*} = \beta_i \frac{Lv'(Q^*)}{\mu(\Phi'(Q^*) - \beta_i)} - u'\big(c^*\big),$$

which implies (21).

8.3 Proof of Lemma 2

Using a traditional argument (see e.g. McKenzie 1986) we can prove that a sequence $(\tilde{s}_{t-1}, \tilde{c}_t, \tilde{m}_t, \tilde{Q}_t)_{t=0}^{\infty}$ given by (22) is a steady-state solution to problem $\mathcal{P}_2$ if and only if there exist q and p such that for $p_t = \beta_i p_{t-1} = \dots = \beta_i^t p$ and $q_{t+1} = \beta_i q_t = \dots = \beta_i^{t+1} q$ the following relationships hold:

$$\beta_i^t u'(\tilde{c}_t) = p_t,$$
$$\beta_i^t v'(\tilde{Q}_t) + q_{t+1}\mu - q_t\mu\Phi'(\tilde{Q}_t) = 0,$$
$$\big(1 + r^*\big)p_t \le p_{t-1}\ (= \text{if } \tilde{s}_{t-1} > 0),$$
$$q_{t+1}L - p_t \ge 0\ (= \text{if } \tilde{m}_t > 0),$$
$$q_{t+1}\tilde{Q}_t + p_t\tilde{s}_{t-1} \to_{t\to\infty} 0,$$

or, equivalently,

$$
\begin{aligned}
&u'(\tilde{c}) = p,\\
&v'(\tilde{Q}) = \mu q\big(\Phi'(\tilde{Q}) - \beta_i\big),\\
&\beta_i \leq \frac{1}{1+r^*} \;(= \text{if } \tilde{s} > 0),\\
&\beta_i Lq - p \geq 0 \;(= \text{if } \tilde{m} > 0).
\end{aligned}
$$

The existence of such q and p is equivalent to conditions (23)–(24).

References

Arrow, K., et al. (1995). Intertemporal equity, discounting, and economic efficiency. In J. P. Bruce & E. F. Haites (Eds.), *Climate change 1995: economic and social dimensions of climate change, contribution to the working group III to the second assessment report of IPCC*. Cambridge: Cambridge University Press.

Baron, D. P. (1996). A dynamic theory of collective goods programs. *American Political Science Review*, *90*, 316–330.

Becker, R. A. (1980). On the long-run steady-state in a simple dynamic model of equilibrium with heterogeneous households. *The Quarterly Journal of Economics*, *94*, 375–383.

Becker, R. A. (2006). Equilibrium dynamics with many agents. In R.-A. Dana, C. Le Van, T. Mitra, & K. Nishimura (Eds.), *Handbook of optimal growth 1. Discrete time*. Berlin: Springer.

Bernheim, B., & Slavov, S. N. (2009). A solution concept for majority rule in dynamic settings. *Review of Economic Studies*, *76*, 33–62.

Caselli, F., & Ventura, J. (2000). A representative consumer theory of distribution. *The American Economic Review*, *90*(4), 909–926.

Cooley, T. F., & Soares, J. (1999). A positive theory of social security based on reputation. *Journal of Political Economy*, *107*, 135–160.

Corbae, D., d'Erasmo, P., & Kuruscu, B. (2009). Politico-economic consequences of rising wage inequality. *Journal of Monetary Economics*, *56*(1), 43–61.

Dasgupta, P. (2008). Discounting climate change. *Journal of Risk and Uncertainty*, *37*, 141–169.

Dasgupta, S., Laplante, B., Wang, H., & Wheeler, D. (2002). Confronting the environmental Kuznets curve. *The Journal of Economic Perspectives*, *16*(1), 147–168.

Gradus, R., & Smulders, S. (1993). The trade-off between environmental care and long-term growth: pollution in three prototype growth models. *Journal of Economics*, *58*, 25–51.

Grandmont, J.-M. (1983). *Money and value: a reconsideration of classical and neoclassical monetary theories. Econometric society of monographs in pure theory: Vol. 5*. Cambridge: Cambridge University Press.

Hicks, J. (1936). *Value and capital*. Oxford: Clarendon press.

John, A., & Pecchenino, R. (1994). An overlapping generations model of growth and the environment. *Econometrics Journal*, *104*, 1393–1410.

Jouvet, P.-A., Michel, Ph., & Pestieau, P. (2008). Public and private environmental spending. A political economy approach. *Environmental Economics & Policy Studies*, *9*, 168–177.

Kempf, H., & Rossignol, S. (2007). Is inequality harmful for the environment in a growing economy? *Economics and Politics*, *19*(1), 53–71.

Kramer, G. H. (1972). Sophisticated voting over multidimensional choice spaces. *The Journal of Mathematical Sociology*, *2*, 165–180.

Krusell, P., Quadrini, V., & Rios-Rull, J. V. (1997). Politico-economic equilibrium and economic growth. *Journal of Economic Dynamics & Control*, *21*, 243–272.

Mankiw, N. G. (2000). The savers-spenders theory of fiscal policy. *The American Economic Review*, *90*, 120–125.
McKenzie, L. (1986). Optimal economic growth, turnpike theorems and comparative dynamics. In K. J. Arrow & M. D. Intriligator (Eds.), *Handbook of mathematical economics* (Vol. 3, pp. 1281–1355).
Meltzer, A. H., & Richard, S. F. (1981). A rational theory of the size of government. *Journal of Political Economy*, *89*, 914–927.
Nordhaus, W. D. (2008). A review of the Stern review on the economics of climate change. *Journal of Economic Literature*, *45*, 686–705.
Prieur, F. (2009). The environmental Kuznets curve in a world of irreversibility. *Economic Theory*, *40*, 57–90.
Rangel, A. (2003). Forward and backward intergenerational goods: why is social security good for the environment? *The American Economic Review*, *93*, 813–834.
Shepsle, K. A. (1979). Institutional arrangements and equilibrium in multidimensional voting models. *American Journal of Political Science*, *23*, 27–59.
Stern, N. (2006). *The Stern review of the economics of climate change*. Cambridge: Cambridge University Press.
Stern, N. (2008). The economics of climate change. *The American Economic Review*, *98*(2), 1–37.
Stokey, N. L. (1998). Are there limits to growth? *International Economic Review*, *39*, 1–31.
Weitzman, M. (2007). A review of the Stern review on the economics of climate change. *Journal of Economic Literature*, *45*(3), 703–724.
Xepapadeas, A. (2005). Economic growth and the environment. In K.-G. Mäler & J. R. Vincent (Eds.), *Handbook of environmental economics* (Vol. 3, pp. 1219–1271).

Optimal Environmental Policy in the Presence of Multiple Equilibria and Reversible Hysteresis

Ben J. Heijdra and Pim Heijnen

Abstract We study optimal environmental policy in an economy-ecology model featuring multiple stable steady-state ecological equilibria. The policy instruments consist of public abatement and a tax on the polluting production input, which we assume to be the stock of capital. The isocline for the stock of pollution features two stable branches, a low-pollution (good) and a high-pollution (bad) one. Assuming that the ecology is initially located on the bad branch of the isocline, the ecological equilibrium is reversibly hysteretic and a suitably designed environmental policy can be used to steer the environment from the bad to the good equilibrium. We study both first-best and second-best social optima. We show that, compared to capital taxation, abatement constitutes a very cheap instrument of environmental policy.

1 Introduction

In this chapter we study optimal environmental policy using a dynamic model featuring interactions between the ecological system and the macro-economy. In line with the recent environmental literature, we assume that the ecological process is nonlinear such that (i) ecosystems do not respond smoothly to gradual changes in dirt flows and abrupt "catastrophic shifts" may be possible in the vicinity of threshold points, (ii) there may be multiple stable equilibria, and (iii) irreversibility and hysteresis are both possible (Scheffer et al. 2001). The nonlinear ecological dynamics described by Scheffer (1998) and employed by us now carries the name Shallow-Lake Dynamics (SLD hereafter).[1]

[1]There is an emerging literature on the SLD approach as it is used in economics—see Heijdra and Heijnen (2013) for an extensive list of references.

B.J. Heijdra (✉) · P. Heijnen
Faculty of Economics & Business, University of Groningen, P.O. Box 800, 9700 Groningen, The Netherlands
e-mail: b.j.heijdra@rug.nl

P. Heijnen
e-mail: p.heijnen@rug.nl

E. Moser et al. (eds.), *Dynamic Optimization in Environmental Economics*, Dynamic Modeling and Econometrics in Economics and Finance 15,
DOI 10.1007/978-3-642-54086-8_3,

To describe the macroeconomic system we use a standard Ramsey-Cass-Koopmans model of a closed economy. Households practice intertemporal consumption smoothing and accumulate capital that is rented out to perfectly competitive firms. Following Bovenberg and Heijdra (1998, 2002), we assume that the capital stock is the polluting production factor. Households enjoy living in a clean environment but act as free riders and thus fail to internalize the external effects caused by their capital accumulation decisions.

We assume that the initial steady-state confronting the policy maker has the following features. First, there is no pre-existing policy regarding the environment, i.e. public abatement activities are absent and there is no externality-correcting tax on capital in place. Second, the flow of dirt is such that there exist two stable ecological steady-state equilibria. Third, the ecological system has settled down at the "bad" equilibrium featuring a high stock of pollution. In this setting the policy maker is in principle able to engineer substantial welfare gains by choosing the appropriate mix of capital taxation and abatement activities.

The chapter is structured as follows. Section 2 presents the model, consisting of an ecological system featuring SLD and an economic system. Section 3 studies the first-best social optimum. The optimal environmental policy can be decentralized with the aid of time-varying abatement and capital taxation. Section 4 studies optimal environmental policy in a second-best setting. In particular we consider the repercussions of two types of constraints on the policy maker's choices, namely the unavailability of instruments and the insufficient flexibility of a given instrument. Finally, in Sect. 5 we offer a brief summary of the main results, whilst the Appendix presents some computational details.

2 The Model[2]

We model the environment as a renewable resource stock, the quality of which depends negatively on the *flow* of dirt, $D(t)$, that is generated in the production process:

$$D(t) \equiv \kappa K(t) - \gamma G(t), \quad \kappa > 0, \gamma > 0, \tag{1}$$

where $K(t)$ is the private capital stock (see below), and $G(t)$ represents abatement activities by the government. Capital is the polluting factor of production, just as in Bovenberg and Heijdra (1998, 2002). By definition the flow of dirt must be non-negative ($D(t) \geq 0$). Denoting the *stock* of pollution at time t by $P(t)$, we write the general form of the emission equation as:

$$\dot{P}(t) = -\Phi\big(P(t)\big) + D(t), \tag{2}$$

[2] Apart from the introduction of a tax on capital, the model used here is identical to the one discussed in more detail in Heijdra and Heijnen (2013).

Fig. 1 Ecological dynamics

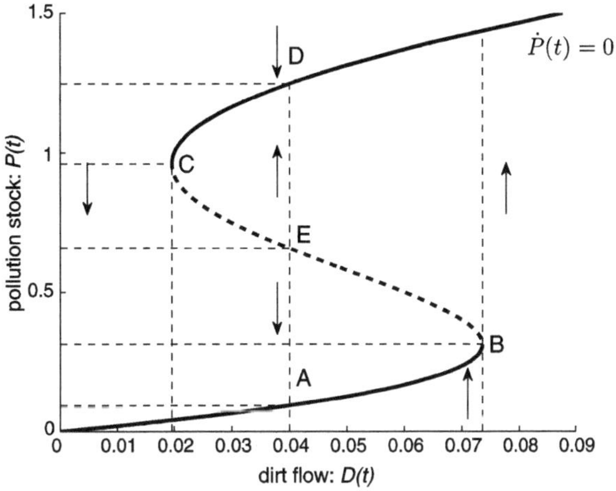

where $\dot{P}(t) \equiv dP(t)/dt$ and $\Phi(P(t))$ is a nonlinear function whose definition and properties are stated in the following Lemma.

Lemma 1 *Let $\Phi(x)$ for $x \geq 0$ be given by:*

$$\Phi(x) \equiv \pi x - \frac{x^2}{x^2+1}, \qquad \frac{1}{2} < \pi < \frac{3\sqrt{3}}{8}.$$

The first- and second derivatives of $\Phi(x)$ are given by:

$$\Phi'(x) \equiv \pi - \frac{2x}{[x^2+1]^2}, \qquad \Phi''(x) \equiv \frac{2[3x^2-1]}{[x^2+1]^3}.$$

The following properties can be established: (i) $\Phi(x) = 0$ *for* $x = 0$ *and* $\Phi(x) > 0$ *for* $x > 0$; (ii) $\Phi(x)$ *attains a local maximum at* x_1 *such that* $\Phi'(x_1) = 0$ *and* $\Phi''(x_1) < 0$ *and a local minimum at* x_2 *such that* $\Phi'(x_2) = 0$ *and* $\Phi''(x_2) > 0$; (iii) $\Phi'(x) > 0$ *for* $0 < x < x_1$ *and* $x > x_2$; (iv) $\Phi'(x) < 0$ *for* $x_1 < x < x_2$.

The isocline for the stock of pollution is depicted in Fig. 1. Given the range of values of π, the pollution isocline is S-shaped, with sharp turns at points C and B. The dirt levels associated with these threshold point are denoted by, respectively, D_L and D_U. The vertical arrows depict the dynamic forces operating on the stock of pollution off the isocline. The upward sloping branches of the isocline are locally stable: Lemma 1(iii) establishes that $\partial \dot{P}(t)/\partial P(t) = -\Phi'(P(t)) < 0$ there. In contrast, the downward sloping (dashed) branch is unstable because Lemma 1(iv) shows that $\partial \dot{P}(t)/\partial P(t) > 0$ for these points. For future reference we state the following Definition.

Definition 1 Define the clean branch of the pollution isocline as $\Phi_C(x) \equiv \Phi(x)$ for $0 \leq x < x_1$ and the dirty branch as $\Phi_D(x) \equiv \Phi(x)$ for $x > x_2$.

Consider a time-invariant dirt flow $\hat{D}$. Depending on its magnitude, three regimes are possible:

- Unique stable and clean steady-state. For $0 \leq \hat{D} < D_L$ there exists a unique steady-state pollution level that is located on the lower branch of the pollution isocline.
- Multiple steady-state pollution levels. For $D_L \leq \hat{D} \leq D_U$ there exist three ecological steady-state equilibria, of which two are stable and one is unstable. For example, if $\hat{D} = 0.04$ the stable equilibria are at points A and D in Fig. 1 whilst the instable one is at point E. Which particular steady state is attained depends on initial conditions, i.e. the ecological model features *reversible hysteresis*.
- Unique stable and polluted steady-state. For $\hat{D} > D_U$ there exists a unique steady-state pollution level that is located on the upper branch of the pollution isocline.

To capture the key features of the economic system we formulate a simple general equilibrium model of the macro-economy. This model describes a closed economy consisting of a government and representative households and firms who are blessed with perfect foresight. The representative household lives forever, and features the following utility functional:

$$\Lambda(t) \equiv \int_t^{\infty} \left[\ln C(\tau) + \varepsilon_E \ln\left[\bar{E} - P(\tau)\right]\right] \cdot e^{-\rho(\tau - t)} d\tau, \tag{3}$$

where $C(\tau)$ denotes consumption of private commodities at time τ, $E(\tau) \equiv \bar{E} - P(\tau) > 0$ measures the quality of the environment, $\bar{E}$ is some pristine value attained in a non-polluting society, ε_E denotes the weight in overall utility attached to environmental amenities, and $\rho > 0$ stands for the pure rate of time preference. Since utility is separable in its two arguments, the quality of the environment does not directly affect household consumption. As the felicity function for private consumption is logarithmic, the model features a unitary intertemporal elasticity of substitution. Without leisure entering utility, labour supply is exogenously fixed.

Households face the following budget identity:

$$\dot{A}(\tau) = r(\tau)A(\tau) + w(\tau) - T(\tau) - C(\tau), \tag{4}$$

where $r(\tau)$ denotes the real rate of interest on financial assets, $w(\tau)$ represents the wage rate, $T(\tau)$ are net lump-sum taxes, and $A(\tau)$ stands for real financial assets owned in period τ.

The representative agent chooses paths for $C(\tau)$ and $A(\tau)$ which maximize (3) subject to (4) and a solvency requirement of the form $\lim_{\tau\to\infty} A(\tau)e^{-\int_t^{\tau} r(s)ds} = 0$. He takes as given the stock of financial assets in the planning period, $A(t)$. The optimal consumption level that the agent chooses at time t is given by:

$$C(t) = \rho\left[A(t) + H(t)\right], \tag{5}$$

where human wealth, $H(t)$, is defined as:

$$H(t) \equiv \int_t^{\infty} \left[w(\tau) - T(\tau)\right] \cdot e^{-\int_t^{\tau} r(s)ds} d\tau. \tag{6}$$

The optimal time profile for consumption is given by the Euler equation:

$$\frac{\dot{C}(\tau)}{C(\tau)} = r(\tau) - \rho, \quad \tau \geq t. \tag{7}$$

The intuitive interpretation of these expressions is as follows. Equation (5) shows that the agent consumes a constant proportion of total wealth in the planning period, whilst Eq. (7) indicates that consumption growth over time is chosen to be equal to the anticipated gap between the interest rate and the rate of time preference. Finally, the expression in (6) implies that human wealth is given by the discounted value of after-tax wage payments using the market rate of interest for discounting purposes. Intuitively it represents the after-tax value of the agent's unitary time endowment.

The production sector of the economy is perfectly competitive. The production function is Cobb-Douglas, with constant returns to scale to the factors capital, $K(t)$, and labour, $L(t)$:

$$Y(t) \equiv F\big(K(t), L(t)\big) = \Omega_0 K(t)^{1-\varepsilon_L} L(t)^{\varepsilon_L}, \quad \Omega_0 > 0,\ 0 < \varepsilon_L < 1, \tag{8}$$

where $Y(t)$ denotes gross output. The value of the firm, $V(t)$, is given by the present value of the after-tax cash flow using the market rate of interest for discounting purposes:

$$V(t) = \int_t^{\infty} \big[\big(1-\theta(\tau)\big)\big[Y(\tau) - w(\tau)L(\tau)\big] - I(\tau)\big] \cdot e^{-\int_t^{\tau} r(s)ds} d\tau, \tag{9}$$

where $\theta(\tau)$ is the capital tax and $I(\tau)$ is gross investment. The capital stock evolves according to:

$$\dot{K}(\tau) = I(\tau) - \delta K(\tau), \tag{10}$$

where $\dot{K}(\tau) \equiv dK(\tau)/d\tau$ denotes the rate of change in the capital stock and δ is the depreciation rate ($\delta > 0$).

The representative firm chooses paths for $Y(\tau)$, $K(\tau)$, $L(\tau)$ and $I(\tau)$ which maximize the value of the firm (9) subject to the production function (8), and the capital accumulation identity (10). The capital stock in the planning period, $K(t)$, is taken as given. The first-order conditions yield the usual marginal productivity conditions:

$$\frac{\partial Y(\tau)}{\partial K(\tau)} = \frac{r(\tau) + \delta}{1 - \theta(\tau)}, \tag{11}$$

$$\frac{\partial Y(\tau)}{\partial L(\tau)} = w(\tau). \tag{12}$$

Since we abstract from adjustment costs in investment, the value of equity corresponds to the replacement value of the capital stock, i.e. $V(t) = K(t)$.

For convenience, the key equations of the core model have been gathered in Table 1. Equation (T1.1) is the Euler equation (7), whilst Eqs. (T1.5) and (T1.8)–(T1.9)

Table 1 The model

$$\frac{\dot{C}(t)}{C(t)} = r(t) - \rho, \quad \rho > 0 \tag{T1.1}$$

$$\dot{K}(t) = Y(t) - C(t) - G(t) - \delta K(t) \tag{T1.2}$$

$$\left[r(t) + \delta\right]K(t) = (1 - \varepsilon_L)\left[1 - \theta(t)\right]Y(t) \tag{T1.3}$$

$$w(t)L(t) = \varepsilon_L Y(t) \tag{T1.4}$$

$$Y(t) = \Omega_0 L(t)^{\varepsilon_L} K(t)^{1-\varepsilon_L}, \quad \Omega_0 > 0, 0 < \varepsilon_L < 1 \tag{T1.5}$$

$$L(t) = 1 \tag{T1.6}$$

$$T(t) = G(t) - \theta(t)\left[Y(t) - w(t)L(t)\right] \tag{T1.7}$$

$$\dot{P}(t) = -\pi P(t) + \frac{P(t)^2}{P(t)^2 + 1} + D(t), \quad \frac{1}{2} < \pi < \frac{3\sqrt{3}}{8} \tag{T1.8}$$

$$D(t) = \kappa K(t) - \gamma G(t), \quad \kappa > 0, \gamma > 0 \tag{T1.9}$$

Endogenous: consumption, $C(t)$, capital stock, $K(t)$, output, $Y(t)$, interest rate, $r(t)$, wage rate, $w(t)$, employment, $L(t)$, pollution stock, $P(t)$, dirt flow, $D(t)$. **Exogenous**: capital tax $\theta(t)$ and government abatement, $G(t)$. **Parameters**: rate of time preference, ρ, depreciation rate of capital, δ, labour coefficient in the technology, ε_L, and scale factor in the technology, Ω_0. **Ecological parameters**: lake resilience, π, capital dirt coefficient, κ, and abatement clean-up coefficient, .γ

just restate, respectively (8), (2), and (1). labour supply is exogenous so $L(t) = 1$—see (T1.6). The factor demand expressions in (11)–(12) have been rewritten by using the production function—see (T1.3) and (T1.4). Equation (T1.2) is obtained by combining (10) with the goods market clearing condition for a closed economy, i.e. $Y(\tau) = C(\tau) + I(\tau) + G(\tau)$. Finally, in the absence of government debt, claims on the capital stock are the only assets available, i.e. $A(t) = K(t)$.

The phase diagram for the economic system is depicted in Fig. 2. The initial equilibrium, by assumption featuring no public abatement, is at point E_0. Steady-state consumption and the capital stock are given by, respectively, $\hat{C}$ and $\hat{K}$. The equilibrium is saddle-point stable, with SP_0 representing the saddle path, and is dynamically efficient, i.e. $\hat{K}$ is strictly less than the golden-rule capital stock, $\hat{K}^{GR}$.

3 First-Best Social Optimum

In the remainder of this chapter we consider optimal environmental policy. The initial situation facing the policy maker is as follows. First, both the economic and ecological systems are in a steady-state equilibrium and environmental abatement is zero. Second, the steady-state dirt flow resulting from the equilibrium capital stock is such that there exist three possible ecological steady-state equilibria. Third, for otherwise unspecified reasons, the ecological system has settled down at the

Fig. 2 Consumption-capital dynamics

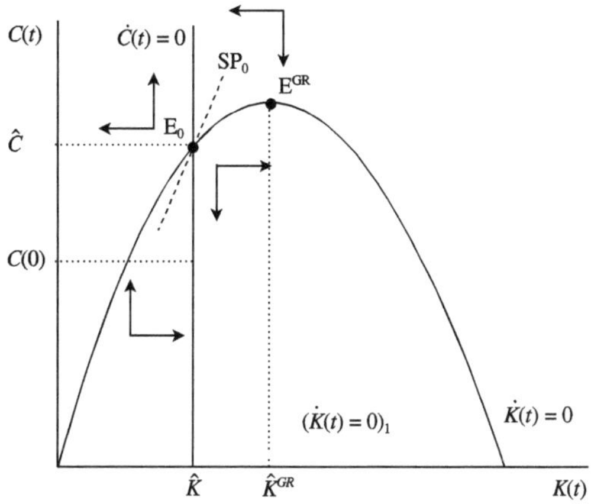

"bad" equilibrium featuring a high stock of pollution. In Fig. 2 the initial economic equilibrium is thus at point E_0. In Fig. 1 the dirt flow equals $\kappa \hat{K}$ and the ecological equilibrium is located at point D. Given this initial condition, can the policy maker bring about substantial welfare gains by choosing the appropriate mix of capital taxation and abatement activities?

In this section we characterize the first-best social optimum, i.e. we study the allocation that would be selected by a benevolent social planner aiming to maximize lifetime utility of the representative agent. In the planning period $t = 0$, the planner chooses paths for $C(t)$, $P(t)$, and $K(t)$ (for $t \geq 0$) in order to maximize (3) subject to the resource constraint (T1.2), the emission equation (2), and the dirt flow definition (1). The initial conditions are:

$$K(0) = \hat{K}, \qquad P(0) = \hat{P}_B = \Phi_D^{-1}(\hat{D}_0), \tag{13}$$

where $\hat{P}_B$ is the steady-state pollution level consistent with the upper branch of the pollution isocline (see Definition 1) and with a dirt flow equal to $\hat{D}_0 = \kappa \hat{K}$—see point D in Fig. 1. Abatement, the dirt flow, and gross investment must remain non-negative:

$$\begin{aligned} &G(t) \geq 0, \quad \big[D(t) \equiv\big]\kappa K(t) - \gamma G(t) \geq 0, \\ &\big[I(t) \equiv\big] F\big(K(t), 1\big) - C(t) - G(t) \geq 0. \end{aligned} \tag{14}$$

Dropping the time index, the current-value Lagrangian can be written as:

$$\begin{aligned} \mathscr{L} \equiv \ln C + \varepsilon_E \ln[\bar{E} - P] + \lambda_K\big[F(K,1) - C - G - \delta K\big] \\ + \lambda_P\big[-\Phi(P) + \kappa K - \gamma G\big] + \eta_D[\kappa K - \gamma G] + \eta_I\big[F(K,1) - C - G\big]. \end{aligned}$$

The control variables for this optimization problem are C and G (and thus implicitly D and I), the state variables are K and P, the co-state variables are λ_K and λ_P, and η_D and η_I are the Lagrange multipliers for, respectively, the dirt and investment constraints. The first-order conditions are:

$$\frac{\partial \mathscr{L}}{\partial C} = \frac{1}{C} - (\lambda_K + \eta_I) = 0, \tag{15}$$

$$\frac{\partial \mathscr{L}}{\partial G} = -(\lambda_K + \eta_I) - \gamma(\lambda_P + \eta_D) \leq 0, \quad G \geq 0, \ G\frac{\partial \mathscr{L}}{\partial G} = 0, \tag{16}$$

$$\frac{\partial \mathscr{L}}{\partial \eta_D} = \kappa K - \gamma G \geq 0, \quad \eta_D \geq 0, \ \eta_D \frac{\partial \mathscr{L}}{\partial \eta_D} = 0, \tag{17}$$

$$\frac{\partial \mathscr{L}}{\partial \eta_I} = F(K,1) - C - G \geq 0, \quad \eta_I \geq 0, \ \eta_I \frac{\partial \mathscr{L}}{\partial \eta_I} = 0, \tag{18}$$

$$\dot{\lambda}_K - \rho\lambda_K = -\frac{\partial \mathscr{L}}{\partial K} = -\kappa(\lambda_P + \eta_D) - \left[F_K(K,1) - \delta\right]\lambda_K - \eta_I F_K(K,1), \tag{19}$$

$$\dot{\lambda}_P - \rho\lambda_P = -\frac{\partial \mathscr{L}}{\partial P} = \frac{\varepsilon_E}{\bar{E} - P} + \lambda_P \Phi'(P). \tag{20}$$

The first-best social optimum is characterized by (2), (T1.2), (14), (15)–(20) and the transversality conditions:

$$\lim_{t\to\infty} e^{-\rho t}\lambda_K(t)K(t) = \lim_{t\to\infty} e^{-\rho t}\lambda_P(t)P(t) = 0.$$

3.1 Long-Run Optimum

We first study the long-run properties of the first-best equilibrium. In terms of notation, hatted variables denote steady-state values and the subscript "f" denotes first-best. In the steady state gross investment is strictly positive, i.e. $\hat{I}_f = \delta\hat{K}_f > 0$ and it follows from (18) that $\hat{\eta}_I = 0$. Depending on the structural parameters and the resulting magnitude of $\hat{G}_f$ two cases are possible.

Case 1: With Long-Run Abatement Assume that $0 < \hat{G}_f < (\kappa/\gamma)\hat{K}_f$ so that $\hat{\eta}_D = 0$ and $\gamma\hat{\lambda}_P = -\hat{\lambda}_K < 0$. It follows that the steady-state first-best equilibrium is given by:

$$F_K(\hat{K}_f, 1) = \rho + \delta + \frac{\kappa}{\gamma}, \tag{21}$$

$$\rho + \Phi_C'(\hat{P}_f) = \gamma\frac{\varepsilon_E \hat{C}_f}{\bar{E} - \hat{P}_f}, \tag{22}$$

$$F(\hat{K}_f, 1) = \hat{C}_f + \hat{G}_f + \delta\hat{K}_f, \tag{23}$$

$$\Phi_C(\hat{P}_f) = \kappa\hat{K}_f - \gamma\hat{G}_f, \tag{24}$$

where $\Phi_C(x)$ is the function representing the lower branch of the P-isocline—see Definition 1. The key thing to note is that a (Pigouvian) capital tax can be used to decentralize the first-best equilibrium. Equation (T1.3) shows that private saving behaviour will result in a steady-state capital stock such that $F_K(\hat{K},1) = (\rho+\delta)/(1-\hat{\theta})$. By comparing this expression to (21) we find that $\hat{K} = \hat{K}_f$ if and only if the steady-state capital tax is set equal to:

$$\hat{\theta}_f = \frac{\kappa/\gamma}{\rho+\delta+\kappa/\gamma}.$$

The optimal Pigouvian capital tax is feasible (as it satisfies $0 < \hat{\theta}_f < 1$) and is increasing in κ/γ. Intuitively, the more polluting is capital (κ up) and the less potent is abatement (γ down), the higher is the optimal environmental tax.

Case 2: Without Long-Run Abatement Assume that $\hat{\lambda}_K > -\gamma\hat{\lambda}_P$ so that $\hat{G}_f = 0$. Since $\hat{K}_f > 0$ it follows that $\hat{D}_f > 0$ and thus $\hat{\eta}_D = 0$ also. The first-best steady-state equilibrium can now be written as:

$$F_K(\hat{K}_f,1) = \rho+\delta-\kappa\hat{\lambda}_P\hat{C}_f, \tag{25}$$

$$\rho+\Phi_C'(\hat{P}_f) = -\frac{1}{\hat{\lambda}_P}\frac{\varepsilon_E}{\bar{E}-\hat{P}_f}, \tag{26}$$

$$F(\hat{K}_f,1) = \hat{C}_f+\delta\hat{K}_f, \tag{27}$$

$$\Phi_C(\hat{P}_f) = \kappa\hat{K}_f. \tag{28}$$

Just as for the previous case, a capital tax is needed to decentralize the first-best optimum:

$$\hat{\theta}_f = \frac{-\kappa\hat{\lambda}_P\hat{C}_f}{\rho+\delta-\kappa\hat{\lambda}_P\hat{C}_f}.$$

Since $\hat{\lambda}_P < 0$ and $\hat{C}_f > 0$ it follows that the optimal Pigouvian capital tax is feasible, i.e. $0 < \hat{\theta}_f < 1$.

3.2 Optimal Dynamic Allocation

In order to avoid having to deal with a taxonomy of possible cases, we use a parameterized version of the model to illustrate its main properties. For reasons of comparison we use the same parameterization as in Heijdra and Heijnen (2013)—see Table 2. For these parameter values we find that $-\hat{\lambda}_K - \gamma\hat{\lambda}_P = -0.9198$, i.e. Case 2 is the relevant one and abatement is not needed in the long run, i.e. $\hat{G}_f = 0$. We furthermore compute $\hat{K}_f = 2.3177$, $\hat{C}_f = 0.7901$, $\hat{Y}_f = 0.9524$, $\hat{P}_f = 0.0766$, and $\hat{D}_f = 0.0340$. For ease of comparison, we report these values in column (b) in Table 3. The long-run Pigouvian capital tax is $\theta_f = 0.1066$ and consumption, output, and the capital stock are all lower than in the initial steady-state equilibrium the key features of which have been reported in column (a) of Table 3.

Table 2 Structural parameters and steady-state features

Economic system:					
$\rho = 0.04$	$\delta = 0.07$	$\varepsilon_L = 0.70$	$\Omega_0 = 0.7401$		
$\hat{r} = 0.04$	$\hat{K} = 2.7273$	$\hat{Y} = 1.000$	$\hat{C} = 0.8091$	$\hat{I} = 0.1909$	$G = 0$
Ecological system:					
$\pi = 0.52$	$\kappa = 0.0147$	$\gamma = 0.302$	$\varepsilon_E = 0.9$	$\bar{E} = 2$	
$D_L = 0.0196$	$D_U = 0.0735$	$\hat{D}_0 = 0.04$	$\hat{P}_B = 1.2482$	$\hat{P}_G = 0.0936$	$P_E = 0.6581$

Table 3 Quantitative effects of taxation and abatement[a]

	BM	FBSO	SBSO			
			Taxation		Abatement	
			TV	TI	TV	TI
	(a)	(b)	(c)	(d)	(e)	(f)
$\hat{Y}$	1.0000	0.9524	0.9524	0.9524	1.0000	1.0000
$C(0)$		0.8677	1.0000	1.0000	0.6798	0.6933
$\hat{C}$	0.8091	0.7901	0.7901	0.7901	0.8091	0.8091
$\hat{K}$	2.7273	2.3177	2.3177	2.3177	2.7273	2.7273
$\hat{P}$	1.2482	0.0766	0.0766	0.0766	0.0936	0.0936
$\Lambda(0)$	−11.9092	−2.7722	−7.7638	−7.9525	−3.2471	−4.3087
$\theta(0)$		0.1234	0.1891	0.8500		
$\hat{\theta}$		0.1077	0.1077	0.1077		
$G(0)$		0.1326			0.1324	0.1166
$\hat{G}$		0.0000			0.0000	0.0000
t_E				39.5	28.2	30.0
$EV(0)$		44.1	17.1	16.2	40.5	34.5

[a]BM: parameterized base model. FBSO: first-best social optimum. SBSO: second-best social optimum. Policy instrument lacking or not sufficiently flexible. TV: time-varying instrument. TI: time-invariant instrument. Notation: $x(0)$ and $\hat{x}$ denote, respectively, the impact- and long-run (steady-state) value of the variable $x(t)$

The dynamic properties of the first-best optimum are illustrated in Fig. 3. Details of the computations are found in the Appendix. There are two critical dates characterizing the optimal solution, namely the earliest time at which the irreversibility constraint on investment ceases to bind, $t_I = 1.27$, and the time at which the dirt constraint becomes slack, $t_D = 27.01$. Together these dates define the three regimes through which the optimal paths evolve.

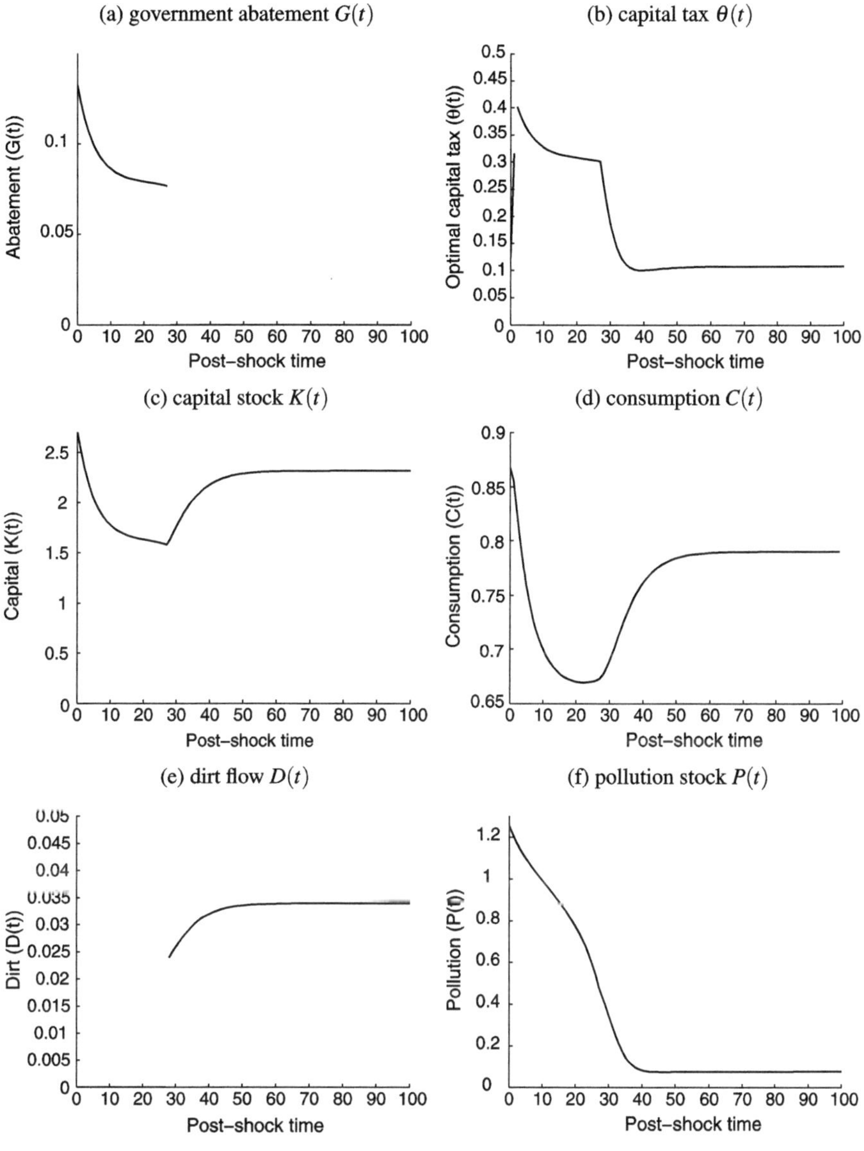

Parameters: see Table 2. The initial ecological equilibrium is at point D in panel (b)

Fig. 3 The first-best optimal policy

3.2.1 Regime 1

For $0 \leq t \leq t_I$ both the dirt flow and gross investment are zero, i.e. $D_f(t) = 0$ and $I_f(t) = 0$. It follows that abatement is at its maximum feasible level given by $G_f(t) = (\kappa/\gamma)K_f(t)$, consumption is described by $C_f(t) = F(K_f(t), 1) -$

$(\kappa/\gamma)K_f(t)$, whilst the capital stock satisfies $\dot{K}_f(t) = -\delta K_f(t)$. By combining these expressions and noting that $K(0) = \hat{K}$ we find:

$$\begin{aligned} K_f(t) &= \hat{K}e^{-\delta t}, \\ C_f(t) &= \hat{Y}e^{-\delta(1-\varepsilon_L)t} - \frac{\kappa}{\gamma}\hat{K}e^{-\delta t}, \\ G_f(t) &= \frac{\kappa}{\gamma}\hat{K}e^{-\delta t}. \end{aligned}$$

The transition paths for $K_f(t)$, $C_f(t)$, and $G_f(t)$ have been depicted in, respectively, panels (c), (d), and (a) of Fig. 3. With the flow of dirt reduced to zero, the stock of pollution falls according to:

$$\dot{P}_f(t) = -\Phi\big(P_f(t)\big).$$

3.2.2 Regime 2

For $t_I < t \le t_D$ the dirt flow is zero but gross investment is strictly positive, i.e. $D_f(t) = 0$ and $I_f(t) > 0$. Abatement remains at its maximum feasible level, $G_f(t) = (\kappa/\gamma)K_f(t)$. Since the capital stock is continuous for all t, it follows that the path of abatement is also continuous throughout this regime. Since the non-negativity constraint for gross investment ceases to be binding for $t > t_I$, the consumption path follows the Euler equation:

$$\frac{\dot{C}_f(t)}{C_f(t)} = F\big(K_f(t), 1\big) - \left(\rho + \delta + \frac{\kappa}{\gamma}\right),$$

whilst the stocks of capital and pollution evolve according to:

$$\begin{aligned} \dot{K}_f(t) &= F\big(K_f(t), 1\big) - C_f(t) - \left(\delta + \frac{\kappa}{\gamma}\right)K_f(t), \\ \dot{P}_f(t) &= -\Phi\big(P_f(t)\big). \end{aligned}$$

Consumption is continuous at time t_I, i.e. $\lim_{t \nearrow t_I} C_f(t) = \lim_{t \searrow t_I} C_f(t) = C_f(t_I)$, so that $C_f(t_I) = \hat{Y}e^{-\delta(1-\varepsilon_L)t_I} - \frac{\kappa}{\gamma}\hat{K}e^{-\delta t_I}$ and $K_f(t_I) = e^{-\delta t_I}\hat{K}$ are the initial conditions for the system of differential equation in $C_f(t)$ and $K_f(t)$.

3.2.3 Regime 3

At time t_D abatement is permanently reduced to zero ($G_f(t) = 0$) and the dirt flow becomes positive (as $D_f(t) = \kappa K_f(t)$). The value of t_D is such that $-\gamma\lambda_P(t_D)C_f(t_D) = 1$. Again, like the stocks of capital and pollution, consumption is continuous at time t_D, i.e. $\lim_{t\nearrow t_D} C_f(t) = \lim_{t\searrow t_D} C_f(t) = C_f(t_D)$. The optimal path for $t > t_D$ is described by:

$$\frac{\dot{C}_f(t)}{C_f(t)} = F_K\big(K_f(t), 1\big) - (\rho + \delta) + \kappa \lambda_P(t) C_f(t),$$

$$\dot{\lambda}_P(t) = \frac{\varepsilon_E}{\bar{E} - P_f(t)} + \big[\rho + \Phi_C'\big(P_f(t)\big)\big]\lambda_P(t),$$

$$\dot{K}_f(t) = F\big(K_f(t), 1\big) - C_f(t) - \delta K_f(t),$$

$$\dot{P}_f(t) = -\Phi\big(P_f(t)\big) + \kappa K_f(t).$$

This system converges to the steady state given in (25)–(28).

In passing through the three regimes, the first-best social optimum is decentralized by means of a tax on capital, $\theta_f(t)$, which is implicitly defined by:

$$\frac{\dot{C}_f(t)}{C_f(t)} = \big(1 - \theta_f(t)\big) F_K\big(K_f(t), 1\big) - (\rho + \delta).$$

As is illustrated in panel (b) of Fig. 3, the tax is quite high during the early phase of the environmental cleanup.

The welfare effect of the first-best optimal policy is considerable. Indeed, as our equivalent variation welfare measure $EV(0)$ in Table 3 reveals, the welfare gain due to the optimal environmental cleanup amounts to 44.1 percent of initial steady-state consumption.[3] Despite the fact that consumption is lower than its initial level during much of the transition, the gradual improvement in environmental quality more than compensates for this.

4 Second-Best Social Optimum

In this section we study optimal environmental policy in a second-best setting. In particular we consider the repercussions of two types of constraints on the policy maker's choices, namely the unavailability of instruments and the insufficient flexibility of a given instrument. In Sects. 4.1 and 4.2 we assume that the policy maker cannot use the abatement instrument and conducts constrained optimal environmental policy with either a time-varying capital tax (in Sect. 4.1) or a time-invariant (step-wise) capital tax (in Sect. 4.2).

In Sects. 4.3 and 4.4 we study the alternative case in which the policy maker cannot use the tax instrument and is constrained to conduct optimal environmental policy with, respectively, a time-varying or time-invariant abatement program. The latter case coincides with the ad hoc policy studied in our earlier paper Heijdra and Heijnen (2013).

[3] See Heijdra and Heijnen (2013) for a further discussion of the equivalent variation measure used here.

4.1 Time-Varying Taxation

The social planner chooses paths for $C(t)$, $P(t)$, and $K(t)$ (for $t \geq 0$) in order to maximize (3) subject to the resource constraint (T1.2), the emission equation (2), and the dirt flow definition (1). The initial conditions are as given in (13) above, and the non-negativity constraint on investment in (14) is still relevant. Compared to the first-best policy, however, the abatement instrument is not available, i.e. $G(t) = 0$ forms an additional constraint. As a result of this, the dirt flow constraint is slack, i.e. $D(t) > 0$ for all t. The second-best optimal plan can be decentralized with the aid of a time-varying tax on capital.

Of course, since abatement is not needed in the long-run first-best social optimum, the steady-state equilibrium under the second-best equilibrium considered here is still as given in (25)–(28) above, i.e. $\hat{K}_s^{TVT} = \hat{K}_f$, $\hat{C}_s^{TVT} = \hat{C}_f$, and $\hat{P}_s^{TVT} = \hat{P}_f$, where the subscript "$s$" denotes second-best and the superscript "*TVT*" indicates that the policy is decentralized with the aid of a time-varying tax. For convenience these quantitative results are reported in column (c) in Table 3.

Whereas the first- and second-best solutions are identical in the long run, the optimal transition paths differ substantially for these two cases. The dynamic properties of the second-best optimum are illustrated in Fig. 4. There is one critical date characterizing the optimal solution, namely $t_I = 21.46$, and there exist two adjustment regimes. Since there is no abatement, the flow of dirt is proportional to the capital stock and environmental pollution evolves in both regimes according to:

$$\dot{P}_s^{TVT}(t) = -\Phi\left(P_s^{TVT}(t)\right) + \kappa K_s^{TVT}(t).$$

4.1.1 Regime 1

For $0 \leq t \leq t_I$ gross investment is zero and the capital stock gradually falls. Since abatement is also absent, consumption is equal to output. To summarize we find for this regime that:

$$K_s^{TVT}(t) = \hat{K}e^{-\delta t},$$
$$C_s^{TVT}(t) = \hat{Y}e^{-\delta(1-\varepsilon_L)t}.$$

These paths have been depicted in panels (b) and (c) in Fig. 4. Because consumption growth in the decentralized equilibrium follows the Euler equation (T1.1) and consumption growth during this social planning regime equals $-\delta(1-\varepsilon_L)$ we find that the second-best social optimum can be decentralized with a time-varying capital tax of the following form:

$$\theta_s^{TVT}(t) = 1 - \frac{\rho + \delta\varepsilon_L}{\rho + \delta}e^{-\delta\varepsilon_L t}.$$

The capital tax is increasing over time in order to ensure that the gap between the equilibrium interest rate and the rate of time preference stays constant despite the fact that the capital stock falls over time. See panel (a) in Fig. 4.

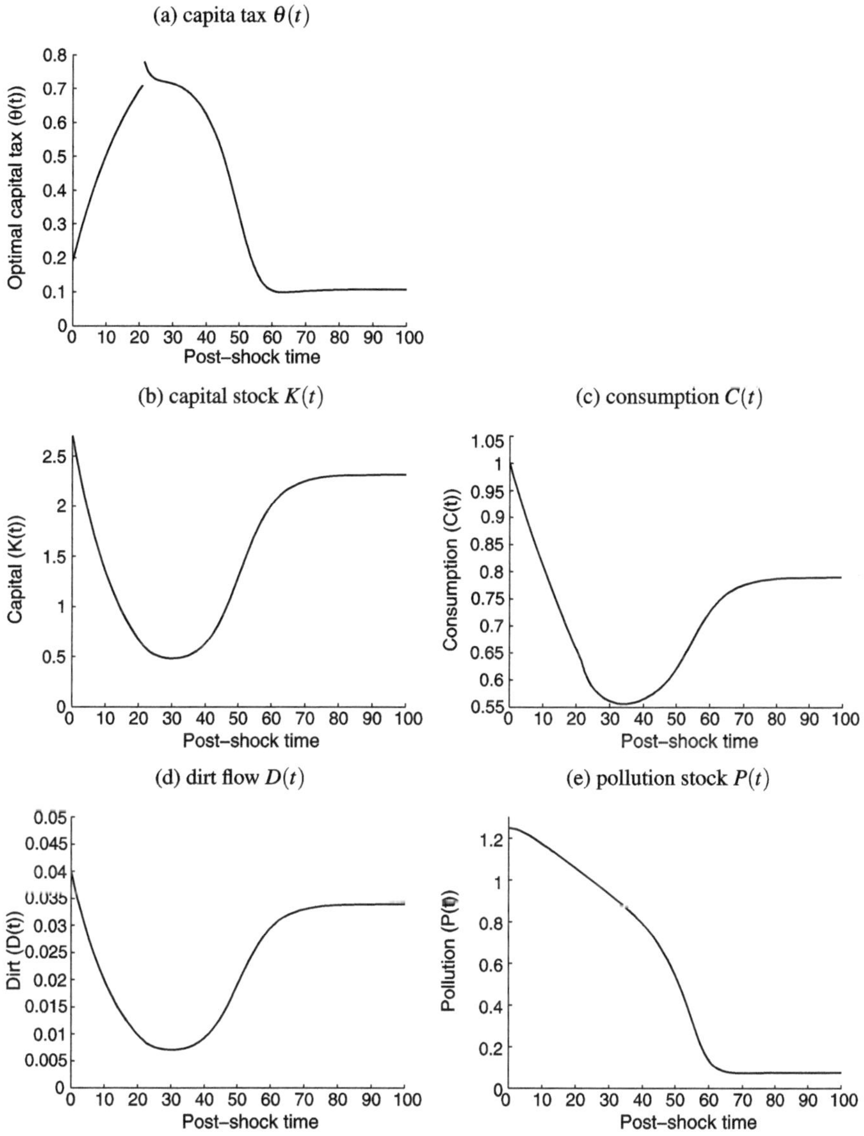

Fig. 4 Second-best optimal policy: time-varying taxation

4.1.2 Regime 2

For $t > t_I$ gross investment is strictly positive ($I_f(t) > 0$) and the consumption path is characterized by:

$$\frac{\dot{C}_s^{TVT}(t)}{C_s^{TVT}(t)} = F_K\left(K_s^{TVT}(t), 1\right) - (\rho + \delta) + \kappa \lambda_P(t) C_s^{TVT}(t),$$

$$\dot{\lambda}_P(t) = \frac{\varepsilon_E}{\bar{E} - P_s^{TVT}(t)} + \left[\rho + \Phi_C'\left(P_f(t)\right)\right]\lambda_P(t)$$

whilst the stock of capital evolves according to:

$$\dot{K}_s^{TVT}(t) = F\left(K_s^{TVT}(t), 1\right) - C_s^{TVT}(t) - \delta K_s^{TVT}(t).$$

Consumption is continuous at time t_I, i.e. $\lim_{t \nearrow t_I} C_s^{TVT}(t) = \lim_{t \searrow t_I} C_s^{TVT}(t) = C_s^{TVT}(t_I)$, so that $C_s^{TVT}(t_I) = \hat{Y}e^{-\delta(1-\varepsilon_L)t_I}$ and $K_s^{TVT}(t_I) = \hat{K}e^{-\delta t_I}$ are the initial conditions for the system of differential equation in $C_s^{TVT}(t)$ and $K_s^{TVT}(t)$. Since the optimal *growth rate* in consumption features a downward jump at $t = t_I$ and the capital stock is a predetermined variable, the optimal capital tax exhibits a discrete increase at that time—see panel (a) in Fig. 4. In the long run the system converges to the steady-state equilibrium discussed above.

Even though steady-state allocations are the same in the first- and second-best social optimum, the "road traveled" to get from the initial (dirty) steady-state to the socially optimal (clean) equilibrium is much more expensive when the policy maker lacks the abatement instrument. Indeed, as is indicated in Table 3 our equivalent variation measure $EV(0)$ falls from 44.1 % to 17.1 % of current consumption when a time-varying capital tax is the sole environmental policy instrument available. The tax is thus a rather blunt instrument in the sense that it must be set at very high (and strongly distortionary) levels during much of the transition in order to sharply reduce the capital stock (and the associated dirt flow) such that the ecology is steered to the basin of attraction of the lower branch of the P-isocline in Fig. 1. In contrast, in the first-best case abatement forms a very cheap instrument to get the pollution dynamics on the right track because it is financed by means of nondistortionary lump-sum taxes.

4.2 Time-Invariant Taxation

In this subsection we further restrict the policy maker's instruments by assuming that the capital tax can only take on two values.[4] In particular, we postulate that $\theta(t)$ is set according to:

$$\theta(t) = \begin{cases} \theta_h & \text{for } 0 \le t \le t_E, \\ \theta_l & \text{for } t > t_E, \end{cases}$$

where θ_h, θ_l, and t_E are chosen optimally by the social planner. Intuitively, in view of the results obtained from the time-varying taxation case (θ_h, t_E) must ensure that the ecology is out on the right track whereas θ_l corrects for the environmental externality in the long run.

[4] See Moser et al. (2013) for a general analysis of multi-stage optimal control techniques in the presence of history dependence.

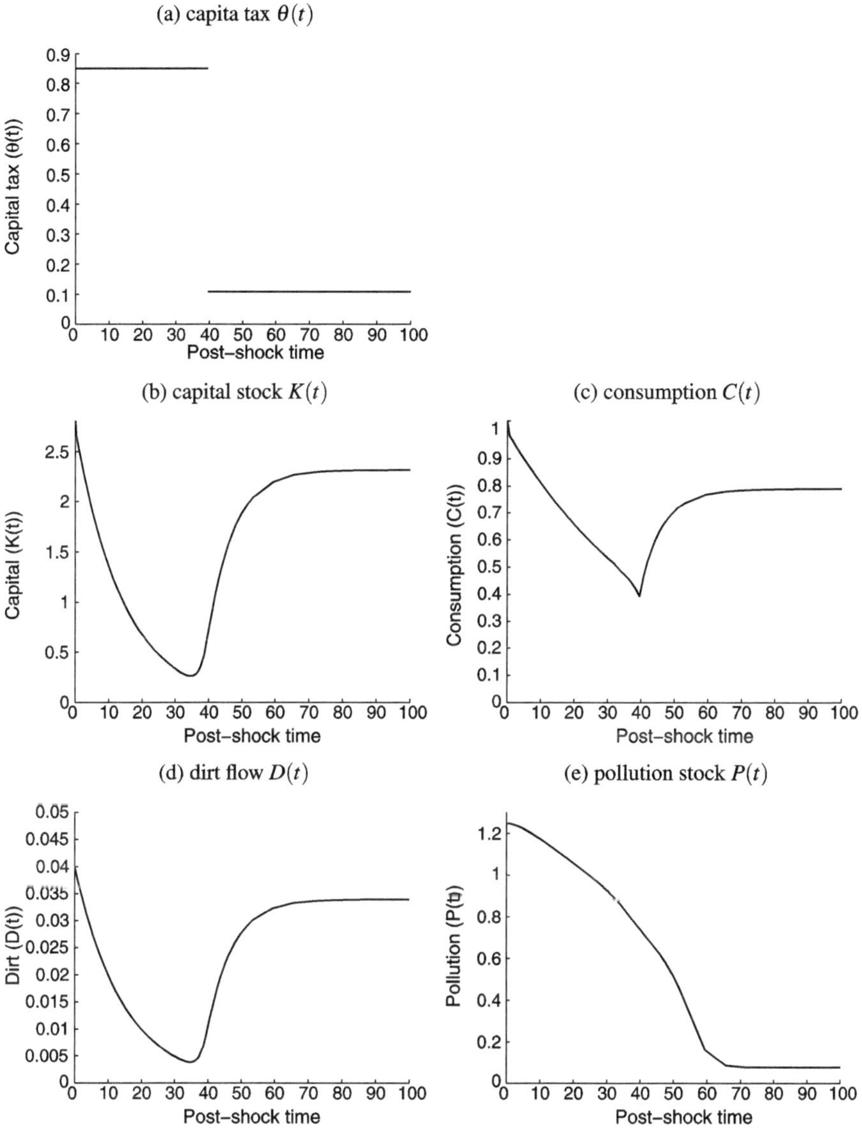

Fig. 5 Second-best optimal policy: time-invariant taxation

Figure 5 depicts the optimal paths for the key variables whilst column (d) in Table 3 presents the quantitative results. Several things are worth noting. First, the long-run allocation is the same under time-varying and time-invariant taxation. Second, during transition the regime configuration is also the same although t_I (the time until which the investment constraint is binding) is highest under time-invariant taxes ($t_I = 32$). Third, the initial capital tax is quite high ($\theta_h = 0.85$) and must be maintained for quite a long time ($t_E = 39.5$) in order to move the ecology to the basin of attraction of the lower branch of the P-isocline in Fig. 1. Fourth, the welfare cost of the instrument inflexibility is modest, i.e. the equivalent variation measure falls from 17.1 % under time-varying taxation to 16.2 % under time-invariant taxes.

4.3 Time-Varying Abatement

In the absence of capital taxation, the policy maker must conduct environmental policy exclusively with the abatement instrument. In order to compute the second-best optimal policy, we follow the approach exposited by Judd (1999). In the determination of the best feasible allocation the social planner faces not only the resource constraint (T1.2), the emission equation (2), and the dirt flow definition (1), but also the following private sector constraints:[5]

$$\lambda_H(t) = \frac{1}{C(t)}, \qquad \dot{\lambda}_H(t) = \big[\rho + \delta - F_K\big(K(t), 1\big)\big]\lambda_H(t). \tag{29}$$

Substituting the dirt constraint into the emission equation and dropping the time index, the current-value Lagrangian can now be written as:

$$\begin{aligned}\mathscr{L} \equiv{}& \ln C + \varepsilon_E \ln[\bar{E} - P] + \lambda_K\big[F(K,1) - C - G - \delta K\big] \\ &+ \lambda_P\big[-\Phi(P) + \kappa K - \gamma G\big] + \eta_\lambda\big[\rho + \delta - F_K(K,1)\big]\lambda_H \\ &+ \eta_C\left[\frac{1}{C} - \lambda_H\right] + \eta_D[\kappa K - \gamma G] + \eta_I\big[F(K,1) - C - G\big].\end{aligned}$$

The control variables are C and G (and thus D and I), the state variables are K, P, and λ_H, the associated co-state variables are λ_K, λ_P, and η_λ, and the Lagrange multipliers are η_C, η_D and η_I. The most relevant first-order conditions are the expressions in (16)–(18), (20), (29) and:

$$\begin{aligned}\frac{\partial \mathscr{L}}{\partial C} &= \frac{1}{C} - (\lambda_K + \eta_I) - \frac{\eta_C}{C^2} = 0, \\ \dot{\lambda}_K - \rho\lambda_K &= -\frac{\partial \mathscr{L}}{\partial K} = -\kappa(\lambda_P + \eta_D) - \big[F_K(K,1) - \delta\big]\lambda_K - \eta_I F_K(K,1) \\ &\quad + F_{KK}(K,1)\eta_\lambda\lambda_H, \\ \dot{\eta}_\lambda - \rho\eta_\lambda &= -\frac{\partial \mathscr{L}}{\partial \lambda_H} = \big[F_K(K,1) - \delta - \rho\big]\eta_\lambda + \eta_C.\end{aligned}$$

[5]Together these give rise to the Euler equation in the decentralized equilibrium, i.e. $\dot{C}(t)/C(t) = r(t) - \rho$, where $r(t) \equiv F_K(K(t), 1) - \delta$.

Since the capital tax is unavailable, the long-run capital stock returns to its initial level:

$$\hat{K}_s^{TVA} = \hat{K},$$

where the superscript "TVA" stands for time-varying abatement. Whilst it is in principle possible for long-run abatement to be positive, the parameter values ensure that this case does not materialize (just as in the first-best social optimum). In summary we find that:

$$\begin{aligned}
G_s^{TVA} &= 0, \\
\hat{C}_s^{TVA} &= F\left(\hat{K}_s^{TVA}, 1\right) - \delta \hat{K}_s^{TVA} = \hat{C}, \\
\hat{P}_s^{TVA} &= \Phi_l^{-1}\left(\kappa \hat{K}_s^{TVA}\right) = \hat{P}_G, \\
\hat{D}_s^{TVA} &= \hat{D}_0.
\end{aligned}$$

In the second-best optimum, the ecology moves from point D to A in Fig. 1. Of course, by construction, the second-best optimum can be decentralized with an abatement policy.

The dynamic properties of the second-best optimum are illustrated in Fig. 6. There is one date characterizing the optimal solution, namely $t_D = 28.2$, and there exist two adjustment regimes. Throughout the two regimes consumption is constrained to follow its decentralized Euler equation:

$$\frac{\dot{C}_s^{TVA}(t)}{C_s^{TVA}(t)} = F_K\left(K_s^{TVA}, 1\right) - (\rho + \delta),$$

whilst gross investment remains non-negative ($I_s^{TVA}(t) \geq 0$) and the dynamic path for capital accumulation for $0 \leq t \leq t_D$ is given in Fig. 6(b).

4.3.1 Regime 1

For $0 \leq t \leq t_D$ abatement is at its maximum feasible level and the dirt flow is reduced to zero ($D_s^{TVA}(t) = 0$). It follows from (1), (T1.2), and (2) that:

$$\begin{aligned}
G_s^{TVA}(t) &= \frac{\kappa}{\gamma} K_s^{TVA}(t), \\
\dot{K}_s^{TVA}(t) &= F\left(K_s^{TVA}(t), 1\right) - C_s^{TVA}(t) - \left(\delta + \frac{\kappa}{\gamma}\right) K_s^{TVA}(t), \\
\dot{P}_s^{TVT}(t) &= -\Phi\left(P_s^{TVT}(t)\right).
\end{aligned}$$

Together with the consumption Euler equation these conditions determine the paths depicted in Fig. 6.

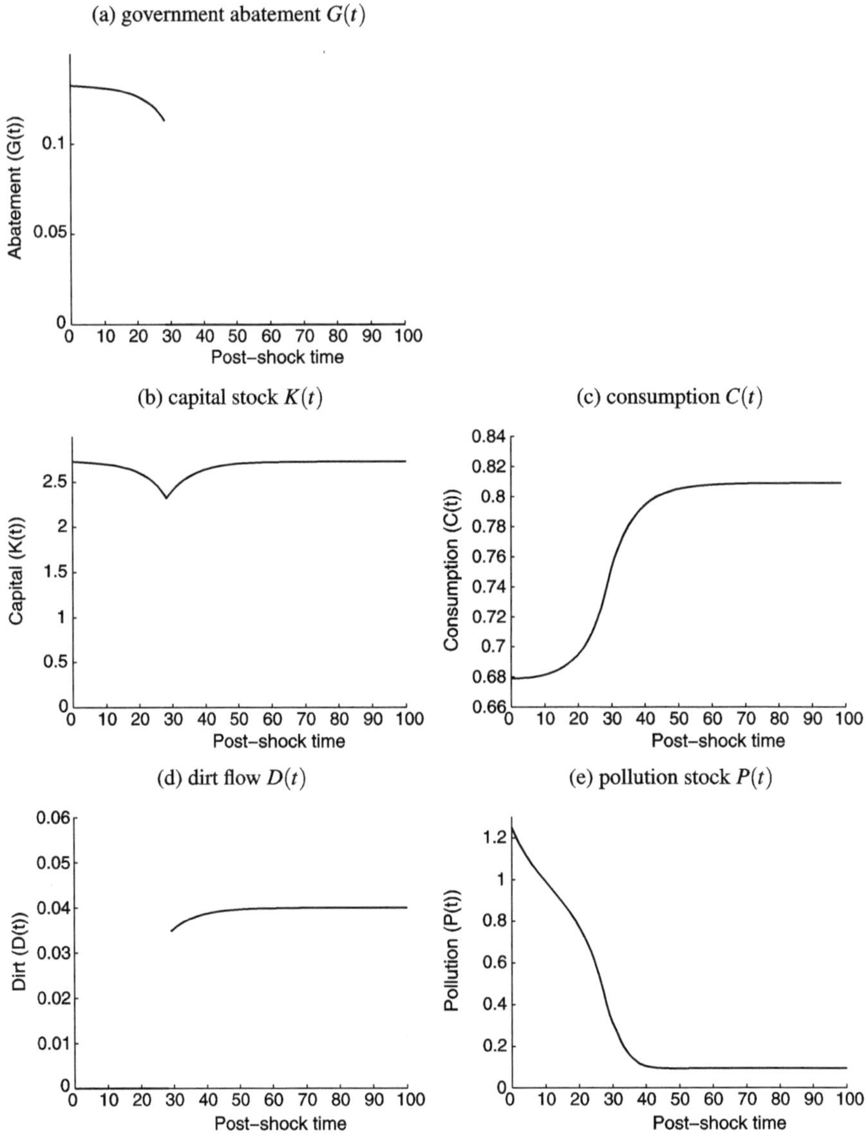

Fig. 6 Second-best optimal policy: time-varying abatement

4.3.2 Regime 2

For $t > t_D$ abatement is reduced to zero and the dirt flow becomes positive. Together with the consumption Euler the paths for the main variables are given by:

$$G_s^{TVA}(t) = 0,$$

$$\dot{K}_s^{TVA}(t) = F\left(K_s^{TVA}(t), 1\right) - C_s^{TVA}(t) - \delta K_s^{TVA}(t),$$
$$\dot{P}_s^{TVA}(t) = -\Phi\left(P_s^{TVA}(t)\right) + \kappa K_s^{TVA}(t).$$

The optimization problem implies that consumption is continuous at time t_D, i.e. $\lim_{t \nearrow t_D} C_s^{TVA}(t) = \lim_{t \searrow t_D} C_s^{TVA}(t) = C_s^{TVT}(t_D)$. This system converges to the steady state discussed above. The quantitative effects of the optimal time-varying abatement policy are reported in column (e) in Table 3. At impact abatement is quite high ($G(0) = 0.13$) and consumption is reduced substantially by about 16 percent. During the early phase of transition the capital stock is crowded out though by a relatively small amount compared to the time-varying taxation case discussed above. The abatement policy is thus a cheap instrument to direct the ecology to the basin of attraction of the lower branch of the P-isocline in Fig. 1. Indeed, as we report in column (e) of Table 3 the welfare gain under time-varying abatement is 40.5 percent of initial consumption which is quite close to the result under the first-best environmental policy.

4.4 *Time-Invariant Abatement*

In Heijdra and Heijnen (2013) we study the case in which the social planner uses an ad hoc abatement policy of the following form:

$$G(t) = \begin{cases} G & \text{for } 0 \le t \le t_E, \\ 0 & \text{for } t > t_E, \end{cases} \tag{30}$$

where G and t_E are chosen optimally by the policy maker. Intuitively, in view of the results obtained from the time-varying abatement case (G, t_E) must ensure that the ecology is put on the right track. The value of G must be chosen such that the non-negativity constraint on the dirt flow is violated *nowhere* along the adjustment path. The optimal policy is demonstrated to possess a "cold turkey" property: within the class of stepwise abatement function (30) the largest feasible G must be chosen for the briefest possible duration.

Figure 7 depicts the optimal paths for the key variables whilst column (f) in Table 3 presents the quantitative results. Several things are worth noting. First, the long-run allocation is the same under time-varying and time-invariant taxation. Second, abatement is set at $G = 0.1166$ which initially is lower than the values it takes under the time-varying policy. As a consequence, abatement must be continued for a slightly longer period ($t_E = 30$ instead of $t_E = t_D = 28.2$). Third, the welfare cost of the instrument inflexibility is relatively small, i.e. the equivalent variation measure falls from 40.3 % under time-varying abatement to 34.5 % under time-invariant abatement.

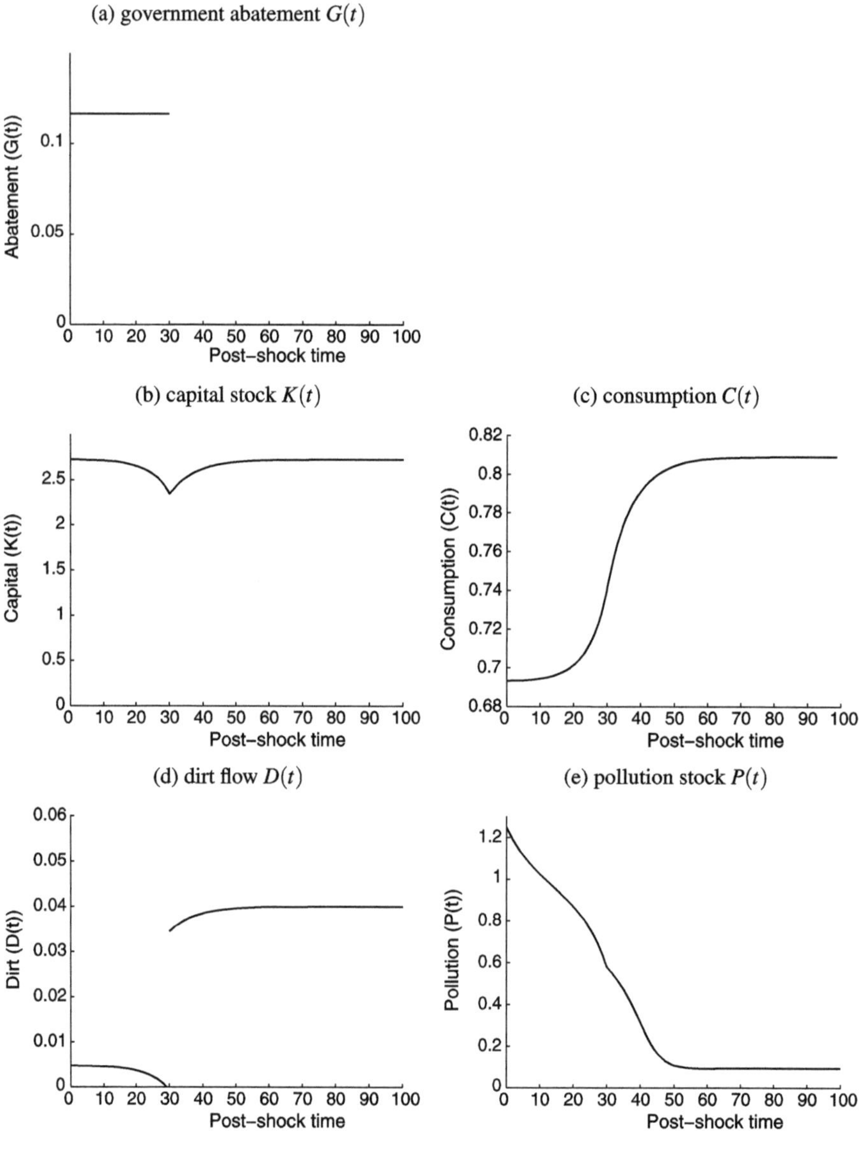

Parameters: see Table 2. The initial ecological equilibrium is at point D in panel (b)

Fig. 7 Second-best optimal policy: time-invariant abatement

5 Conclusions

In this paper we have studied optimal environmental policy in the presence of an ecological process featuring multiple stable steady-state ecological equilibria and reversible hysteresis. Assuming that the ecological steady-state equilibrium is initially located on the high-pollution (low-welfare) branch of the pollution isocline, the policy maker is in principle able to engineer substantial welfare gains by choosing the appropriate mix of Pigouvian taxation and abatement activities.

In the first-best social optimum the two available policy instruments each play a very distinct role. During the initial phase of the policy, abatement is used to choke off the flow of dirt as much as is feasible whereas the tax is employed to bring down the stock of the polluting capital input as quickly as possible. In the long run, however, abatement is no longer needed and the capital tax settles down at its externality-correcting Pigouvian level.

Interestingly, in a second-best setting it matters very much which additional constraint is faced by the policy maker. In the case where capital taxation is unavailable as an instrument for environmental policy, a suitably designed abatement policy can achieve a social outcome that is only marginally worse than the first-best result. Intuitively, lump-sum tax financed abatement is a cheap instrument to steer the ecology from the high- to the low-pollution equilibrium.

In contrast, if the abatement instrument is not available and the tax must be used to clean up the environment then the "road traveled" is a very expensive one. Intuitively, because of the distorting nature of the capital tax, using it to get out of the hysteretic equilibrium is a high-price option. Indeed, we show that in that case it is only marginally welfare improving to steer the ecology from the high- to the low-pollution equilibrium and to correct for the environmental externality.

Acknowledgements A previous draft of this paper was presented at the 12th Viennese Workshop on Optimal Control, Dynamic Games and Nonlinear Dynamics, held at the Vienna University of Technology in May–June 2012. We thank the Editors, an anonymous referee, Jochen Mierau, Laurie Reijnders, and various conference and seminar participants for their useful comments.

Appendix: Computational Details

First-Best (FB) We use a continuation method to compute the first-best. Let $X_s = (K, P)$ denote the state variables, $X_c = (\lambda_K, \lambda_P)$ the costates, and $X = (X_s, X_c)$. The controls and the Lagrange-multipliers are denoted by $U = (C, G, \eta_I, \eta_D)$. From Pontryagin's maximum principle, we get $U = U^*(X)$: the state- and costate-variables determine consumption, abatement and the multipliers for the investment- and dirt constraints. Recall that the other first-order conditions can be written as follows:

$$\dot{X}(t) = H\big(X(t), U^*\big(X(t)\big)\big).$$

The optimal path is determined by constraints on $X_s(0)$ and $X(\infty)$. In particular $X_s(0) = (\hat{K}, \hat{P}_B)$ and $X(\infty) = X^*$, where X^* is a root of $H(\cdot)$. The end condition is replaced by the requirement that at time $T = 200$, the trajectory is orthogonal to the stable manifold of X^*. We approximate the first-order condition as follows. First, we discretize the time grid $t \in \{0, 1, \ldots, 200\}$ and at time t we replace the differential equation by a fourth-order Runge-Kutta approximation. This leads to a system of equations of which we have to find the root.

For the continuation method, we need a trivial solution. Note that $X(t) \equiv X^*$ is a solution for the initial condition $X_s(0) = X_s^*$. We slowly change this initial condition into the direction of the actual initial condition, using a simple predictor-correction algorithm. See Grass (2012) for details.

The time at which the investment constraint stops being binding is calculated in the following manner. Suppose that for $t \leq t^*$, we have $\eta_I(t) > 0$ (and $\eta_I(t) = 0$ for $t > t^*$). This means that the investment constraint is binding until $t_I \in [t^*, t^* + 1]$. Using cubic extrapolation, we determine the value of t_I. It turns out that $t_I = 1.27$. For the dirt constraint, we use a similar method and it turns out that $t_D = 27.01$.

Time-Varying Tax (TVT) In principle, in this case we should be able to use a similar algorithm as for the first-best. However, the continuation algorithm fails to terminate (the path "bends back" to the $X_s(0) = X_s^*$). We note that at some point the investment constraint becomes binding. Therefore, we postulate that the optimal path first goes through a regime where the investment constraint is binding. If the investment constraint is binding until $t = t_I$, then we can calculate the value of capital and pollution at $t = t_I$. We take these as the initial value for capital and pollution and solve for the optimal time-varying tax from that point onward. Then we choose t_I such that this is the point where the investment constraint stops being binding (i.e. $\lambda_K(t_I)F(K(t_I), 1) = 1$). It turns out that this is the case for $t_I = 21.46$.

Please note that in both FB and TVT the long-run tax rate is $\hat{\theta} = 0.1066$.

Time-Invariant Tax (TIT) In the long-run, we set the tax rate equal to $\hat{\theta}$, but we start with a higher tax rate to move the system towards a lower pollution level. It turns out that the initial tax rate θ_0 is high enough to make the investment constraint binding. This means that we have to determine t_I (the time at which the investment constraint stops being binding) and t_E (the time at which the tax rate shifts from θ_0 to $\hat{\theta}$). Since the consumption path cannot jump, we can only switch from θ_0 to $\hat{\theta}$ if we are on the stable saddle path leading to the clean equilibrium. Hence, the free variables are θ_0 and t_I. We somewhat crudely search for the lowest values that can force the system to the clean steady state by increasing θ_0 with step size 0.05 and t_I with step size 1. We end up with $\theta_0 = 0.85$ and $t_I = 32$. Since the EV under TIT is close to the EV under TIA (time-invariant abatement), we are confident that these values are close to the optimal tax of this form.

Time-Varying Abatement and Time-Invariant Abatement (TVA and TIA) See Heijdra and Heijnen (2013). We have added a bit of accuracy for the case with TVA: full abatement until $t_E = 28.2$, increases the EV to 40.5 %.

Calculation of Utility Levels In the FB, we calculate utility level by calculating the Lagrangian at time zero and dividing this value by ρ. In all other cases, we use the following method to calculate the utility of the representative consumer. Given paths for consumption and pollution, this amounts to evaluating an integral of the form

$$W = \int_0^{\infty} u\big(C(s), P(s)\big)\mathrm{e}^{-\rho s} ds.$$

As inputs we have the levels of consumption and pollution at discrete points in time $t \in \{t_0, t_1, t_2, \ldots, t_n\}$, where t_n is sufficiently large for consumption and pollution to be close to the steady state values. Then, as is also noted by Heijnen and Wagener (2013), W is approximately equal to:

$$W \approx \frac{1}{2}\sum_{i=1}^{n}\big[u\big(C(t_i), P(t_i)\big)\mathrm{e}^{-\rho t_i} + u\big(C(t_{i-1}), P(t_{i-1})\big)\mathrm{e}^{-\rho t_{i-1}}\big](t_i - t_{i-1})$$
$$+ u\big(C(t_n), P(t_n)\big)\frac{\mathrm{e}^{-\rho t_n}}{\rho}.$$

Since our grid is not very dense, this gives a rather rough approximation, limiting the accuracy with which we can calculate the optimal policy.

References

Bovenberg, A. L., & Heijdra, B. J. (1998). Environmental tax policy and intergenerational distribution. *Journal of Public Economics*, *67*, 1–24.

Bovenberg, A. L., & Heijdra, B. J. (2002). Environmental abatement and intergenerational distribution. *Environmental & Resource Economics*, *23*, 45–84.

Grass, D. (2012). Numerical computation of the optimal vector field: exemplified by a fishery model. *Journal of Economic Dynamics & Control*, *36*, 1626–1658.

Heijdra, B. J., & Heijnen, P. (2013). Environmental abatement policy and the macroeconomy in the presence of ecological thresholds. *Environmental & Resource Economics*, *55*, 47–70.

Heijnen, P., & Wagener, F. O. O. (2013). Avoiding an ecological regime shift is sound economic policy. *Journal of Economic Dynamics & Control*, *37*, 1322–1341.

Judd, K. L. (1999). Optimal taxation and spending in general competitive growth models. *Journal of Public Economics*, *71*, 1–26.

Moser, E., Seidl, A., & Feichtinger, G. (2013). History-dependence in production-pollution-trade-off models: a multi-stage approach. *Annals of Operations Research*. doi:10.1007/s10479-013-1349-9.

Scheffer, M. (1998). *Ecology of shallow lakes*. Dordrecht: Kluwer Academic.

Scheffer, M., Carpenter, S., Foley, J. A., Folke, C., & Walker, B. (2001). Catastrophic shifts in ecosystems. *Nature*, *413*, 591–596.

Modeling the Dynamics of the Transition to a Green Economy

Stefan Mittnik, Willi Semmler, Mika Kato, and Daniel Samaan

Abstract Recent academic work argues for a greater urgency to implement effective climate policies to combat global warming. Concrete policy proposals for reducing CO_2 emissions have been developed by the IPCC. Yet, it has not been sufficiently explored to what extent mitigation policies, such as cap-and-trade, carbon tax or the phasing in of green technology, will entail structural change in an economy. Here, we explore the transition to a green economy using a growth model with structural change resulting from three types of policies: (1) shifting preferences, (2) taxing high-carbon intensive goods, or (3) imposing a carbon tax while subsidizing low-carbon intensive economic activities. We also will consider a strategy of imposing a carbon tax and subsidizing labor cost. Our focus will be on two questions: What impact do the policies under consideration have on employment and output, and whether resulting growth paths will be stable. We also indicate how the effects of carbon policies can be assessed empirically.

1 Introduction

Given the recent scientific evidence on global warming and its consequences, as documented in the numerous reports by the IPCC, the importance of climate change mitigation policies has been sufficiently demonstrated. The need for climate actions

---

S. Mittnik
Department of Statistics, Ludwig-Maximilians-Universität München, Munich, Germany
e-mail: finmetrics@stat.uni-muenchen.de

W. Semmler (✉)
New School for Social Research, New York, USA
e-mail: semmlerw@newschool.edu

M. Kato
Department of Economics, Howard University, Washington, DC, USA
e-mail: mkato@howard.edu

D. Samaan
Department of Economics, New School for Social Research, New York, USA
e-mail: samd01@newschool.edu

E. Moser et al. (eds.), *Dynamic Optimization in Environmental Economics*,
Dynamic Modeling and Econometrics in Economics and Finance 15,
DOI 10.1007/978-3-642-54086-8_4,

becomes particularly urgent if—as recent research indicates—some threshold has already been reached.[1] This produces the danger of doing "too little too late." In Greiner and Semmler (2008) and Greiner et al. (2009), models are presented that motivate a great urgency of actions. As shown there, discount rates, which have heavily been stressed in previous academic studies (see Nordhaus 2008; Weitzmann 2009; Stern 2007), are not as important as the danger of delaying actions. Delaying action can mean that a high temperature (and low growth path) is approached that becomes irreversible.

Along those lines, IPCC reports have been published that propose a large number of policy measures to prevent further emission of Greenhouse Gases and a further rise of global temperature. The IPCC 4th Assessment Report urgently suggests a broad range of mitigation policies, such as coordinated and integrated climate policies, broader development policies, regulations and standards, voluntary agreements, information instruments and financial incentives to control and reduce Greenhouse Gas emission. As measures toward a green economy, it emphasizes the role of technology policies to achieve lower CO_2 stabilization levels, a greater need for more efficient R&D efforts, and higher investment in new technologies over the next few decades. It also recommends government initiatives for funding or subsidizing alternative energy sources (solar energy, ocean power, windmills, biomass, and nuclear fusion). Overall, the IPCC stresses the fact that there are a number of effective policy measures available now that can reduce Greenhouse Gas emission.

Yet, the major instruments that the IPCC and a number of economists propose are two specific tools in order to fulfill the agreements of the Kyoto protocol and other international agreements. These two tools are decentralized market trading of emission right (tradeable permits) and carbon taxation—in the public discussion often called "cap-and-trade" and "carbon tax" (see Uzawa 2003; Nordhaus 2008; Mankiw 2007; IPCC 2007). Both measures have a long-standing history in economic theory, originating in the works of Pigou and Coase. Most economists seem to agree now that an emission (or carbon) tax is preferable to a cap&trade system.[2] Here, we

---

[1]This is, for example, occurring through some albedo effect. This effect refers to the reflection of incoming energy from the sun. As scientists have found out, the amount of energy reflected back to space is decreasing as the earth becomes warmer. The melting of the arctic ice, for example, leads to an absorption of a higher fraction of energy by the earth and the amount of energy reflected back to space, the albedo, falls. This in turn heats up the earth faster. For details of such a canonical model of climate change with threshold effects, see Greiner et al. (2009).

[2]*Cap-and-Trade System*: Cap-and-trade system requires that the actual polluter can be identified, for example firms. Enforcement of the cap is difficult and trading of emission certificates are exposed to speculative investments, generating a high volatility of the carbon price as the European example shows. According to an estimate by Nell and Semmler (2009), the carbon price, in case of emission trading, is even ten times more volatile than stock prices, which is already about seven times more volatile than the GNP.

Carbon Tax: Carbon tax, on the other hand, allows for a broader application, including energy supply, major polluting industries, the service sector, transport system and households. Furthermore, the generated tax revenue can be employed to reduce other taxes, and tax funds can to be used to compensate developing economies, or can be used to induce climate-friendly investment behavior as argued in Uzawa (2003).

mainly focus on tax schemes as regulatory instruments. Yet, whatever measure will be pursued, it will have a major impact on growth and the structure of the economy.

Below, we propose a growth model that, in contrast to much recent work, allows for structural change due to policy influences. In economics, models that allow to study simultaneously growth and structural change have been developed by Kuznets (1957), Kaldor (1957), and Pasinetti (1981) in the context of traditional Keynesian-oriented growth models. In Pasinetti (1981), for example, structural change occurs through a change in final demand—driven by the income elasticity of demand (Engel curves). The Kuznets-Kaldor-Pasinetti ideas are incorporated into a recent optimal growth model that allows for structural change (see Kongsamut et al. 2001). In the latter, three types of preferences are driving structural change: preferences for agricultural goods, manufactured goods, and services. The authors call such a path a generalized balanced growth path. Since models of this type allow one to trace the impact of climate policies on structural change along the growth path, we use them as a starting point to model the impact of climate policies on growth and also the structure of output and employment.

One can generally distinguish different factors that have an effect on total CO_2 emissions of an economy (see Proops et al. 1993): the carbon intensity of the energy sources used, C/E, the energy intensity of the production, E/TO, and the total output, TO.[3] Total output effects can be further decomposed into the structure of intermediate goods, the structure of final demand and the volume of final demand. Thus all climate policies have to influence ultimately at least one of those factors. It is important to note that TO can affect CO_2 emissions through its level *and* through its structure. Furthermore, we can distinguish a share of total output that directly satisfies final demand and another share that is used to produce intermediate goods. However, the volume of intermediate goods produced will directly depend on the volume of final goods demanded. We will propose a type of carbon tax which emphasizes a dependency of the structure of intermediate goods on the structure of final demand. For lasting and long-term results, climate policies have to induce structural changes in an economy, and this can be best achieved through a change of the demand structure.

We, thus, can perceive three main ways of how carbon intensity of an economy can be reduced:

Although economists seem to lean toward a carbon tax, policy makers appear to tend toward a market-based cap-and-trade system, as it is unpopular to announce the increase of tax rates when running for public offices.

[3]This can be seen from the following identity:

$$C \equiv \frac{C}{E} \cdot \frac{E}{TO} \cdot TO,$$

with C denoting the absolute CO_2 emissions, E the energy use, and TO the total output of the economy. We can, therefore, identify the five listed factors that affect CO_2 emissions.

- Preferences can shift over time, away from carbon-intensive and towards less-carbon-intensive industries. This implies that the final demand will change.
- Carbon-intensive sectors can be taxed and the tax revenue can be used to reduce wage cost. This is often seen as a policy measure that generates a double dividend.
- Carbon-intensive sectors and products can be taxed and less-carbon intensive sectors and activities can be subsidized.

Yet, capital goods are delivered and used in final-goods sectors, so one also has to consider the capital goods sector and its generation of CO_2. We will show that also the carbon intensity of the capital goods sector that delivers inputs to the final goods sector can be reduced which will finally reduce the carbon intensity of the economy. Thus, although we are considering a capital goods sector, our model makes preferences for final goods central. What we are claiming is that preferences shape the final demand of goods and, hence, the CO_2 emissions of the economy. If preferences shift from high- to low-carbon intensive goods, overall carbon intensity will be reduced. The same holds for an industry: If, for an industry, preferences are redirected to a product variety with less carbon intensity, an industry's carbon intensity will also be reduced. As we will show below, this implies a change of final demand which will entail a change of the carbon intensity of the capital goods sector as mentioned above. We will show that even if the capital goods sector is not given direct incentive to reduce carbon intensity, a reduction will indirectly occur through changed preferences for final goods.[4]

As to the first point, however, the question is how to impact preferences through policies in such a way that they change in the desired direction. Preferences are often evolving historically and are impacted by sociological and cultural factors, such as through role models and adapting the behavior of others. Conspicuous consumption in the sense of Veblen is another example which leads to a certain preference adoption. Population segments may copy the behavior of other population segments as Veblen suggested, or activities of green political movements may influence the behavior of households purchasing less-carbon intensive goods and services.

Preferences may also change through regulations and standards, norms and conventions. development policies, voluntary agreements, and information instruments. If standards and rules are set for construction of housing and for fuel efficiency of cars, then preferences are likely to change over time. Moral persuasion is another way to change preferences. Moreover, there exist policy tools to affect final demand. Final demand may, for example, be changed through financial incentives, such as taxes, emission certificates and subsidies, and, thus, help to redirect demand to less-carbon incentive sectors and goods.

In our subsequent model, we first will consider indirect influences on preferences by looking at the way how private preferences are shaped over time. This represents our baseline dynamic model, which is presented in Sect. 2. In the second version

[4]Note that one could also introduce a financial penalty, for example, a tax on the use of capital goods or intermediate goods with high direct and indirect carbon emission. A VAT on consumption whose rates are based on the cumulated carbon intensity could achieve this.

of our model, in Sect. 3, we study the issue when a carbon tax (or tradeable emission permits) are imposed. In a third version of our model, in Sect. 4, we consider a further extension where we allow for both carbon taxes on emission-intensive industries and subsidization of low-carbon intensive products. The latter model will exhibit what has been called a generalized steady state. Although there are some common growth rates, there will be a structural change and thus a change of the composition of sectors over time. Thus, the output composition of the economy and the fraction of labor and capital employed will change over time. In Sect. 5, we briefly summarize how one can take the model to the data, to empirically estimate employment and output effects of mitigation policies, as suggested by Mittnik et al. (2013). Empirical implications will be spelled out, but for a detailed econometric treatment of those issues, using a double-sided VAR, we refer the reader to Mittnik et al. (2013).

2 A Baseline Dynamic Model with Preferences

In the baseline model we explain the main mechanism of how our model works.[5] We also show here what role preferences play for the sectoral composition of the economy and how preferences can evolve over time.

2.1 Allocation of Labor and Capital

Consider an economy with three (high-CO_2-emitting, low-CO_2-emitting, and capital-goods) sectors. There are two factors of production, capital K_t and labor N_t. The total amount of labor available in the economy is normalized to 1.

Denote the fraction of capital devoted to sector i as $\phi_t^i K_t$ for $i \in \{H$ (high-CO_2-emitting sector), L (low-CO_2-emitting sector), K (capital-goods sector)$\}$ and

$$\sum_{i \in \{H,L,K\}} \phi_t^i = 1,$$

for all t.

Denote the fraction of labor devoted to sector i as N_t^i for $i \in \{H, L, K\}$ where

$$\sum_{i \in \{H,L,K\}} N_t^i = 1, \tag{1}$$

for all t.

[5] Our baseline model is related to Kongsamut et al. (2001).

2.2 Production and Technology

The level of technological progress is denoted by X_t and the path of X_t is given. We, however, assume that the growth rate of technology gradually converges to a constant rate g, i.e.,

$$\dot{X}_t = X_t g_t,$$

with a given $X_0 > 0$ and

$$\lim_{t \to \infty} g_t = g \geq 0.$$

Moreover, we assume that technological progress is labor augmenting.

Then the output of the high-CO_2-emitting sector is

$$H_t = B_H F\big(\phi_t^H K_t, N_t^H X_t\big), \tag{2}$$

the output of the low-CO_2-emitting sector is

$$L_t = B_L F\big(\phi_t^L K_t, N_t^L X_t\big), \tag{3}$$

and the output of the capital-goods sector is

$$\dot{K}_t + \delta K_t = B_K F\big(\phi_t^K K_t, N_t^K X_t\big), \tag{4}$$

where B_i measures the efficiency of production in sector $i \in \{H, L, K\}$.

For reasons of simplicity, the functions, $F(\phi_t^i K_t, N_t^i X_t)$, in (2), (3), and (4) are identical to all three sectors and assumed to have constant returns to scale. Denote the partial derivatives using subscripts as

$$F_1 \equiv \frac{\partial F(\phi_t^i K_t, N_t^i X_t)}{\partial(\phi_t^i K_t)} \quad \text{and} \quad F_2 \equiv \frac{\partial F(\phi_t^i K_t, N_t^i X_t)}{\partial(N_t^i X_t)},$$

so $B_i F_1$ and $B_i F_2$ are the marginal products of capital and labor in sector $i \in \{H, L, K\}$, respectively.

Note that the outputs of the high-CO_2- and the low-CO_2-emitting sectors can only be consumed while the output of the capital-goods sector can only be invested.

2.3 Efficient Factor Allocation

We assume that factor markets are competitive and that both factors are fully mobile across sectors. As the three sectors have identical F, efficiency of production in each sector should be measured by B_i only. Then an efficient factor allocation requires the fraction of capital employed in sector i and the fraction of labor employed in sector i be equalized, i.e., $\phi_t^i = N_t^i$. That is, capital-labor ratios of all three production sectors are

$$\frac{\phi_t^H}{N_t^H} = \frac{\phi_t^K}{N_t^K} = \frac{\phi_t^L}{N_t^L} = 1, \tag{5}$$

for all t.

Similarly, the price P_i should reflect efficiency of production in sector i, i.e., the higher the efficiency is, the lower the price is. Competitive factor markets imply that the marginal products of capital, $B_i F_1$, and marginal products of labor, $B_i F_2$, have to be equalized in all three production sectors. Then, using the price of capital goods as numeraire, the relative prices of high-CO_2-emitting products and low-CO_2-emitting products in terms of capital goods are

$$P_H = \frac{B_K}{B_H} \quad \text{and} \quad P_L = \frac{B_K}{B_L} \quad \text{when } P_K = 1. \tag{6}$$

2.4 *Investment and Capital Accumulation*

Since production functions F have constant returns to scale, from (2), (5), and (6), the total revenue from the high-CO_2-emitting sector is

$$\begin{aligned} P_H H_t &= P_H B_H N_t^H F(K_t, X_t) \\ &= B_K N_t^H F(K_t, X_t), \end{aligned} \tag{7}$$

and from Eqs. (3), (5), and (6), the total revenue from the low-CO_2-emitting sector is

$$\begin{aligned} P_L L_t &= P_L B_L N_t^L F(K_t, X_t) \\ &= B_K N_t^L F(K_t, X_t). \end{aligned} \tag{8}$$

Then, from Eqs. (1), (4), (5), (7), and (8), the accumulation of capital, K_t, should follow the law of motion,

$$\begin{aligned} \dot{K}_t + \delta K_t &= B_K N_t^K F(K_t, X_t) \\ &= B_K \big(1 - N_t^H - N_t^L\big) F(K_t, X_t) \\ &= B_K F(K_t, X_t) - P_H H_t - P_L L_t, \end{aligned} \tag{9}$$

with a given $K_0 = K(0) > 0$.

2.5 Preferences

We use a constant relative risk aversion (CRRA) type utility function,

$$U_t = \int_0^{\infty} e^{-\rho t} \frac{(H_t^{\beta} L_t^{\theta})^{1-\sigma} - 1}{1-\sigma} dt, \tag{10}$$

where parameters ρ, β, θ, σ are all strictly positive and $\beta + \theta = 1$.

2.6 Equilibrium

Let us define capital per unit of efficiency labor as $k_t \equiv K_t/X_t$. Since production functions F have constant returns to scale and from (5), we may rewrite the output, (4), of the capital-goods sector as

$$B_K F\big(\phi_t^K K_t, N_t^K X_t\big) = B_K N_t^K X_t F(k_t, 1),$$

and similarly the cost of capital in the capital-goods sector as

$$(r_t + \delta)\phi_t^K K_t = (r_t + \delta) N_t^K X_t k_t.$$

Competitive equilibrium requires that the marginal products of capital, $B_K N_t^K X_t F_1(k_t, 1)$, should be equal to the marginal cost of capital, $(r_t + \delta) N_t^K X_t$. Therefore, at an equilibrium, we should have the interest rate, r_t, that satisfies

$$r_t = B_K F_1(k_t, 1) - \delta, \tag{11}$$

which implies that capital is paid the marginal products of capital net of depreciation.

2.7 Consumption

Let us define output per unit of efficiency labor as $h_t \equiv H_t/X_t$ and $l_t \equiv L_t/X_t$. Using k_t, h_t, and l_t, the law of motion of k_t, Eq. (9), can be rewritten as[6]

$$\dot{k}_t + (g_t + \delta)k_t + P_H h_t + P_L l_t = B_K F(k_t, 1). \tag{12}$$

---

[6]Dividing both sides of (9) by X_t gives

$$\frac{\dot{K}_t}{X_t} + \delta k_t + P_H h_t + P_L l_t = B_K F(k_t, 1).$$

As $\dot{k}_t = \dot{K}_t/X_t - g k_t$, inserting this fact into the above equation derives the law of motion of k_t.

Similarly, the utility function, Eq. (10), can be rewritten as

$$U_t = \int_0^{\infty} e^{-\rho t} \frac{[(h_t X_t)^{\beta} (l_t X_t)^{\theta}]^{1-\sigma} - 1}{1-\sigma} dt. \tag{13}$$

The optimal consumption can be obtained by solving the following problem:

$$\max_{h_t, l_t} \ (13)$$

subject to (12), for an initial value of capital per unit of efficiency labor, $k_0 \equiv k(0) = K(0)/X(0)$, with $K(0) > 0$ and $X(0) > 0$ given.

To solve this dynamic problem, we use Pontryagin's maximum principle. The current-value Hamiltonian function, Φ_t, is defined as

$$\Phi_t \equiv \frac{[(h_t X_t)^{\beta} (l_t X_t)^{\theta}]^{1-\sigma} - 1}{1-\sigma} + q_t \big[B_K F(k_t, 1) - P_H h_t - P_L l_t - (\delta + g_t) k_t \big],$$

where q_t is the costate variable.

Then an optimal solution must satisfy the following first-order necessary conditions:

$$\partial \Phi_t / \partial h_t = 0 \Leftrightarrow [\cdot]^{-\sigma} \beta (h_t X_t)^{\beta-1} (l_t X_t)^{\theta} X_t = P_H q_t, \tag{14}$$

and

$$\partial \Phi_t / \partial l_t = 0 \Leftrightarrow [\cdot]^{-\sigma} \theta (h_t X_t)^{\beta} (l_t X_t)^{\theta-1} X_t = P_L q_t, \tag{15}$$

for all t. The optimal solution must also satisfy the law of motion of k_t, (12), and the following law of motion of q_t,

$$\begin{aligned} \dot{q}_t &= \rho q_t - \partial \Phi_t / \partial k_t \\ &= \rho q_t - \big[B_K F_1(k_t, 1) - (\delta + g_t) \big] q_t, \end{aligned} \tag{16}$$

for all t. The transversality condition, $\lim_{t \to \infty} k_t q_t e^{-\rho t} = 0$, must be also satisfied.

From (14) and (15), we get

$$\frac{\beta}{\theta} \left(\frac{l_t}{h_t} \right) = \frac{P_H}{P_L}. \tag{17}$$

Equation (17) implies that along the optimal path, l_t and h_t must grow at the same rate (denoted by κ), i.e.,

$$\frac{\dot{l}_t X_t + l_t \dot{X}_t}{l_t X_t} = \frac{\dot{h}_t X_t + h_t \dot{X}_t}{h_t X_t} \equiv \kappa. \tag{18}$$

By taking the natural log of (14) and then taking its time-derivative, we get

$$\begin{aligned} &-\sigma \left(\beta \frac{\dot{h}_t X_t + h_t \dot{X}_t}{h_t X_t} + \theta \frac{\dot{l}_t X_t + l_t \dot{X}_t}{l_t X_t} \right) \\ &\quad + (\beta - 1) \frac{\dot{h}_t X_t + h_t \dot{X}_t}{h_t X_t} + \theta \frac{\dot{l}_t X_t + l_t \dot{X}_t}{l_t X_t} + \frac{\dot{X}_t}{X_t} = \frac{\dot{q}_t}{q_t}, \end{aligned}$$

but from (18) and $\beta + \theta = 1$, it can reduce to

$$\kappa = -\frac{1}{\sigma}\left(\frac{\dot{q}_t}{q_t} - g_t\right). \tag{19}$$

As competitive equilibrium requires (11), we may rewrite the law of motion of q_t, (16), as

$$\frac{\dot{q}_t}{q_t} = \rho - r_t + g_t. \tag{20}$$

Thus, from (19) and (20), we have

$$\kappa = \frac{r_t - \rho}{\sigma}. \tag{21}$$

2.8 Balanced Growth

The growth rate of technology, g_t, by assumption, should approach a constant rate g. Thus (18) becomes

$$\frac{\dot{l}_t}{l_t} = \frac{\dot{h}_t}{h_t} = \kappa - g.$$

The law of motion of k_t, (12), suggests that the only path along which all variables grow at constant rate is that L_t, H_t, K_t, and X_t grow at rate g, i.e.,

$$\kappa - g = 0. \tag{22}$$

Let k^*, h^*, and l^* be the steady-state values. Then in balanced growth,

$$B_K F(k^*, 1) = (g + \delta)k^* + P_H h^* + P_L l^*. \tag{23}$$

From (21) and (22), the equilibrium interest rate r_t is

$$r_t = \rho + \sigma g. \tag{24}$$

Inserting (24) into (11), the steady-state value k^* solves

$$\rho + \sigma g + \delta = B_K F_1(k^*, 1).$$

We can also obtain, from (17) and (23), the steady-state values h^* and l^* as

$$h^* = \frac{\beta}{P_H}\left[B_K F(k^*, 1) - (g + \delta)k^*\right], \tag{25}$$

and

$$l^* = \frac{\theta}{P_L}\left[B_K F(k^*, 1) - (g + \delta)k^*\right]. \tag{26}$$

Stability of the dynamics in the vicinity of these steady-state values is analyzed in the Appendix.

2.8.1 Output and Employment

In balanced growth, the output in high-CO_2-emitting sector and the output in low-CO_2-emitting sector grow at the given long-run technological growth rate g, i.e.,

$$H_t^* = \frac{\beta}{P_H}\big[B_K F\big(k^*, 1\big) - (g+\delta)k^*\big]X_t,$$

and

$$L_t^* = \frac{\theta}{P_L}\big[B_K F\big(k^*, 1\big) - (g+\delta)k^*\big]X_t.$$

From (2) and (25), employment of the high-CO_2-emitting sector is

$$N^{H^*} = \frac{\beta}{P_H B_H F(k^*, 1)}\big[B_K F\big(k^*, 1\big) - (g+\delta)k^*\big], \tag{27}$$

and similarly, from (3) and (26), employment of the low-CO_2-emitting sector is

$$N^{L^*} = \frac{\theta}{P_L B_L F(k^*, 1)}\big[B_K F\big(k^*, 1\big) - (g+\delta)k^*\big]. \tag{28}$$

Finally, from (1), (27), and (28), employment of the capital goods sector is

$$N^{K^*} = 1 - N^{H^*} - N^{L^*}. \tag{29}$$

3 Preferences and Carbon Tax

Policy measures to change preferences has already been discussed in Sect. 1. Next we want to study the effects of a carbon tax on preferences.

3.1 A Carbon Tax

For such a version of a dynamic decision model with preferences we could rewrite the above baseline model with the utility function of (13), but including a tax on high-CO_2-emitting sector, in our model aggregated as $h_t X_t$.

We introduce a form of carbon tax policy that imposes a negative endowment of high-CO_2-emitting goods, $-\bar{H} < 0$, on each household. By reducing endowments, such a policy changes households' preferences (see Kongsamut et al. 2001). Then the utility function, (13), can be rewritten as

$$U_t = \int_0^{\infty} e^{-\rho t} \frac{[(h_t X_t - \overline{H})^{\beta}(l_t X_t)^{\theta}]^{1-\sigma} - 1}{1-\sigma} dt, \tag{30}$$

with $\beta + \theta = 1$. We may interpret $\bar{H} > 0$ as a type of carbon tax imposed on high-CO_2-emitting goods.[7]

3.2 Consumption Under Carbon Tax

In order to obtain an optimal consumption, we solve the following problem:

$$\max_{h_t, l_t} (30)$$

subject to (12) for a given value of $k_0 = k(0) > 0$.

As before, we use Pontryagin's maximum principle to solve this problem. The current-value Hamiltonian function, Ψ_t, is newly defined as

$$\Psi_t \equiv \frac{[(h_t X_t - \overline{H})^{\beta}(l_t X_t)^{\theta}]^{1-\sigma} - 1}{1-\sigma} + q_t\big[B_K F(k_t, 1) - P_H h_t - P_L l_t - (\delta + g_t) k_t\big],$$

where q_t is the costate variable.

The optimal solution must satisfy the following first-order necessary conditions:

$$\partial \Psi_t / \partial h_t = 0 \Leftrightarrow [\cdot]^{-\sigma} \beta (h_t X_t - \overline{H})^{\beta - 1} (l_t X_t)^{\theta} X_t = P_H q_t, \tag{31}$$

and

$$\partial \Psi_t / \partial l_t = 0 \Leftrightarrow [\cdot]^{-\sigma} \theta (h_t X_t - \overline{H})^{\beta} (l_t X_t)^{\theta - 1} X_t = P_L q_t, \tag{32}$$

for all t. The optimal solution must also satisfy the law of motion of k_t, (12), the law of motion of the costate variable, q_t, (16), for all t and the transversality condition, $\lim_{t \to \infty} k_t q_t e^{-\rho t} = 0$.

From (31) and (32), we get

$$\frac{\beta}{\theta}\left(\frac{l_t X_t}{h_t X_t - \overline{H}}\right) = \frac{P_H}{P_L}. \tag{33}$$

---

[7] We could also assume that the price P_H is raised in imperfectly competitive markets and then the reallocation of production and employment from high to low-carbon-intensive goods could occur through relative prices: As the price P_H is raised the relative share of high-carbon-intensive industries would decline. In this latter case the derivation of our baseline model of Sect. 3.1 would hold. The relative prices of the two sectors could also change through an environmental tax reform. In the latter case one can think, as Boehringer et al. (2008) suggest, that one has some tax incidence effect, where the tax is shifted forward and the demand for those products will react—reducing demand as a result of higher prices.

Equation (33) implies that along the optimal path, consumption of low-CO_2-emitting goods, $l_t X_t$, and consumption of high-CO_2-emitting goods, $h_t X_t - \bar{H}$, must grow at the same rate (denoted by κ), i.e.,

$$\frac{\dot{h}_t X_t + h_t \dot{X}_t}{h_t X_t - \overline{H}} = \frac{\dot{l}_t X_t + l_t \dot{X}_t}{l_t X_t} \equiv \kappa. \tag{34}$$

By taking the natural log of (31) and then taking its time-derivative, we get

$$-\sigma \left\{ \beta \frac{\dot{h}_t X_t + h_t \dot{X}_t}{h_t X_t - \overline{H}} + \theta \frac{\dot{l}_t X_t + l_t \dot{X}_t}{l_t X_t} \right\}$$
$$+ (\beta - 1) \frac{\dot{h}_t X_t + h_t \dot{X}_t}{h_t X_t - \overline{H}} + \theta \frac{\dot{l}_t X_t + l_t \dot{X}_t}{l_t X_t} + \frac{\dot{X}_t}{X_t} = \frac{\dot{q}_t}{q_t},$$

but from Eq. (34) and $\beta + \theta = 1$, it can reduce to

$$\kappa = -\frac{1}{\sigma} \left(\frac{\dot{q}_t}{q_t} - g_t \right). \tag{35}$$

The competitive equilibrium is described as (11). Thus the law of motion of the costate, q_t, (16), that describes the competitive equilibrium is the same as (20). Inserting this fact into (35), we have

$$\kappa = \frac{r_t - \rho}{\sigma}.$$

3.3 Balanced Growth Under Carbon Tax

We attempt to find a trajectory along which the real interest rate r_t is constant. Kongsamut et al. (2001) call such a trajectory a "Generalized Balanced Growth Path".

From (11), we know that k_t has to be constant in order for the real interest rate r_t to be constant. Moreover, by assumption, g_t converges to a constant rate g in the long run.

Then, from (12), along the GBG path where $\dot{k}_t = 0$, we have

$$P_H h_t + P_L l_t = B_K F(k, 1) - (g + \delta)k. \tag{36}$$

The right side of (36) is constant. The left side of (36), on the other hand, is not constant because from (34),

$$\dot{h}_t = (\kappa - g) h_t - \frac{\overline{H}}{X_t} \kappa, \tag{37}$$

and

$$\dot{l}_t = (\kappa - g)l_t, \tag{38}$$

and therefore,

$$P_H \dot{h}_t + P_L \dot{l}_t = (P_H h_t + P_L l_t)(\kappa - g) - \frac{P_H \bar{H}}{X_t}\kappa. \tag{39}$$

Thus, for $\overline{H} > 0$, there exits no κ that makes the right side of (39) zero unless $g = 0$. There exists no competitive equilibrium path along which the requirement of a constant real interest rate is satisfied.[8] This implies that a GBG path does not exists under the given carbon tax policy.

3.4 Output and Employment

When $\kappa = g$ and $g > 0$, consumption of high-CO_2-emitting goods, $h_t X_t - \overline{H}$, and low-CO_2-emitting goods, $l_t X_t$, grow at the rate g. Inserting this fact into (37), we get

$$\dot{h}_t = -\frac{\overline{H}}{X_t} g \leq 0. \tag{40}$$

Thus, the growth rate of output in high-CO_2-emitting sector is

$$\begin{aligned} \frac{\dot{H}_t}{H_t} &= \dot{h}_t \frac{X_t}{H_t} + g \\ &= g \frac{H_t - \overline{H}}{H_t}. \end{aligned}$$

Similarly, from (38), when $\kappa = g$,

$$\dot{l}_t = 0. \tag{41}$$

Thus, the growth rate of output in low-CO_2-emitting sector is

$$\frac{\dot{L}_t}{L_t} = g.$$

From (2), the output per unit of efficiency labor is $h_t = B_H N_t^H F(k_t, 1)$ in the high-CO_2-emitting sector. Taking a time derivative of this gives

$$\dot{N}_t^H = \frac{\dot{h}_t}{B_H F(k_t, 1)}, \tag{42}$$

[8] The effect of a carbon tax case on the price P_H in an environment of imperfect competition is discussed in a later subsection.

but from (40), we find the evolution of the employment in the high-CO_2-emitting sector as

$$\dot{N}_t^H = -\frac{\overline{H}}{X_t B_H F(k_t, 1)} g \leq 0. \tag{43}$$

Similarly, from (3), the output per unit of efficiency labor is $l_t = B_L N_t^L F(k_t, 1)$ in the low-CO_2-emitting sector. Taking a time derivative of this gives

$$\dot{N}_t^L = \frac{\dot{l}_t}{B_L F(k_t, 1)}, \tag{44}$$

but from (41), we find the evolution of the employment in the low-CO_2-emitting sector as

$$\dot{N}_t^L = 0. \tag{45}$$

Finally, from (1), (43), and (45), the evolution of the employment in the capital-goods sector is

$$\begin{aligned} \dot{N}_t^K &= -\dot{N}_t^H - \dot{N}_t^L \\ &= \frac{\overline{H}}{X_t B_H F(k_t, 1)} g \geq 0. \end{aligned} \tag{46}$$

So, due to taxation of the high carbon intensive sector, its output and employment will shrink, whereas the low carbon intensive sector stays unchanged and output and employment for the capital goods sector is rising.

Note however, as mitigation policy instrument, we might also assume that instead of a carbon tax a cap-and-trade system with emission permits, that have to be bought and traded, can be established that would imply a cost on the carbon-intensive activities.

4 Preferences, Carbon Tax and Product Subsidies

Next we consider the case of a double sided action, namely a carbon tax on high carbon intensive sectors and subsidies on low carbon intensive sectors.

4.1 Tax and Product Subsidies

In the next version we allow for taxation and subsidies of the sectors simultaneously. We consider an environmental policy through which each household is imposed a negative endowment of high-CO_2-emitting goods, $-\bar{H} < 0$, and a positive endowment of low-CO_2-emitting goods, $\overline{L} > 0$. We may express such an endowment-effect, as in Kongsamut et al. (2001), in the utility function. Due to this policy change, the utility function of the baseline model, (13), can be rewritten as

$$U_t = \int_0^{\infty} e^{-\rho t} \frac{[(h_t X_t - \overline{H})^{\beta} (l_t X_t + \overline{L})^{\theta}]^{1-\sigma} - 1}{1-\sigma} dt. \tag{47}$$

We may interpret the two policy parameters, $\bar{H}$ and $\bar{L}$, as a type of carbon tax imposed on high-CO_2-emitting goods and subsidies in the form of low-CO_2-emitting goods, respectively.

We also introduce a policy requirement that the levels of $\bar{H}$ and $\bar{L}$ are chosen by a policy maker such that they satisfy

$$P_L \overline{L} - P_H \overline{H} = 0. \tag{48}$$

An interpretation of (48) is that the market value of these endowments be equal to zero.

4.2 Consumption Under Carbon Tax and Product Subsidies

The household's problem is

$$\max_{h_t, l_t} \ (47)$$

subject to (12) for a given $k_0 = k(0) > 0$.

The current-value Hamiltonian function, Θ_t, is defined as

$$\Theta_t \equiv \frac{[(h_t X_t - \overline{H})^{\beta} (l_t X_t + \overline{L})^{\theta}]^{1-\sigma} - 1}{1-\sigma} + q_t \big[B_K F(k_t, 1) - P_H h_t - P_L l_t - (\delta + g_t) k_t\big].$$

The optimal solution must satisfy the following first-order necessary conditions:

$$\partial \Theta_t / \partial h_t = 0 \Leftrightarrow [\cdot]^{-\sigma} \beta (h_t X_t - \overline{H})^{\beta - 1} (l_t X_t + \overline{L})^{\theta} X_t = P_H q_t, \tag{49}$$

and

$$\partial \Theta_t / \partial l_t = 0 \Leftrightarrow [\cdot]^{-\sigma} \theta (h_t X_t - \overline{H})^{\beta} (l_t X_t + \overline{L})^{\theta - 1} X_t = P_L q_t, \tag{50}$$

for all t. As before, the optimal solution must also satisfy the law of motion of k_t, (12), the law of motion of the costate variable, q_t, (16), for all t and the transversality condition, $\lim_{t\to\infty} k_t q_t e^{-\rho t} = 0$.

From (49) and (50),

$$\frac{\beta}{\theta} \left(\frac{l_t X_t + \overline{L}}{h_t X_t - \overline{H}} \right) = \frac{P_H}{P_L}. \tag{51}$$

Equation (51) implies that along the optimal path, $l_t X_t + \bar{L}$ and $h_t X_t - \bar{H}$ must grow at the same rate (denoted by κ), i.e.,

$$\frac{\dot{l}_t X_t + l_t \dot{X}_t}{l_t X_t + \overline{L}} = \frac{\dot{h}_t X_t + h_t \dot{X}_t}{h_t X_t - \overline{H}} \equiv \kappa. \tag{52}$$

By taking the natural log of (49) and then its time derivative, we get

$$-\sigma\left(\beta\frac{\dot{h}_t X_t + h_t \dot{X}_t}{h_t X_t - \overline{H}} + \theta\frac{\dot{l}_t X_t + l_t \dot{X}_t}{l_t X_t + \overline{L}}\right)$$
$$+ (\beta - 1)\frac{\dot{h}_t X_t + h_t \dot{X}_t}{h_t X_t - \overline{H}} + \theta\frac{\dot{l}_t X_t + l_t \dot{X}_t}{l_t X_t + \overline{L}} + \frac{\dot{X}_t}{X_t} = \frac{\dot{q}_t}{q_t},$$

but from (52) and $\beta + \theta = 1$, it reduces to

$$\kappa = -\frac{1}{\sigma}\left(\frac{\dot{q}_t}{q_t} - g_t\right). \tag{53}$$

Inserting (20) into (53), we have

$$\kappa = \frac{r_t - \rho}{\sigma}.$$

4.3 Balanced Growth Under Carbon Tax and Product Subsidies

We find again a GBG path along which the real interest rate r_t is constant. From (11), k_t has to be constant in order for r_t to be constant. Moreover, in balanced growth, g_t is constant at g.

Then, from (12), along the GBG path where $\dot{k}_t = 0$, we have

$$P_H h_t + P_L l_t = B_K F(k, 1) - (g + \delta)k. \tag{54}$$

The right side of (54) is constant. On the other hand, the left side of (54) is not constant without additional assumptions.

However, recall our environmental policy requirement, (48). Then we may rewrite (51) as

$$P_L l_t X_t = \frac{\theta}{\beta} P_H h_t X_t - \frac{1}{\beta} P_H \overline{H}. \tag{55}$$

Inserting (55) into (54) gives

$$\frac{1}{\beta} P_H (h_t X_t - \overline{H}) = \left[B_K F(k, 1) - (g + \delta)k\right] X_t. \tag{56}$$

The right side of (56) grows at a given constant rate g. On the left side, from (52), $h_t X_t - \bar{H}$ also grows at a constant rate κ. Thus there exists a competitive equilibrium path at $\kappa = g$ along which the real interest rate is constant. This implies that a GBG path exists under our environmental policy with a requirement, (48).

4.4 Output and Employment

In the generalized balanced growth, consumption of both high-CO_2-emitting goods, $h_t X_t - \overline{H}$, and low-CO_2-emitting goods, $l_t X_t + \overline{L}$, grow at rate g, i.e., $\kappa = g$. Inserting this fact into (52) gives

$$\dot{h}_t = -\frac{\overline{H}}{X_t} g \leq 0. \tag{57}$$

Thus, the growth rate of output in high-CO_2-emitting sector is

$$\begin{aligned} \frac{\dot{H}_t}{H_t} &= \dot{h}_t \frac{X_t}{H_t} + g \\ &= g \frac{H_t - \overline{H}}{H_t}. \end{aligned}$$

Similarly, from (52),

$$\dot{l}_t = \frac{\overline{L}}{X_t} g \geq 0. \tag{58}$$

Thus, the growth rate of output in low-CO_2-emitting sector is

$$\begin{aligned} \frac{\dot{L}_t}{L_t} &= \dot{l}_t \frac{X_t}{L_t} + g \\ &= g \frac{L_t + \overline{L}}{L_t}. \end{aligned}$$

From (42) and (57), the evolution of the employment in the high-CO_2-emitting sector is

$$\dot{N}_t^H = -\frac{\overline{H}}{X_t B_H F(k_t, 1)} g \leq 0. \tag{59}$$

Similarly, from (44) and (58), the evolution of the employment in the low-CO_2-emitting sector is

$$\dot{N}_t^L = \frac{\overline{L}}{X_t B_L F(k_t, 1)} g \geq 0. \tag{60}$$

Finally, from (1), (59), and (60),

$$\begin{aligned} \dot{N}_t^K &= -\dot{N}_t^H - \dot{N}_t^L \\ &= \frac{\overline{H}}{X_t B_H F(k_t, 1)} g - \frac{\overline{L}}{X_t B_L F(k_t, 1)} g. \end{aligned} \tag{61}$$

However, recall our environmental policy requirement, (48). Then we may rewrite (51) as

$$\frac{\overline{H}}{\overline{L}} = \frac{B_H}{B_L}.$$

Thus, plugging this into (61), we obtain the evolution of the employment in the low-CO_2-emitting sector as

$$\dot{N}_t^K = 0. \tag{62}$$

The theoretical analysis in this section leads to the prediction that, due to the two-track policy—i.e., taxation of high-carbon intensity goods and subsidies for low-intensity goods—output and employment will shrink in the high-carbon sector and expand in the low-carbon sector. Next, we examine to what extent this prediction can be examined empirically.

5 Empirics of the Model

The effects of mitigation policies on output and employment can be estimated through econometric techniques. This is undertaken in Mittnik et al. (2013).

First one has to construct appropriate data sets. One can employ input-output tables in order to identify high- and low-carbon intensive industries. In Mittnik et al. (2013) German input-output tables, which are available for 71 sectors, are aggregated into two sectors: a high-carbon intensive sector (HCIS), and a low-carbon intensive sector (LCIS). In addition to traditional input-output tables, the German Federal Statistical Office provides industry specific data on CO_2 emission in kilo tons. With these data, one can calculate the CO_2-intensity of each industry measured in kilo tons over gross output in million euros (direct CO_2 intensity). This ratio describes how many kilo tons of CO_2 emissions a specific sector in the economy creates in order to generate one million euros of gross output. With the help of these key figures, we can rank different industries according to their CO_2 intensity and classify industries in the two sectors (HCIS and LCIS). Industries whose carbon intensity per unit of output is above (below) the median are classified as belonging to the high carbon intensity (low carbon intensity) sector.

Note that this grouping can be done on basis of one country's CO_2 intensity data and this information can be used for other countries as well by assuming that the ranking of industries is identical for other countries analyzed. The absolute level of CO_2 emissions as well as the absolute CO_2 intensity in a particular sector may of course differ among countries. This depends on the size of the industry, the technology used, the energy mix, and possibly on other factors. However, the relative position of an industry within a country can be expected to be roughly the same, especially among industrialized countries. Thus, energy intensive manufacturing industries like metals, coke, and mechanical wood can be expected to be relatively high-carbon intensive in any country. Since we aim at aggregating the industries into just two sectors (HCIS and LCIS), only changes in CO_2 intensities of industries around the median have an effect on the composition of the HCIS and LCIS in a country. As a next step, we use the industry time series data from EU KLEMS, available from the OECD, to determine the past growth of output and employment in the HCIS and the LCIS for our countries under study. The countries examined in Mittnik et al. (2013) are Germany, Australia, France, Hungary, Japan, South Korea, Sweden, UK, and the United States.

Next given the two sectors aggregation with respect to HCIS and LCIS, one can implement—following the sub-division of the economy in the two sectors in the theoretical model—a vector autoregression (VAR) and estimate concrete policy effects. Mittnik et al. (2013) consider three types of policies: (1) imposing a carbon tax on carbon intensive industries, (2) imposing a carbon tax and subsidizing the less carbon intensive industries, and (3) imposing a carbon tax and subsidizing wage cost of industries (reducing overhead cost for labor).[9] The policy effects of (1) and (2) correspond to our model variants of Sects. 3 and 4. Also the effects of the policy (3) were also estimated empirically in Mittnik et al. (2013), employing impulse response analysis involving individual and simultaneous policy shocks, and led to interesting results. The least favorable outcome was obtained when only a carbon tax was imposed on carbon-intensive industries and the revenue not used for other purposes, such as reducing other tax rates, subsidizing wages or the development of other (less carbon intensive) products. Since the proposed double-sided VAR setup employed allows one to impose budgetary neutrality one can study the cases when the revenue is used for other purposes. The empirical results show that in particular the second policy measure, i.e., carbon tax revenues are used to subsidize the development of other products, has the greatest net gains in terms of output and employment.

In summary, Mittnik et al. (2013) find that the specified simultaneous policy shocks do not have a huge impact on the level of aggregate output and employment. For the most part, structural adjustments are triggered and not reductions of economic activity as a whole. In several countries, like the US or Hungary, Mittnik et al. (2013) observe somewhat positive effects on total economic activity. For Australia, the effects are on the negative side. The reasons for these differences on an individual country level need to be studied further. It is likely, however, that also the initial conditions (for example, sizes of the HCIS and LCIS in terms of output and employment) at the time of the shock as well as differing sample periods play a role.

---

[9]This was a key element of the German ecological tax reform which was implemented in 1999, see Boehringer et al. (2008). If wages are subsidized in all sectors, high- and low-carbon intensive sectors, then the firms can employ labor at a lower cost. In this version of mitigation policy, we would have a reduction of employment in the carbon-intensive sector and an overall increase in employment due to the use of the tax revenue to subsidize wages across the board. Yet, the overall output and employment effects are ambiguous for such a mitigation policy, see Mittnik et al. (2013).

6 Conclusions

Recently, academic work has been put forward that argues for a greater urgency to implement effective climate policies to combat global warming. Concrete policy proposals for reducing CO_2 emissions have been developed by the IPCC. So far, however, it has not been sufficiently explored to what extent mitigation policies, such as cap-and-trade, carbon tax or phasing in of green technology, will entail a structural change of the economy, and what implication this may have for the stability of the growth path. Another essential issue that had not been sufficiently studied—done here using a growth model with structural change—is the question of the consequences of a transition to a green economy with respect to output and employment. Here, we have considered four types of policies: (1) changing households' preferences, (2) imposing a carbon tax, (3) imposing a carbon tax and subsidizing low-carbon-intensive economic activities, and (4) imposing carbon tax and subsidizing wage costs. Further research on these questions is needed.

Acknowledgements This paper has been presented at the following occasions: 2nd International Wuppertal Colloquium on "Sustainable Growth, Resource Productivity and Sustainable Industrial Policy"; International Institute for Labour Studies at ILO, Geneva, the Eastern Economic Association, Philadelphia, "The Economics of Climate Change" conference at the New School for Social Research, and a workshop at the University of Technology, Vienna. We thank all participants for the comments provided. The paper was also discussed at the "Meeting of the Task Force on Global Warming," at Manchester University, organized by Joseph Stiglitz.

Appendix

Stability Analysis

From the equilibrium market condition, (11), and the first-order necessary conditions, (12), (16), (18), and (21), we may summarize our model's dynamics by the following 3×3 dynamic system:

$$\dot{l}_t = \left(\frac{B_K F_1(k_t, 1) - \delta - \rho}{\sigma} - g_t \right) l_t,$$

$$\dot{h}_t = \left(\frac{B_K F_1(k_t, 1) - \delta - \rho}{\sigma} - g_t \right) h_t,$$

$$\dot{k}_t = B_K F(k_t, 1) - P_H h_t - P_L l_t - (g_t + \delta) k_t.$$

Then the Jacobian matrix, J, of our 3×3 dynamic system is

$$J = \begin{bmatrix} \frac{\partial \dot{l}_t}{\partial l_t} & \frac{\partial \dot{l}_t}{\partial h_t} & \frac{\partial \dot{l}_t}{\partial k_t} \\ \frac{\partial \dot{h}_t}{\partial l_t} & \frac{\partial \dot{h}_t}{\partial h_t} & \frac{\partial \dot{h}_t}{\partial k_t} \\ \frac{\partial \dot{k}_t}{\partial l_t} & \frac{\partial \dot{k}_t}{\partial h_t} & \frac{\partial \dot{k}_t}{\partial k_t} \end{bmatrix}$$

$$= \begin{bmatrix} \frac{B_K F_1(k_t,1)-\delta-\rho}{\sigma} - g_t & 0 & \frac{1}{\sigma} B_K F_{11}(k_t,1) l_t \\ 0 & \frac{B_K F_1(k_t,1)-\delta-\rho}{\sigma} - g_t & \frac{1}{\sigma} B_K F_{11}(k_t,1) h_t \\ -P_L & -P_H & B_K F_1(k_t,1) - (g_t+\delta) \end{bmatrix}.$$

Let J^* be J evaluated at the vicinity of the steady state where $l_t = l^*$, $h_t = h^*$, $k_t = k^*$, $g_t = g$, and $B_K F_1(k_t, 1) = \rho + \sigma g + \delta$.

Then the associated characteristic equation is

$$\det(J^* - \lambda I) = \begin{vmatrix} -\lambda & 0 & \frac{1}{\sigma} B_K F_{11}(k^*,1) l^* \\ 0 & -\lambda & \frac{1}{\sigma} B_K F_{11}(k^*,1) h^* \\ -P_L & -P_H & B_K F_1(k^*,1) - (g+\delta) - \lambda \end{vmatrix}$$

$$= -\lambda \Big\{ \lambda^2 - \big[B_K F_1(k^*,1) - (g+\delta)\big]\lambda + \frac{1}{\sigma} F_{11}(k^*,1)\big(P_H B_K h^* + P_L B_K l^*\big) \Big\},$$

where λ is the eigenvalue of J.

Thus the eigenvalues are

$$\lambda_1 = 0 \quad \text{and} \quad \lambda_{2,3} = \frac{B_K F_1(k^*,1) - (g+\delta) \pm \sqrt{\Delta}}{2},$$

where

$$\Delta \equiv \big[B_K F_1(k^*,1) - (g+\delta)\big]^2 - \frac{4}{\sigma} F_{11}(k^*,1)\big(P_H B_K h^* + P_L B_K l^*\big) > 0.$$

Therefore, we have a zero eigenvalue, a real positive eigenvalue, and a real negative eigenvalue.

References

Boehringer, C., Loeschel, A., & Welsch, H. (2008). Environmental taxation and induced structural change in an open economy: the role of market structure. *German Economic Review*, *9*(1), 17–40.

Greiner, A., & Semmler, W. (2008). *The global environment, natural resources and economic growth*. London: Oxford University Press.

Greiner, A., Gruene, L., & Semmler, W. (2009). Growth and climate change: thresholds and multiple equilibria. In J. Crespo Cuaresma, T. Palokangas, & A. Tarasyev (Eds.), *Dynamic systems, economic growth, and the environment* (pp. 63–78). Berlin: Springer.

IPCC (2007). *Climate change 2007*. International Panel on Climate Change.

Kaldor, N. (1957). A model of economic growth. *The Economic Journal*, *67*(268), 591–624.

Kongsamut, P., Rebelo, S., & Xie, D. (2001). Beyond balanced growth. *The Review of Economic Studies*, *68*, 869–892.

Kuznets, S. (1957). Quantitative aspects of the economic growth of nations: II. *Economic Development and Cultural Change*, *5*, 3–111.

Mankiw, N. G. (2007). One answer to global warming: a new tax. The New York Times, 07/16/09.
Mittnik, S., Semmler, W., Kato, M., & Samaan, D. (2013). Employment and output effects of climate policies. In *Oxford University handbook on the macroeconomics of climate change*. Forthcoming.
Nell, E., & Semmler, W. (2009). Economic growth and climate change: cap-and-trade or emission tax? Discussion paper. SCEPA, New York.
Nordhaus, W. (2008). *A question of balance*. Princeton: Princeton University Press.
Pasinetti, L. (1981). *Structural change and economic growth: a theoretical essay on the dynamics of the wealth of nations*. Cambridge: Cambridge University Press.
Proops, J. L., Faber, M., & Wagenhals, G. (1993). *Reducing CO_2 emissions: a comparative input-output study for Germany and the UK*. Berlin: Springer.
Stern, N. (2007). *The economics of climate change: the Stern review*. Cambridge: Cambridge University Press.
Uzawa, H. (2003). *Economic theory and global warming*. Cambridge: Cambridge University Press.
Weitzmann, M. (2009). On modeling and interpreting the economics of catastrophic climate change. *Review of Economics and Statistics*, *91*(1), 1–19.

One-Parameter GHG Emission Policy with R&D-Based Growth

Tapio Palokangas

Abstract This document examines the GHG emission policy of regions which use land, labor and emitting inputs in production and enhance their productivity by devoting labor to R&D, but with different endowments and technology. The regions also have different impacts on global pollution. The problem is to organize common emission policy, if the regions cannot form a federation with a common budget and the policy parameters must be uniform for all regions. The results are the following. If a self-interested central planner allocates emission caps in fixed proportion to past emissions (i.e. grandfathering), then it establishes the Pareto optimum, decreasing emissions and promoting R&D and economic growth.

1 Introduction

This document examines regions that produce the final good from land, labor and an emitting input and enhance their productivity by devoting labor to R&D. There is no limit to how much a region can emit, but because local emissions harm local production, there is an optimal level of emissions for a region. There is a central planner that decides how much each region can emit greenhouse gases (GHGs). Because the regions cannot form a federation, the central planner is *self-interested* (i.e. subject to lobbying) with no budget of its own. Furthermore, the central planner can use only one policy parameter that must be uniformly applied to all regions. In this framework, it is instructive to compare the cases of laissez-faire, Pareto optimum and lobbying.

It has been common in environmental economics to consider abatement in a two-sector framework where one sector produces a final good, but the other sector alleviates the use of natural resources (cf. Xepapadeas 2005, Chap. 4.3). The problem of environmental policy is then basically static: it answers the question of how resources could be optimally allocated between the sectors. Because that approach ig-

T. Palokangas (✉)
University of Helsinki, HECER and IIASA, Economicum Building, Arkadiankatu 7, 00014 Helsinki, Finland
e-mail: tapio.palokangas@helsinki.fi

E. Moser et al. (eds.), *Dynamic Optimization in Environmental Economics*, Dynamic Modeling and Econometrics in Economics and Finance 15,
DOI 10.1007/978-3-642-54086-8_5, © Springer-Verlag Berlin Heidelberg 2014

nores the long traces that environmental policy may cause for the economic growth of countries, this document examines emissions in a R&D-based growth model.

Haurie et al. (2006) examine a negotiation game where the regions talk over an international agreement on their use of GHGs to foster their economic development. They show that if GHGs in the atmosphere are exogenously constrained, then there is a Pareto optimum in these talks. Böhringer and Lange (2005) and Mackenzie et al. (2008) consider emissions-based allocation rules for which the basis of allocation is updated over time. They show that if the emission cap is absolute, then *grandfathering* schemes—which allocate allowances proportionally to past emissions—lead to the first-best. This document extends the analysis of these papers as follows. First, the policy maker in the coalition is self-interested, being subject to lobbying from the regions. Second, the international emission cap is endogenously determined by the same bargaining between the coalition members and the policy maker.

Jouvet et al. (2008) incorporate externality through pollution in an overlapping-generations (OLG) model, showing that the optimal growth path can be decentralized only with lump-sum transfers and a market for GHG permits. All permits should then be auctioned, which rules out all grandfathering practises. Jouvet et al. (2008) explain these results as follows: grandfathering practices cause a distortion by raising the return on investment, but the lump-sum provision of pollution rights to households does not distort anything. In contrast, this document considers the coordination of environmental policy through the design of a policy maker with no budget. It is instructing to see whether grandfathering schemes distort in that setting.

Palokangas (2009) considers emission policy with a self-interested central planner in a coalition of identical regions. That paper however assumes, rather unrealistically, that technology and primary resources are similar in all regions and that the central planner can negotiate over different emission caps with different regions. In this document, that assumption is relaxed: the central planner has only one policy parameter—the proportion of grandfathering in allocating emissions caps—that must be uniformly applied to heterogeneous regions. Section 2 presents the structure of the economy and Sect. 3 constructs the model for a single region. Sections 4, 5 and 6 examine the cases of laissez-faire, the Pareto optimum and lobbying, respectively. It is shown that a one-parameter grandfathering agreement is *self-enforcing* (cf. Haurie et al. 2006): no region has incentives to break it.

2 The Economy

The economy contains a large number (a "continuum") of regions that are placed evenly in the limit [0, 1]. The regions have different endowments of labor and land, different production functions in manufacturing and different technology in R&D. Their emissions have different impacts on global pollution. All regions produce the same consumption good from land, labor and energy. That good is chosen as the numeraire in the model, for convenience.

Each region $j \in [0, 1]$ supplies land A_j and labor L_j inelastically, and devotes l_j units of labor to production and the remainder,

$$z_j = L_j - l_j, \tag{1}$$

to R&D. There exists an emitting input called energy the extraction costs of which are ignored, for simplicity. It is assumed that emissions are proportional to the use of energy, m_j, in each region j. Pollution m is a linearly homogeneous function M of the emissions of all regions $j \in [0, 1]$:

$$m = M\big(m_j \mid j \in [0, 1]\big), \quad M \text{ homogeneous of degree one.} \tag{2}$$

In global warming problems, it is the stock of GHGs that causes damages and not the flow. In this document, however, the flow is used instead to simplify the dynamics. In the model, pollution affects the economy in two ways. First, pollution decreases utility globally. Second, local pollution harms local production. Except realism, there is also a technical reason to introduce the "local" effect: it enables the existence of the laissez-faire equilibrium in the case there is no international agent running emission policy.

To enable that the regions can increase their efficiency and consequently grow at different rates in a stationary-state equilibrium, we eliminate

- the terms-of-trade effect by the assumption that all regions produced the same internationally-traded good, and
- international capital movements by the assumption that all regions share the same constant rate of time preference, ρ.

On the assumption of perfect markets, each region $j \in [0, 1]$ behaves as if there were a single agent (hereafter called *region* j) that controls fully the resources in that region. This document ignores free riding, for simplicity: all regions $j \in [0, 1]$ are committed to common emission policy.

3 Single Region $j \in [0, 1]$

3.1 Production

When region j develops a new technology, it increases its productivity by constant proportion $a_j > 1$. The level of productivity in region j is then equal to $a_j{}^{\gamma_j}$, where γ_j is its serial number of technology. The innovation of new technology in region j increases γ_j by one.

Region j produces its output y_j from land A_j, labor l_j and energy m_j. It is assumed that local emissions, which are proportional to energy input m_j, harm local production.[1] It is furthermore assumed that labor l_j and energy m_j form a composite input $\phi^j(l_j, m_j)$ through CES technology, but otherwise there is Cobb-Douglas technology:[2]

$$y_j = a_j{}^{\gamma_j} f^j(l_j, m_j) m_j^{-\beta}, \quad f^j(l_j, m_j) \doteq A_j^{1-\alpha_j} \phi^j(l_j, m_j)^{\alpha_j}, \ 0 < \alpha_j < 1, \ \beta > 0,$$
$$f_l^j > 0, \ f_m^j > 0, \ \phi_l^j > 0, \ \phi_m^j > 0, \ \phi_{ll}^j < 0, \ \phi_{mm}^j < 0, \ \phi_{lm}^j > 0, \tag{3}$$

where the subscripts l and m denote the partial derivative of the function with respect to l_j and m_j, respectively, $a_j{}^{\gamma_j}$ is total factor productivity, α_j a parameter and β is the constant elasticity of output with respect to emissions m_j. The higher β, the more local emissions m_j harm local production.

When the markets are perfect in region j, one can interpret $1 - \alpha_j$ as the expenditure share of land and α_j that of labor and energy taken together. Noting (3), the expenditure shares of energy and labor in production are

$$\begin{aligned}
\frac{m_j f_m^j(l_j, m_j)}{f^j(l_j, m_j)} &= \alpha_j \frac{m_j \phi_m^j(l_j, m_j)}{\phi^j(l_j, m_j)} = \alpha_j \frac{\phi_m^j(l_j/m_j, 1)}{\phi^j(l_j/m_j, 1)} \doteq \xi^j\left(\frac{l_j}{m_j}\right) \in (0, \alpha_j), \\
\frac{l_j f_l^j(l_j, m_j)}{f^j(l_j, m_j)} &= \alpha_j \frac{l_j \phi_l^j(l_j, m_j)}{\phi^j(l_j, m_j)} = \alpha_j\left[1 - \frac{m_j \phi_m^j(l_j, m_j)}{\phi^j(l_j, m_j)}\right] \\
&= \alpha_j - \xi^j\left(\frac{l_j}{m_j}\right) \in (0, \alpha_j).
\end{aligned} \tag{4}$$

Because the composite input $\phi^j(l_j, m_j)$ is a CES function, one obtains

$$(\xi^j)'\left(\frac{l_j}{m_j}\right) = \frac{d\xi^j}{d(l_j/m_j)} \begin{cases} > 0 & \text{for } 0 < \sigma_j < 1, \\ < 0 & \text{for } \sigma_j > 1, \end{cases} \tag{5}$$

where σ_j is the constant elasticity of substitution between inputs l_j and m_j.

3.2 *Research and Development (R&D)*

An increase in productivity in region j, $a_j{}^{\gamma_j}$ (cf. the production function (3)), depends on labor devoted to R&D, z_j, in that region: the probability that input z_j leads to development of a new technology with a jump from γ_j to $\gamma_j + 1$ in a small period

---

[1] Without this assumption, region j would use an indefinitely large amount of energy in the case of laissez-faire (cf. Sect. 4).

[2] The use of a general production function $y_j = a_j^{\gamma_j} F(A_j, l_j, m_j)$ would excessively complicate the analysis.

of time $d\theta$ is given by $\lambda_j z_j d\theta$, while the probability that input z_j remains without success is given by $1 - \lambda_j z_j d\theta$, where $\lambda_j > 0$ is a constant. Noting (1), this defines a Poisson process χ_j with

$$d\chi_j = \begin{cases} 1 & \text{with probability } \lambda_j z_j d\theta, \\ 0 & \text{with probability } 1 - \lambda_j z_j d\theta, \end{cases} \qquad z_j = L_j - l_j, \tag{6}$$

where $d\chi_j$ is the increment of the process χ_j.

3.3 Preferences

All regions have the same preferences: the expected utility of region $j \in [0, 1]$ starting at time T is given by

$$E \int_T^\infty c_j m^{-\delta} e^{-\rho(\theta - T)} d\theta, \quad \delta > 0, \ \rho > 0, \tag{7}$$

where E is the expectation operator, θ time, c_j consumption in region j, ρ the constant rate of time preference and δ the constant elasticity of temporary utility with respect to economy-wide emissions m. The lower ρ, the more patient the regions are. Total pollution m decreases welfare in all regions $j \in [0, 1]$, but a single region is so small that it ignores this dependence. The higher δ, the more pollution m is disliked.

4 Laissez-Faire

Because all regions $j \in [0, 1]$ produce the same consumption good, then, without GHG emissions management, each region j consumes what it produces, $c_j = y_j$. Noting (3) and $c_j = y_j$, the expected utility of the region starting at time T, (7), becomes

$$\Upsilon^j = E \int_T^\infty y_j m^{-\delta} e^{-\rho(\theta - T)} d\theta = E \int_T^\infty a_j^{\gamma_j} f^j(l_j, m_j) m_j^{-\beta} m^{-\delta} e^{-\rho(\theta - T)} d\theta. \tag{8}$$

Assume for a while that energy input m_j is held constant. Region j then maximizes its expected utility (8) by its labor devoted to production, l_j, subject to its technological change (6), given pollution m. The solution of this maximization is the following (cf. Appendix 1):

Proposition 1 *The expected utility of region j is*

$$\Upsilon^j = m^{-\delta} \Pi^j(\gamma_j, m_j, T), \quad \textit{for which } \frac{\partial \Pi^j}{\partial m_j} = \frac{\Pi^j}{m_j}\left[\xi^j\left(\frac{l_j}{m_j}\right) - \beta\right]. \tag{9}$$

Region j chooses its labor input l_j so that

$$\frac{(a_j-1)\lambda_j l_j}{\rho+(1-a_j)\lambda_j(L_j-l_j)}=\alpha_j-\xi^j\left(\frac{l_j}{m_j}\right). \tag{10}$$

In the presence of laissez-faire, region j can optimally determine its energy input m_j as well: it maximizes the value of its program, Υ^j, by m_j. Given (9), this leads to the first-order condition

$$\frac{\partial\Upsilon^j}{\partial m_j}=m^{-\delta}\frac{\partial\Pi^j}{\partial m_j}=m^{-\delta}\frac{\Pi^j}{m_j}\left[\xi^j\left(\frac{l_j}{m_j}\right)-\beta\right]=0 \quad\text{and}\quad \xi^j\left(\frac{l_j}{m_j}\right)=\beta. \tag{11}$$

The second-order condition of the maximization is given by

$$\frac{\partial^2\Upsilon^j}{\partial m_j^2}=-\underbrace{m^{-\delta}\frac{\Pi^j}{m_j}}_{+}(\xi^j)'\underbrace{\frac{l_j}{m_j^2}}_{+}<0 \quad\text{and}\quad (\xi^j)'>0.$$

Given this and (5), labor and energy are gross complements, $0<\sigma_j<1$, and $(\xi^j)'>0$ holds true everywhere. From this, (10) and (11) it follows that

$$\frac{(a_j-1)\lambda_j l_j^L}{\rho+(1-a_j)\lambda_j(L_j-l_j^L)}=\alpha_j-\beta, \qquad \xi^j\left(\frac{l_j^L}{m_j^L}\right)=\beta \quad\text{with } (\xi^j)'>0, \tag{12}$$

where the superscript L denotes the laissez-faire equilibrium.

Finally, the following result is proven in Appendix 2:

Proposition 2 *The more emissions harm locally* (*i.e. the higher* β), *the less there are emissions* m_j^L, $dm_j^L/d\beta<0$, *and the more there is R&D* (*i.e. the higher* z_j^L), $dz_j^L/d\beta>0$.

Because technological change generated by R&D decreases the need for polluting energy, there are incentives to perform R&D.

5 The Pareto Optimum

Grandfathering means that emission caps have a base that is determined by the history, but updated over time. In models with discrete time, that base would be calculated by a moving average of past emissions. In the quality-ladders model of this document where time is continuous, the base is specified as follows. The central planner sets the pollutant caps in fixed proportion ε to the energy input of that region under previous technology, $\widehat{m}_j$:

$$m_j\le\varepsilon\widehat{m}_j \quad\text{for } j\in[0,1] \text{ and } \varepsilon>0. \tag{13}$$

If the current number of technology is γ_j, then the allocation base $\widehat{m}_j$ is calculated by energy input under previous technology $\gamma_j - 1$ (cf. Sect. 3.1). If the central planner tightens emission policy by decreasing ε below one, then the constraint (13) becomes binding for all regions $j \in [0,1]$. Because the function M in (2) is linearly homogeneous, one then obtains:

$$m_j = \varepsilon \widehat{m}_j \quad \text{for } j \in [0,1], m = \varepsilon \widehat{m},\ \widehat{m} \doteq M\big(\widehat{m}_j \mid j \in [0,1]\big). \tag{14}$$

In the grandfathering scheme, there is thus only one policy parameter ε.

Because all regions $j \in [0,1]$ produce the same consumption good, total consumption is equal to total production, $\int_0^1 c_j dj = \int_0^1 y_j dj$. To construct the Pareto optimum, let us introduce a benevolent central planner that maximizes the welfare of the representative agent of the economy, $\mathscr{W}$. Given (7), (8), (9) and $\int_0^1 c_j dj = \int_0^1 y_j dj$, that welfare is

$$\begin{aligned}\mathscr{W} &\doteq \int_0^1 \left[E \int_T^\infty c_j m^{-\delta} e^{-\rho(\theta - T)} d\theta \right] dj = E \int_T^\infty \left(\int_0^1 c_j dj \right) m^{-\delta} e^{-\rho(\theta - T)} d\theta \\ &= E \int_T^\infty \left(\int_0^1 y_j dj \right) m^{-\delta} e^{-\rho(\theta - T)} d\theta = E \int_0^1 \left(\int_T^\infty y_j m^{-\delta} e^{-\rho(\theta - T)} d\theta \right) dj \\ &= \int_0^1 \Upsilon^j dj = m^{-\delta} \int_0^1 \Pi^j(\gamma_j, m_j, T) dj\end{aligned}$$

which should be maximized by the policy parameter ε. Given (9) and (14), this leads to the first-order conditions

$$\begin{aligned}0 = \frac{d\mathscr{W}}{d\varepsilon} &= m^{-\delta} \int_0^1 \frac{\partial \Pi^j}{\partial m_j} \underbrace{\frac{\partial m_j}{\partial \varepsilon}}_{=\widehat{m}_j} dj - \delta m^{-\delta - 1} \underbrace{\frac{\partial m}{\partial \varepsilon}}_{=\widehat{m}} \int_0^1 \Pi^j dj \\ &= m^{-\delta} \left[\int_0^1 \frac{\partial \Pi^j}{\partial m_j} \widehat{m}_j dj - \delta \frac{\widehat{m}}{m} \int_0^1 \Pi^j dj \right] \\ &= m^{-\delta} \left\{ \int_0^1 \Pi^j \left[\xi^j \left(\frac{l_j}{m_j} \right) - \beta \right] \frac{\widehat{m}_j}{m_j} dj - \delta \frac{\widehat{m}}{m} \int_0^1 \Pi^j dj \right\} \\ &= m^{-\delta} \int_0^1 \Pi^j \left\{ \left[\xi^j \left(\frac{l_j}{m_j} \right) - \beta \right] \frac{\widehat{m}_j}{m_j} - \delta \frac{\widehat{m}}{m} \right\} dj.\end{aligned} \tag{15}$$

In the stationary state, all inputs (l_j, m_j) for all regions $j \in [0,1]$ must be constant. Once the economy attains the stationary state, the emissions under the previous and current technology become equal: $\widehat{m} = m$ and $\widehat{m}_j = m_j$ for $j \in [0,1]$. Plugging these conditions and into (15) yields

$$0 = m^{-\delta} \int_0^1 \Pi^j \left[\xi^j \left(\frac{l_j}{m_j} \right) - \beta - \delta \right] dj. \tag{16}$$

Because the expected utilities Π^j for $j \in [0, 1]$ are random variables, then, given (16), the only possible stationary state is

$$\xi^j\left(\frac{l_j}{m_j}\right) = \beta + \delta \quad \text{for } j \in [0, 1]. \tag{17}$$

The equilibrium conditions (10) for the regions $j \in [0, 1]$ as well as those (17) for the central planner can be written as

$$\xi^j\left(\frac{l_j^P}{m_j^P}\right) = \beta + \delta, \qquad \frac{(a-1)\lambda_j l_j^P}{\rho + (1-a)\lambda_j(L_j - l_j^P)} = \alpha_j - \beta - \delta, \tag{18}$$

where the superscript P denotes the Pareto optimum equilibrium.

The comparison of (18) with (12) shows that the introduction of a benevolent central planner increases the parameter β up to $\beta + \delta$ in the system. Thus, Proposition 2 has the following corollary:

Proposition 3 *A shift from laissez-faire to the Pareto optimum decreases emissions, $m_j^P < m_j^L$, and increases R&D, $z_j^P > z_j^L$.*

The introduction of a benevolent central planner internalizes the negative externality through emissions. This increases incentives to perform R&D. With the uniform proportionality rule ε, all regions face the same marginal benefits from pollutants via allocation in subsequent periods. In contrast to Böhringer and Lange (2005), the regulatory cap m^P is not exogenous but endogenously determined.

6 Regulation

In this section, regions $j \in [0, 1]$ lobby the central planner over the policy parameter ε. Following Grossman and Helpman (1994), it is assumed that the central planner has its own interests and collects political contributions R_j from regions $j \in [0, 1]$. This is a common agency game, the order of which is then the following (cf. Grossman and Helpman 1994, and Dixit et al. 1997). First, the regions $j \in [0, 1]$ set their political contributions R_j conditional on the central planner's prospective policy ε. Second, the central planner sets its policy ε and collects the contributions from the regions. Third, the regions maximize their utilities. This game is solved in reverse order: Sect. 6.1 considers the equilibrium of the regions and Sect. 6.2 the political equilibrium.

6.1 Optimal Program

Region j pays its political contributions R_j to the central planner. It is assumed, for simplicity, that the central planner consists of civil servants who inhabit regions

$j \in [0, 1]$ evenly. Thus, the regions gets an equal share R of total contributions,

$$R \doteq \int_0^1 R_j dj \Big/ \int_0^1 dk = \int_0^1 R_j dj. \tag{19}$$

Noting the production function (3), consumption in region j is then

$$c_j = y_j + R - R_j = a_j^{\gamma_j} f^j(l_j, m_j) m_j^{-\beta} + R - R_j, \tag{20}$$

where y_j is income from production and $R - R_j$ net revenue from political contributions in region j. Noting (20), the expected utility of region j starting at time T, (7), becomes

$$\Theta^j = E \int_T^\infty \left[a_j^{\gamma_j} f^j(l_j, m_j) m_j^{-\beta} + R - R_j\right] m^{-\delta} e^{-\rho(\theta - T)} d\theta. \tag{21}$$

Region j maximizes its expected utility (21) by its labor devoted to production, l_j, subject to technological change in the region, (6), given the emission cap m_j, pollution m and political contributions R_j and R. The solution for this optimal program is the function (cf. Appendix 3)

$$\begin{aligned} &\Theta^j(m_j, m, R, R_j, \gamma_j), \qquad \frac{\partial \Theta^j}{\partial m_j} = m^{-\delta} \frac{\Gamma^j(\gamma_j, m_j, T)}{m_j} \left[\xi^j\left(\frac{l_j^*}{m_j}\right) - \beta\right], \\ &\frac{\partial \Theta^j}{\partial m} = -\delta m^{-\delta-1}\left(\Gamma^j + \frac{R - R_j}{\rho}\right), \qquad -\frac{\partial \Theta^j}{\partial R_j} = \frac{\partial \Theta^j}{\partial R} = \frac{m^{-\delta}}{\rho}, \end{aligned} \tag{22}$$

where the random variable Γ^j is the expected value of the flow of output for region j and l_j^* is the optimal labor input in production for which

$$\frac{(a_j - 1)\lambda_j l_j^*}{\rho + (1 - a_j)\lambda_j (L_j - l_j^*)} = \alpha_j - \xi^j\left(\frac{l_j^*}{m_j}\right). \tag{23}$$

6.2 The Political Equilibrium

Because each region j affects the central planner by its contributions R_j, its contribution schedule depends on the central planner's policy ε (cf. (19)):

$$R_j(\varepsilon) \quad \text{for } j \in [0, 1], \qquad R(\varepsilon) \doteq \int_0^1 R_k(\varepsilon) dk. \tag{24}$$

The central planner maximizes present value of the expected flow of the political contributions R from all regions $j \in [0, 1]$:

$$G(R) \doteq E \int_T^\infty R e^{-\theta(\theta - T)} d\theta = \frac{R}{\rho}. \tag{25}$$

Each region j maximizes its expected utility Θ^j (cf. (22)).

According to Dixit et al. (1997), a subgame perfect Nash equilibrium for this lobbying game is a set of contribution schedules $R_j(\varepsilon)$ and a policy ε such that the following conditions (i)–(iv) hold:

(i) Contributions R_j are non-negative but no more than the contributor's income, $\Theta^j \geq 0$.

(ii) The policy ε maximizes the central planner's welfare (25) taking the contribution schedules $R_j(\varepsilon)$ as given,

$$\varepsilon = \arg\max_{\varepsilon} G\big(R(\varepsilon)\big) = \arg\max_{\varepsilon \in [0,1]} R(\varepsilon). \tag{26}$$

(iii) Region j cannot have a feasible strategy $R_j(\varepsilon)$ that yields it a higher level of utility than in equilibrium, given the central planner's anticipated decision rule (14),

$$\varepsilon = \arg\max_{\varepsilon} \Theta^j\big(m_j, m, R, R_j(\varepsilon), \gamma_j\big) \quad \text{with } m_j = \varepsilon \widehat{m}_j \text{ and } m = \varepsilon \widehat{m}. \tag{27}$$

Because the region is small, it takes the total contributions of all regions, R, as given. However, the region observes the dependency of pollution m on environmental policy ε (cf. (14)).

(iv) Region j provides the central planner at least with the level of utility than in the case it offers nothing ($R_j = 0$), and the central planner responds optimally given the other regions contribution functions,

$$G\big(R(\varepsilon)\big) \geq \max_{\varepsilon} G\big(R(\varepsilon)\big)\big|_{R_j=0}.$$

6.3 The Stationary State

Noting (22), the conditions (27) for regions $j \in [0, 1]$ is equivalent to

$$0 = \frac{d\Theta^j}{d\varepsilon} = \frac{\partial \Theta^j}{\partial R_j}\frac{dR_j}{d\varepsilon} + \frac{\partial \Theta^j}{\partial m_j}\underbrace{\frac{\partial m_j}{\partial \varepsilon}}_{=\widehat{m}_j} + \frac{\partial \Theta^j}{\partial m}\underbrace{\frac{\partial m}{\partial \varepsilon}}_{=\widehat{m}} = \frac{\partial \Theta^j}{\partial R_j}\frac{dR_j}{d\varepsilon} + \frac{\partial \Theta^j}{\partial m_j}\widehat{m}_j + \frac{\partial \Theta^j}{\partial m}\widehat{m}$$

$$= -\frac{m^{-\delta}}{\rho}\frac{dR_j}{d\varepsilon} + m^{-\delta}\Gamma^j\left[\xi^j\left(\frac{l_j}{m_j}\right) - \beta\right]\frac{\widehat{m}_j}{m_j} - \delta m^{-\delta}\left(\Gamma^j + \frac{R - R_j}{\rho}\right)\frac{\widehat{m}}{m}$$

and

$$\frac{1}{\rho}\frac{dR_j}{d\varepsilon} = \Gamma^j\left[\xi^j\left(\frac{l_j}{m_j}\right) - \beta\right]\frac{\widehat{m}_j}{m_j} - \delta\left(\Gamma^j + \frac{R - R_j}{\rho}\right)\frac{\widehat{m}}{m} \quad \text{for } j \in [0, 1]. \tag{28}$$

Once the economy attains the stationary state, the emissions under the previous and current technology become equal: $\widehat{m} = m$ and $\widehat{m}_j = m_j$ for $j \in [0, 1]$. Plugging these conditions into (28) yields

$$\frac{1}{\rho}\frac{dR_j}{d\varepsilon} = \left[\xi^j\left(\frac{l_j}{m_j}\right) - \beta\right]\Gamma^j - \delta\left(\Gamma^j + \frac{R - R_j}{\rho}\right) \quad \text{for } j \in [0, 1].$$

Noting these equations and (24), the central planner's equilibrium condition (26) is equivalent to

$$\begin{aligned} 0 = \frac{dR}{d\varepsilon} &= \int_0^1 \frac{dR_j}{d\varepsilon}dj = \rho\int_0^1 \left\{\left[\xi^j\left(\frac{l_j}{m_j}\right) - \beta\right]\Gamma^j - \delta\left(\Gamma^j + \frac{R - R_j}{\rho}\right)\right\}dj \\ &= \rho\left\{\int_0^1 \left[\xi^j\left(\frac{l_j}{m_j}\right) - \beta - \delta\right]\Gamma^j dj - \frac{\delta}{\rho}\underbrace{\int_0^1 (R - R_j)dj}_{=0}\right\} \\ &= \rho\int_0^1 \left[\xi^j\left(\frac{l_j}{m_j}\right) - \beta - \delta\right]\Gamma^j dj. \end{aligned} \tag{29}$$

In the stationary state, all inputs (l_j, m_j) for all regions $j \in [0, 1]$ must be constant. Because the expected value of the flow of output, Γ^j, is a random variable for all regions $j \in [0, 1]$, then, given (29), the only possible stationary state in the economy of regions $j \in [0, 1]$ is

$$\xi^j\left(\frac{l_j}{m_j}\right) = \beta + \delta \quad \text{for } j \in [0, 1]. \tag{30}$$

This means that if region $j \in [0, 1]$ has confidence on stable development, then it expects that its expenditure share of energy, ξ^j, will be equal to $\beta + \delta$ in the long run. From the equilibrium conditions (23) of the regions $j \in [0, 1]$ as well as those (30) of the central planner, one obtains

$$\xi^j\left(\frac{l_j^G}{m_j^G}\right) = \beta + \delta, \qquad \frac{(a_j - 1)\lambda_j l_j^G}{\rho + (1 - a_j)\lambda_j(L_j - l_j^G)} = \alpha_j - \beta - \delta, \tag{31}$$

where the superscript G denotes grandfathering of emissions.

Comparing the systems (18) and (31) yields the following result:

Proposition 4 *Regulation leads to the Pareto optimum,* $(l_j^G, m_j^G) = (l_j^P, m_j^P)$ *for* $j \in [0, 1]$.

The introduction of a self-interested central planner has the same impact as that of a benevolent central planner: it internalizes the externality of emissions through pollution, leading to the Pareto optimum. This means that an agreement on a self-interested policy maker is self-enforcing: no region has incentives to break it.

7 Conclusions

This document examines the design of emission policy for a large number of regions which use land, labor and emitting inputs in production, but which can increase their total factor productivity by allocating labor to R&D. The use of emitting inputs pollutes, decreasing welfare everywhere. The regions can agree on a central planner and authorize it to grant them GHG emission caps. Because the regions do not form a federation with a budget of its own, the central planner is non-benevolent, self-interested and subject to lobbying. It is plausible to assume that the policy parameter of the central planner is uniform throughout all regions.

By the use of grandfathering schemes with one policy parameter only, the central planner internalizes the negative externality through GHG emissions. When emission caps are set in proportion to past emissions, all regions face the same marginal benefits from emissions via allocation in subsequent periods. Because the basis for allocation is updated over time, the central planner has the full control of resources. Thus, an agreement on the central planner, benevolent or self-interested, leads to the first-best allocation of resources (i.e. the Pareto optimum). Consequently, that agreement is self-enforcing.

Appendix 1

Region j maximizes (21) by (l_j, m_j) subject to (6), given m. It is equivalent to maximize

$$E\int_T^\infty a_j^{\gamma_j} f^j(l_j, m_j) m_j^{-\beta} e^{-\rho(t-T)} dt$$

by (l_j, m_j) subject to (6).

Assume for a while that energy input m_j is kept constant. The value of this maximization is

$$\Pi^j(\gamma_j, m_j, T) = \max_{l_j \text{ s.t. } (6)} E\int_T^\infty a_j^{\gamma_j} f^j(l_j, m_j) m_j^{-\beta} e^{-\rho(t-T)} dt. \tag{32}$$

Let us denote $\Pi^j = \Pi^j(\gamma_j, m_j, T)$ and $\widetilde{\Pi}^j = \Pi^j(\gamma_j + 1, m_j, T)$. The Bellman equation corresponding to the optimal program (32) is given by (cf. Dixit and Pindyck 1994)

$$\rho\Pi^j = \max_{l_j, m_j} \Psi(l_j, m_j, \gamma_j, T), \quad \text{where}$$

$$\Psi(l_j, m_j, \gamma_j, T) = a_j^{\gamma_j} f^j(l_j, m_j) m_j^{-\beta} + (\widetilde{\Pi}^j - \Pi^j)\lambda_j(L_j - l_j). \tag{33}$$

Noting (4), this leads to the first-order condition

$$\frac{\partial\Psi}{\partial l_j} = a_j^{\gamma_j} f_l^j(l_j, m_j) m_j^{-\beta} - \lambda_j(\widetilde{\Pi}^j - \Pi^j)$$

$$= \frac{1}{l_j} a_j^{\gamma_j} f^j(l_j, m_j) m_j^{-\beta}\left[1 - \xi^j\left(\frac{l_j}{m_j}\right)\right] - \lambda_j(\widetilde{\Pi}^j - \Pi^j) = 0. \tag{34}$$

To solve the dynamic program (32), assume that the value of the program, Π^j, is in fixed proportion $\vartheta_j > 0$ to instantaneous utility at the optimum. Noting (4), this implies

$$\begin{aligned} &\Pi^j(\gamma_j, m_j, T) = \vartheta_j a_j^{\gamma_j} f^j(l_j^*, m_j) m_j^{-\beta} \quad \text{with} \\ &\frac{\partial \Pi^j}{\partial m_j} = \Pi^j \left[\frac{f_m^j(l_j, m_j)}{f^j(l_j, m_j)} - \frac{\beta}{m_j} \right] = \frac{\Pi^j}{m_j} \left[\xi^j \left(\frac{l_j}{m_j} \right) - \beta \right], \end{aligned} \tag{35}$$

where l_j^* is the optimal value of the control variable l_j. This implies

$$(\widetilde{\Pi}^j - \Pi^j)/\Pi^j = a_j - 1. \tag{36}$$

Inserting (35) and (36) into the Bellman equation (33) yields

$$1/\vartheta_j = \rho + (1 - a_j)\lambda_j (L_j - l_j^*) > 0. \tag{37}$$

Inserting (35), (36) and (37) into (34), and noting $(\xi^j)' > 0$ yield (12):

$$\begin{aligned} 0 &= \vartheta_j \frac{l_j}{\Pi^j} \frac{\partial \Psi}{\partial l_j} = \underbrace{a_j^{\gamma_j} f^j(l_j, m_j) m_j^{-\beta} \frac{\vartheta_j}{\Pi^j}}_{=1} \left[\alpha_j - \xi^j \left(\frac{l_j}{m_j} \right) \right] - \Big(\underbrace{\frac{\widetilde{\Pi}^j}{\Pi_j^j}}_{=a_j} - 1 \Big) \lambda_j l_j \vartheta_j \\ &= \alpha_j - \xi^j \left(\frac{l_j}{m_j} \right) - \frac{(a_j - 1)\lambda_j l_j}{\rho + (1 - a_j)\lambda_j (L_j - l_j^*)}. \end{aligned} \tag{38}$$

From (8), (32) and (37) it follows that

$$\begin{aligned} \Upsilon^j &= \max_{l_j \text{ s.t. } (6)} E \int_T^\infty a_j^{\gamma_i} f^j(l_j, m_j) m_j^{-\beta} m^{-\delta} e^{-\rho(\theta - T)} d\theta \\ &= m^{-\delta} E \int_T^\infty a_j^{\gamma_j} f^j(l_j, m_j) m_j^{-\beta} e^{-\rho(\theta - T)} d\theta = m^{-\delta} \Pi^j(\gamma_j, m_j, T). \end{aligned} \tag{39}$$

Results (35), (38) and (39) lead to Proposition 1.

Appendix 2

Given (1), (3), (4) and (12), it then holds true that

$$\begin{aligned} &\rho + \underbrace{(1 - a_j)}_{-} \lambda_j \underbrace{(L_j - l_j^L)}_{+} \underbrace{\xi^j}_{\in(0,1)} > \rho + (1 - a_j)\lambda_j (L_j - l_j^L) > 0, \\ &\frac{(a_j - 1)\lambda_j l_j^L}{\rho + (1 - a_j)\lambda_j (L_j - l_j^L)} < \alpha_j - \beta < \alpha_j < 1, \quad \rho + (1 - a_j)\lambda_j L_j > 0. \end{aligned} \tag{40}$$

Noting (1), (12) and (40) yield

$$\frac{d}{dl_j^L}\log\left[\frac{(a_j-1)\lambda_j l_j^L}{\rho+(1-a_j)\lambda_j(L_j-l_j^L)}\right]=\frac{1}{l_j^L}\Bigg[1-\underbrace{\frac{(a_j-1)\lambda_j l_j^L}{\rho+(1-a_j)\lambda_j(L_j-l_j^L)}}_{\in(0,1)}\Bigg]>0$$

and

$$\frac{d}{dl_j^L}\left[\frac{(a_j-1)\lambda_j l_j^L}{\rho+(1-a_j)\lambda_j(L_j-l_j^L)}\right]>0.$$

Noting this and differentiating the left-hand equation in (12), one obtains

$$\underbrace{\frac{d}{dl_j^L}\left[\frac{(a_j-1)\lambda_j l_j^L}{\rho+(1-a_j)\lambda_j(L_j-l_j^L)}\right]}_{+}dl_j^L+d\beta=0,$$

and $dl_j^L/d\beta<0$. Given (1), this implies $dz_j^L/d\beta=-dl_j^L/d\beta>0$. Finally, differentiating the right-hand equation in (12), and noting (12), one obtains

$$\frac{dm_j^L}{d\beta}=\frac{m_j^L}{l_j^L}\Bigg[\underbrace{\frac{dl_j^L}{d\beta}}_{-}-\underbrace{\frac{m_j^L}{(\xi^j)'}}_{+}\Bigg]<0.$$

Appendix 3

Region j maximizes (21) by l_j subject to (6), given (m,m_j,R,R_j). It is equivalent to maximize *the expected value of the flow of output for region* j,

$$E\int_T^\infty a_j^{\gamma_j}f^j(l_j,m_j)m_j^{-\beta}e^{-\rho(\theta-T)}d\theta,$$

by l_j subject to (6), given m_j. The value of this maximization is

$$\Gamma_j^j(\gamma_j,m_j,T)=\max_{l_j\text{ s.t. }(6)}E\int_T^\infty a_j^{\gamma_j}f^j(l_j,m_j)m_j^{-\beta}e^{-\rho(\theta-T)}d\theta. \tag{41}$$

Denote $\Gamma^j=\Gamma^j(\gamma_j,m_j,T)$ and $\widetilde{\Gamma}^j=\Gamma^j(\gamma_j+1,m_j,T)$. The Bellman equation corresponding to the optimal program (41) is

$$\rho\Gamma^j=\max_{l_j}\Psi(l_j,\gamma_j,m_j,R-R_j,T),\quad\text{where}$$

$$\Psi(l_j,\gamma_j,m_j,T)=a_j^{\gamma_j}f^j(l_j,m_j)m_j^{-\beta}+\lambda_j(L_j-l_j)\big(\widetilde{\Gamma}^j-\Gamma^j\big). \tag{42}$$

Noting (4), this leads to the first-order condition

$$\frac{\partial \Psi}{\partial l_j} = a_j^{\gamma_j} f_l^j(l_j, m_j) m_j^{-\beta} - \lambda_j(\widetilde{\Gamma}^j - \Gamma^j)$$

$$= \frac{1}{l_j} a_j^{\gamma_j} f^j(l_j, m_j) m_j^{-\beta}\left[\alpha_j - \xi^j\left(\frac{l_j}{m_j}\right)\right] - \lambda_j(\widetilde{\Gamma}^j - \Gamma^j) = 0. \qquad (43)$$

To solve the dynamic program (41), assume that the value of the program, Γ^j, is in fixed proportion $\vartheta_j > 0$ to instantaneous utility:

$$\Gamma^j(\gamma_j, m_j, T) = \vartheta_j a_j^{\gamma_j} f^j(l_j^*, m_j) m_j^{-\beta}, \qquad (44)$$

where l_j^* is the optimal value of the control variable l_j. This implies

$$(\widetilde{\Gamma}^j - \Gamma^j)/\Gamma^j = a_j - 1. \qquad (45)$$

Inserting (44) and (45) into the Bellman equation (42) yields

$$1/\vartheta_j = \rho + (1 - a_j)\lambda_j(L_j - l_j) > 0. \qquad (46)$$

Plugging this (46) into (44), one obtains

$$\Gamma^j(\gamma_j, m_j, T) = \frac{a_j^{\gamma_j} f^j(l_j, m_j) m_j^{-\beta}}{\rho + (1 - a_j)\lambda_j(L_j - l_j^*)}, \qquad (47)$$

where l_j^*—the optimal value of the control variable l_j—is taken as given.

Inserting (47), (45) and (46) into (43), one obtains (23):

$$0 = \vartheta_j \frac{l_j}{\Gamma^j}\frac{\partial \Psi}{\partial l_j} = \underbrace{a_j^{\gamma_j} f^j(l_j, m_j) m_j^{-\beta}\frac{\vartheta_j}{\Gamma^j}}_{=1}\left[\alpha_j - \xi^j\left(\frac{l_j}{m_j}\right)\right] - \Big(\underbrace{\frac{\widetilde{\Gamma}^j}{\Gamma^j}}_{=a_j} - 1\Big)\lambda_j l_j \vartheta_j$$

$$= \alpha_j - \xi^j\left(\frac{l_j}{m_j}\right) - \frac{(a_j - 1)\lambda_j l_j}{\rho + (1 - a_j)\lambda_j(L_j - l_j)}.$$

Noting (41) and (47), the expected utility (21) becomes (22):

$$\Theta(m_j, m, R_j, R)$$

$$= m^{-\delta} E \int_T^\infty \left[a_j^{\gamma_j} f^j(l_j, m_j) m_j^{-\beta} + R - R_j\right] e^{-\rho(\theta - T)} d\theta$$

$$= m^{-\delta}\left[E \int_T^\infty a_j^{\gamma_j} f^j(l_j, m_j) m_j^{-\beta} e^{-\rho(\theta - T)} d\theta + \int_T^\infty (R - R_j) e^{-\rho(\theta - T)} d\theta\right]$$

$$= m^{-\delta}\left[E \int_T^\infty a_j^{\gamma_j} f^j(l_j, m_j) m_j^{-\beta} e^{-\rho(\theta - T)} d\theta + \frac{R - R_j}{\rho}\right]$$

$$= m^{-\delta}\big[\Gamma^j(\gamma_j, m_j, T) + (R - R_j)/\rho\big],$$

$$\frac{\partial\Theta}{\partial m_j} = \frac{\Gamma^j}{m^\delta}\left[\frac{f^j_m(l_j, m_j)}{f^j(l_j, m_j)} - \frac{\beta}{m_j}\right] = \frac{\Gamma^j}{m^\delta m_j}\left[\xi^j\left(\frac{l_j}{m_j}\right) - \beta\right],$$

$$\partial\Theta/\partial M = -\delta m^{-\delta-1}\big[\Gamma^j + (R - R_j)/\rho\big], \qquad -\partial\Theta/\partial R_j = \partial\Theta/\partial R = m^{-\delta}/\rho.$$

References

Böhringer, C., & Lange, A. (2005). On the design of optimal grandfathering schemes for emission allowances. *European Economic Review*, *49*, 2041–2055.

Dixit, A., & Pindyck, K. (1994). *Investment under uncertainty*. Princeton: Princeton University Press.

Dixit, A., Grossman, G. M., & Helpman, E. (1997). Common agency and coordination: general theory and application to management policy making. *Journal of Political Economy*, *105*, 752–769.

Grossman, G. M., & Helpman, E. (1994). Protection for sale. *The American Economic Review*, *84*, 833–850.

Haurie, A., Moresino, F., & Viguier, L. (2006). A two-level differential game of international emissions trading. Advances in dynamic games. *Annals of the International Society of Dynamic Games*, *8*, 293–307. Part V.

Jouvet, P.-A., Michel, P., & Rotillon, G. (2008). The optimal initial allocation of pollution permits: a relative performance approach. *Journal of Economic Dynamics & Control*, *29*, 1597–1609.

Mackenzie, I. A., Hanley, N., & Kornienko, T. (2008). Optimal growth with pollution: how to use pollution permits? *Environmental & Resource Economics*, *39*, 265–282.

Palokangas, T. (2009). International emission policy with lobbying and technological change. In J. Crespo Cuaresma, T. Palokangas, & A. Tarasyev (Eds.), *Dynamic systems, economic growth and the environment*, Heidelberg: Springer.

Xepapadeas, A. (2005). Economic growth and the environment. In K.-G. Mähler & J. R. Vincent (Eds.), *Handbook of environmental economics* (Vol. III, pp. 1219–1271). Amsterdam: Elsevier.

Pollution, Public Health Care, and Life Expectancy when Inequality Matters

Andreas Schaefer and Alexia Prskawetz

Abstract We analyze the link between economic inequality in terms of wealth, life expectancy, health care and pollution. The distribution of wealth is decisive for the number of households investing in human capital. Moreover, the willingness to invest in human capital depends on agents' life expectancy which determines the length of the amortization period of human capital investments. Life expectancy is positively affected by public health care expenditures but adversely affected by the pollution stock generated by aggregate production. Our model accounts for an endogenous take-off in terms of human capital investments. Higher initial inequality delays the take-off because a given set of policies (abatement measures and public health care) is less effective in improving agents' survival probabilities. We compare a change in taxes to a change in expenditure shares on health and abatement given different amounts of (initial) inequality. The advantage of the latter as compared to the former is the achieved increase in the tax base which induces more expenditures on health care and abatement measures, such that an even higher economic activity is compatible with a similar level of long-run pollution.

1 Introduction

Beginning with the work of John and Pecchenino (1994) and John et al. (1995) several authors have argued that one of the difficulties in the interaction of the en-

A. Schaefer (✉)
Institute of Theoretical Economics/Macroeconomics, University of Leipzig, Grimmaische Strasse 12, 04109 Leipzig, Germany
e-mail: schaefer@wifa.uni-leipzig.de

A. Prskawetz
Wittgenstein Centre for Demography and Global Human Capital (IIASA, VID/ÖAW, WU), Schloßplatz 1, 2361 Laxenburg, Austria
e-mail: afp@econ.tuwien.ac.at

A. Prskawetz
Institute of Mathematical Methods in Economics, Research Unit Economics, Vienna University of Technology (TU), Argentinierstrasse 8/4/105–3, 1040 Vienna, Austria

E. Moser et al. (eds.), *Dynamic Optimization in Environmental Economics*, Dynamic Modeling and Econometrics in Economics and Finance 15,
DOI 10.1007/978-3-642-54086-8_6, © Springer-Verlag Berlin Heidelberg 2014

vironment and economic activity is the different life span of both systems. While the lifetime of the environment is infinite, the lifetime of economic agents is finite. Hence, the incentive to invest into the environment might be limited by the lifetime of the individuals. Recently a paper by Mariani et al. (2010) has extended this literature considering the two way interaction between pollution and life expectancy, i.e. it is assumed that life expectancy and environmental quality are jointly determined. In an extension to the model by Mariani et al. (2010), Raffin and Seegmuller (2012) studied the path of pollution and economic growth when households' longevity is endogenously determined not only by environmental quality but also by health policy. While economic growth may induce negative externalities on the environment, it may also be the engine of growth for investment into health and thereby enhance life expectancy. As argued in Raffin and Seegmuller (2012) the tax base will be positively associated with higher economic growth and hence more resources will be available to finance investments such as health expenditures and abatement measures. These models allow for multiple steady states, with a low level trap of high pollution and low life expectancy and a high level equilibrium with low pollution and high life expectancy. None of these models has so far considered the role of inequality in the process of economic growth. However, the initial distribution of wealth in a society may limit the possibilities for economic growth and the effectiveness of economic policy in terms of public health care and abatement measures.

In this paper, we analyze the link between economic inequality in terms of wealth, life expectancy, health care and pollution based on the work by Galor and Zeira (1993). In our framework, life expectancy is positively affected by public health care expenditures but adversely affected by the pollution stock generated by aggregate production. Life expectancy plays a key role in our model since it determines the level of human capital investment and therefore aggregate output. If households expect to live longer, they are more inclined to invest in human capital as the returns to human capital will accrue over a longer period and borrowers' credit costs shrink. We assume that the government levies taxes on households' income (where we distinguish between skilled and unskilled households) and uses taxes to finance health care and abatement measures. As earlier stages of economic development are characterized by low life expectancy, human capital investments are zero. However, tax financed health care and abatement measures may improve life expectancy such that agents start to invest in skills once the level of the life expectancy has passed a certain threshold. Therefore, our framework takes account for an endogenous take-off in terms of human capital investment. Higher initial inequality delays the take-off because a given set of policies reflected by income taxes and expenditure shares on public health care and abatement measures is less effective.

Moreover, we compare a change in taxes to a change in expenditure shares on health and abatement given different amounts of (initial) inequality. Our results show that an increase in the tax rate (hence, the government budget) benefits skilled and unskilled agents in terms of wealth as long as the marginal cost of taxes in terms of foregone lifetime earnings are smaller than the marginal increase in lifetime net-earnings generated through the improvement in life expectancies. However, since the marginal benefit of an increase in tax revenues for the skilled group exceeds

Fig. 1 Expectation of life at birth (London excluded) (Szreter 1997)

the corresponding level of the marginal benefit of the unskilled population group, economic policy increases long-run inequality in terms of wealth. An increase in health expenditures compared to investments in abatement always raises the wealth of skilled and unskilled, but again the gain is greater for skilled workers. Moreover, we find that an increase in the expenditure share on public health care increases the tax base which induces more expenditures on health care and abatement measures, such that an even higher economic activity is compatible with a similar level of long-run pollution as compared to the levels resulting from an increase in the income tax.

The initially adverse impact of economic development on individuals' health is mirrored in the evolution of life expectancy at birth as shown in Fig. 1. Average life expectancy at birth stagnated during the second phase of the industrial revolution and started to increase only in the last four decades of the 19th century. In cities, life expectancies at birth started even to decline and reached a level passed in the 15th century already, although per capita output was already growing. It is well documented that the gap in mortality rates between cities and rural areas can be explained by environmental degradation and pollution. In this line of argumentation the significance of water as an industrial raw material has been documented by Hassan (1985): fresh water was used for commercial purposes while the new entrepreneurial class saw no point in spending money for sanitation and sewage treatment plants. In addition, Haines (2004) and Komlos (1998) provide evidence for the adverse impacts of economic development during the Industrial Revolution, in the sense that physical height of soldiers declined during the 19th century in the US as well as England and the Netherlands indicating an increase in morbidity over the same period of time. Adverse effects of economic growth on the environment in earlier stages of economic development are even today of greatest concern, for example the combined health and non-health cost of outdoor air and water pollution for China's economy comes to around 5.8 % of the GDP per year (World Bank 2007). Moreover, as regards later stages of economic development, Chay and Greenstone (2003) provide evidence for the impact of air pollution on infant mortality in the US during the recession period 1981–1982 and conclude that a 1—percent reduction in total suspended particulates results in a 0.35—percent decline in infant mortality at the county level.

The remainder of the paper is organized as follows: in Sect. 2, we introduce the model. Section 3 explores the evolution of wealth and Sect. 4 discusses the set of economic policies in terms of income taxes and expenditures shares on public health care and abatement measures. In order to capture the entire evolution of the economy towards its steady state and in order to illustrate our analytical findings, we perform numerical experiments in Sect. 5. In Sect. 6, we provide a critical discussion of our results and, finally, Sect. 7 concludes.

2 The Model

Households live for two periods and decide in their first period whether or not to invest in skills. The amount of inherited wealth by the parental household determines whether agents acquire skills since the human capital investment is subject to indivisibilities and capital market imperfections in the sense that borrowers' interest rate exceeds lenders' opportunity costs. Moreover, the willingness to invest in human capital depends on agents' life expectancy as it triggers the amortization period of human capital investments. The long-run performance of the economy depends on the initial distribution of wealth which determines the number of agents investing in skills.

2.1 *Production*

Consider a small open economy which produces a homogeneous good Y_t in two sectors, an unskilled and a skilled sector denoted by superscripts u, s in the following. Output of the unskilled sector, Y_t^u, is subject to a linear production function employing unskilled labor, L_t^u, only

$$Y_t^u = aL_t^u, \quad a > 0,$$

with a denoting a positive scaling factor. The high skilled sector produces Y_t^s subject to a neoclassical production function of Cobb-Douglas type and employs skilled labor, L_t^s, as well as physical capital, K_t, such that

$$Y_t^s = b(K_t)^{\gamma}\left(L_t^s\right)^{1-\gamma}, \quad b > 0, \; \gamma \in (0, 1),$$

with b denoting a positive scaling factor and γ representing the output elasticity of capital. Aggregate output is given by

$$Y_t = Y_t^s + Y_t^u.$$

The small open economy assumption implies an exogenous interest rate that equals the international interest rate $\bar{r}$. Markets are assumed to be perfectly competitive. Given $\bar{r}$ and profit maximizing behavior of firms, the capital intensity, k_t,

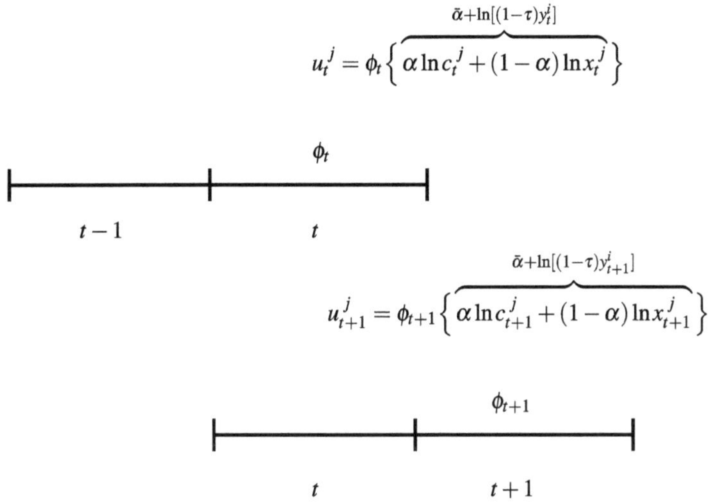

Fig. 2 Demographics

is determined by:

$$\bar{r} = \gamma b k_t^{\gamma-1} - \delta,$$

with $k_t = K_t/L_t^s$, and $0 \leq \delta \leq 1$ representing the rate of depreciation of physical capital. In addition the wage rates for skilled and unskilled labor are given as

$$w_t^s = (1-\gamma) b k_t^{\gamma},$$
$$w_t^u = a.$$

Hence, the small-open economy assumption switches off any dynamics with respect to k and factor prices, such that k, w^s and w^u are constant for all t, since

$$k_t = \left(\frac{\gamma b}{\bar{r}+\delta}\right)^{\frac{1}{1-\gamma}}.$$

Thus, in period t, the level of output in the skilled sector depends only on the amount of skilled labor and exogenously fixed parameters:

$$Y_t^s = b\left(k_t L_t^s\right)^{\gamma}\left(L_t^s\right)^{1-\gamma} = b^{\frac{1}{1-\gamma}}\left(\frac{\gamma}{\bar{r}+\delta}\right)^{\frac{\gamma}{1-\gamma}} L_t^s.$$

2.2 Demographics and Households' Decisions

An individual born in $t-1$ expects to live for $1+\phi_t$ periods with $0 \leq \phi_t \leq 1$ representing the probability to reach the end of period t (see also Fig. 2). Hence, the terms life expectancy and survival probability can be used interchangeably. The probability to reach the end of the second period of life is determined by the level of public health expenditures, H_t, and the exposure to pollutants, P_t, i.e. $\phi_t = \phi(H_t, P_t)$.

Definition 1 Life expectancy $0 \leq \phi_t \leq 1$ is a non-decreasing function in public health care expenditures, H_t, and a non-increasing function in the pollution stock P_t, such that

$$\frac{\partial \phi(H_t, P_t)}{\partial H_t} \geq 0,$$
$$\frac{\partial \phi(H_t, P_t)}{\partial P_t} \leq 0.$$

Moreover, the cross-derivative is non-positive, i.e.

$$\frac{\partial^2 \phi(H_t, P_t)}{\partial H_t \partial P_t} \leq 0.$$

A non-positive cross-derivative of the life expectancy with respect to H and P means that an increase in pollution may reduce the effectiveness of public health expenditures on ϕ.

At this point it is worth to notice that public health expenditures improve the life expectancy of those generations which are taxed while an improvement of environmental quality reflected by a decline of the pollution stock in the subsequent period, P_{t+1}, benefits only those generations which are alive from $t + 1$ onwards. Moreover, in Sect. 2.3, we will see that public health expenditures depend on aggregate tax revenues, i.e. aggregate income. Since we abstract from population growth an increase in aggregate income corresponds to an increase in per capita income and thus to an improvement in health expenditures and life expectancy. Thus, our theory is compatible to the well-known Preston-curve suggesting a positive association between per capita income and life expectancy.

Note that we only consider public expenditures on health and ignore private health expenditures. Our approach is similar to Aisa and Pueyo (2006). Different to Aisa and Pueyo we assume that the level of health expenditures and not the share of health expenditures in total GDP positively affects life expectancy.[1] By only focusing on public health care we aim to emphasize the role of the allocation of public expenditures between health care and pollution abatement on economic growth (see also Agenor and Neanidis 2011). On one hand health expenditures reduce income through taxes thereby also reducing spending on pollution abatement, on the other hand higher taxes induce a higher life expectancy and thereby foster human capital accumulation and economic growth.

[1]But remember that pollution adversely affects agents' life expectancy in our framework. Moreover, we will assume further below a logistic functional form of $\phi(H_t, P_t)$. Thus, our model is not more optimistic with respect to the effectiveness of public health expenditures on improvements of ϕ compared to existing literature.

Agents, j, work either as unskilled workers, $j = u$, in both periods or invest in their first period of life in human capital and become a skilled worker, $j = s$, in their second period of life. Fertility is exogenous in the sense that each household has exactly one descendant which replaces him after she dies. Agents born in $t-1$ derive utility out of consumption, c_t^j, and out of bequests to their offspring, x_t^j, in their second period of life. Lifetime utility in t of an agent j born in period $t-1$ is specified as

$$u_t^j = E\left[\bar{u}_t^j\right] = \phi_t^j \bar{u}_t^j, \tag{1}$$

with $\phi_t^j = \phi_t \; \forall j$ and $\bar{u}_t^j = \alpha \ln c_t^j + (1-\alpha) \ln x_t^j$. Hence, we assume the same life expectancy for skilled and unskilled people.

An agent j born in period $t-1$ maximizes lifetime utility (1) subject to lifetime earnings net of taxes, $\tau \in (0, 1)$, resulting in

$$c_t^j = \alpha(1-\tau)y_t^j, \tag{2}$$

$$x_t^j = (1-\alpha)(1-\tau)y_t^j, \tag{3}$$

with y_t^j denoting agents' second period's income depending on life expectancy ϕ_t. Before we specify life time earnings further below, we obtain the indirect utility function from (1)–(3) as

$$\bar{u}_t^j = \bar{\alpha} + \ln\left[(1-\tau)y_t^j\right], \tag{4}$$

with $\bar{\alpha} = \alpha \ln \alpha + (1-\alpha)\ln[1-\alpha]$.

2.3 The Government

The government raises income taxes $\tau \in (0, 1)$ in order to finance public health expenditures H_t and abatement measures A_t. In period t the government taxes unskilled households working in their first period of life, L_t^u, and skilled and unskilled households born in $t-1$ that survived to period t, i.e. $\phi_t(L_{t-1}^u + L_{t-1}^s)$. Hence, tax revenues in period t are

$$G_t = \tau\left(\phi_t\left(y_t^s L_{t-1}^s + y_t^u L_{t-1}^u\right) + w^u L_t^u\right).$$

Abstracting from intertemporal debts and assuming constant expenditure shares for public health, ν, and abatement measures, $1-\nu$, a balanced budget in each period requires

$$H_t = \nu G_t,$$

$$A_t = (1-\nu)G_t, \quad \nu \in (0, 1).$$

In $t+1$, the stock of pollutants, P_{t+1}, increases by current emissions, E_t, generated by the production process. We assume for simplicity $E_t = E(Y_t) = \varepsilon_0 Y_t$. On the other hand, the impact of emissions on the pollution stock can be reduced through tax financed abatement measures, A_t. Moreover, the environment regenerates at rate $0 < \eta < 1$, such that the pollution stock evolves over time according to

$$P_{t+1} = (1-\eta)P_t + \varepsilon_0 Y_t - \varepsilon_1 A_t, \quad 0 < \varepsilon_1 < \varepsilon_0,\ \eta \in (0,1),$$

with ε_0 denoting the impact of one unit of output on the pollution stock and ε_1 reflecting the productivity of abatement measures.

2.4 The Credit Market

The credit market is subject to imperfections as in Galor and Zeira (1993), in the sense that borrowers' interest rate, i_t, exceeds the world market interest rate, $\bar{r}$. In contrast to Galor and Zeira (1993), i_t depends inversely on agents' life expectancy, ϕ_t. Moreover, as $\phi_t = \phi(H_t, P_t)$, borrowers' credit costs are not time invariant and affected by public health expenditures and abatement measures triggering the wedge between i_t and $\bar{r}$. Thus, economic policy affects the incentive to invest in skills (by determining life expectancy), but the effectiveness of economic policy will depend on the amount of economic inequality. Before we come back to this issue further below, we elaborate more on the mechanisms on the credit market.

Human capital investments are (see Galor and Zeira 1993) subject to indivisibilities, in the sense that it requires an amount $h > 0$ to become a skilled worker. Workers born in $t-1$ with inherited wealth $x^j_{t-1} < h$ can borrow $h - x^j_{t-1}$ at the capital market, but since human capital investments are unobservable and the transition to the end of the second period of life is uncertain, moral hazard and mortality risks induce a wedge between the equilibrium interest rate $\bar{r}$ and the interest rate i_t at which lenders are willing to lend money to borrowers. Hence, credits are subject to monitoring costs z, such that the zero profit condition is given by

$$(1+i_t)\phi_t\big(h - x^j_{t-1}\big) = z + (1+\bar{r})\big(h - x^j_{t-1}\big).$$

The left hand side denotes the lender's credit costs, i.e. the interest rate that the lender faces times the probability to survive to the end of the next period and times the amount of investment to be borrowed. The right hand side indicates the costs that accrue to the borrower. These are the monitoring costs, z, plus the value of the borrowed investment if it would be invested at the international interest rate.

As lenders can still evade repayment by spending βz with $\beta > 1$, borrowers set monitoring effort, z, such that lenders are indifferent between repayment and evasion

$$(1+i_t)\phi_t\big(h - x^j_{t-1}\big) = \beta z.$$

From the last two equations we obtain i_t

$$(1+i_t)\phi_t\big(h - x^j_{t-1}\big) = \frac{(1+i_t)\phi_t(h - x^j_{t-1})}{\beta} + (1+\bar{r})\big(h - x^j_{t-1}\big)$$

$$\Rightarrow i_t = \frac{\beta}{(\beta-1)}\frac{(1+\bar{r})}{\phi_t} - 1.$$

The following proposition summarizes the association between borrower's and lender's interest rates as well as life expectancy.

Proposition 1 *Since $\beta > 1$ it follows that $i_t > \bar{r}$.*[2] *Moreover, the interest rate for credits, i_t, is inversely related to life expectancy, i.e. $\frac{\partial i_t}{\partial \phi_t} < 0$, such that higher health risks increase lenders' credit costs.*

3 The Evolution of Wealth

Lifetime utility (4) of agents born in t depends positively on lifetime earnings y^j_{t+1} which in turn depends on human capital investment. Whether or not to invest in human capital depends on the level of inherited bequests, i.e. $x^j_t \gtreqless h$, and life expectancy ϕ_{t+1}, with h representing an exogenous fixed cost of human capital investment. Households with $x^j_t \geq h$ invest in human capital, if lifetime utility of becoming a skilled worker is at least as high as lifetime utility from remaining unskilled, i.e. $u^s_{t+1} \geq u^u_{t+1}$ which implies in light of (4) that $y^s_{t+1} \geq y^u_{t+1}$, such that[3]

$$\underbrace{\phi_{t+1} w^s + \left(x^j_t - h\right)(1+\bar{r})}_{y^s_{t+1}} \geq \underbrace{\phi_{t+1} w^u + \left((1-\tau)w^u + x^j_t\right)(1+\bar{r})}_{y^u_{t+1}}.$$

In contrast, households with $x^j_t < h$ wish to invest in human capital, if $u^{u,s}_{t+1} \geq u^u_{t+1}$.[4] These households borrow $h - x^j_t$ at an interest rate i_{t+1} from the capital market, such that

$$(1-\tau)y^{u,s}_{t+1} = (1-\tau)\left(w^s\phi_{t+1} + \left(x^j_t - h\right)(1+i_{t+1})\right), \quad \text{with } x^j_t < h.$$

The requirement of $u^{u,s}_{t+1} \geq u^u_{t+1}$ implies again in light of (4) that the last expression holds with equality, if

$$y^{u,s}_{t+1} = y^u_{t+1}. \tag{5}$$

Condition (5) determines the minimum level of inherited wealth necessary to become a skilled worker, $x^j_t = x^{crit}_t$, conditional on the survival probability ϕ_{t+1}:

---

[2] $i_t > \bar{r}$ implies $\frac{(1+\bar{r})\beta}{\phi_t(\beta-1)} > 1 + \bar{r}$ and therefore $\frac{\beta}{(\beta-1)} > \phi_t$ which is valid as long as $\beta > 1$ since $0 < \phi_t \leq 1$.

[3] The left-hand side of the last expression captures the lifetime income of a skilled household in her second period of life, i.e. labor income, w^s, multiplied by the corresponding level of life expectancy, ϕ_{t+1} plus wealth net human capital investment times accrued interests, $(x^j_t - h)(1+r)$. The right-hand side captures lifetime income of an unskilled household that does not invest in human capital, i.e. labor income, w^u, multiplied by ϕ_{t+1}, plus the sum out of first-period labor income net of taxes, $(1-\tau)w^u$, and wealth, x^j_t, times accrued interests.

[4] The superscript u, s denotes agents that are born in unskilled households and decide to invest in skills.

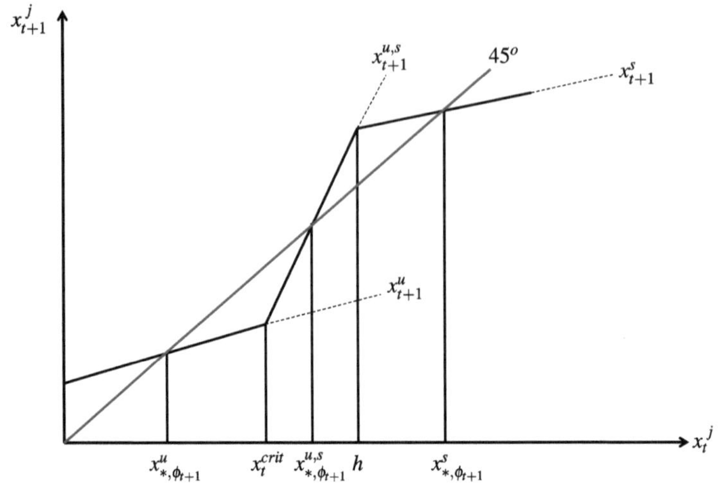

Fig. 3 Evolution of wealth conditional on life expectancies ϕ_{t+1} and taxes τ

$$x_t^{crit} = \frac{1}{i_{t+1} - \bar{r}}\big[(1-\tau)w^u(1+\bar{r}) + h(1+i_{t+1}) - \phi_{t+1}\big(w^s - w^u\big)\big].$$

Since $x^j_{t+1} = (1-\alpha)(1-\tau)y^j_{t+1}$ and given that $y^s_{t+1} \geq y^u_{t+1}$, wealth of agents born in t given life expectancy, ϕ_{t+1}, evolves according to:

1. Agents born in unskilled households with $x^u_t < x^{crit}_t$ remain unskilled

$$\begin{aligned} x^u_{t+1} &= (1-\alpha)(1-\tau)y^u_{t+1} \\ &= (1-\alpha)(1-\tau)\big[\big(x^u_t + (1-\tau)w^u\big)(1+\bar{r}) + w^u\phi_{t+1}\big]. \end{aligned} \tag{6}$$

2. Agents born in skilled households with $x^s_t \geq h$ invest in skills

$$\begin{aligned} x^s_{t+1} &= (1-\alpha)(1-\tau)y^s_{t+1} \\ &= (1-\alpha)(1-\tau)\big[w^s\phi_{t+1} + \big(x^s_t - h\big)(1+\bar{r})\big]. \end{aligned} \tag{7}$$

3. Agents born in unskilled households with $h > x^u_t \geq x^{crit}_t$ invest in skills

$$\begin{aligned} x^{u,s}_{t+1} &= (1-\alpha)(1-\tau)y^{u,s}_{t+1} \\ &= (1-\alpha)(1-\tau)\big[w^s\phi_{t+1} + \big(x^u_t - h\big)(1+i_{t+1})\big]. \end{aligned} \tag{8}$$

The system (6)–(8) can be presented graphically in the (x_{t+1}, x_t)—plane for a given stock of pollution and public health expenditures, i.e. a given life expectancy ϕ_{t+1}. Thus, the loci depicted in Fig. 3 are conditional on the state of ϕ_{t+1}. This is the reason why we refer to conditional loci and steady states. The positions and the slope of these conditional loci and steady states are important as they trigger the dynamics of the economy to their long-run values and the composition of the population in terms of skilled and unskilled agents. We will describe the behavior of the conditional steady states and x_t^{crit} in detail further below (see also Fig. 4). Noting that i_{t+1} is a function of ϕ_{t+1}, the conditional stationary solutions read as follows

$$x^s_{*,\phi_{t+1}} = \frac{(1-\alpha)(1-\tau)(w^s\phi_{t+1} - h(1+\bar{r}))}{1-(1-\alpha)(1-\tau)(1+\bar{r})}, \tag{9}$$

$$x^u_{*,\phi_{t+1}} = \frac{(1-\alpha)(1-\tau)w^u((1+\bar{r})(1-\tau)+\phi_{t+1})}{1-(1-\alpha)(1-\tau)(1+\bar{r})}, \tag{10}$$

$$x^{u,s}_{*,\phi_{t+1}} = \frac{(1-\alpha)(1-\tau)(h(1+i_{t+1}) - w^s\phi_{t+1})}{(1-\alpha)(1-\tau)(1+i_{t+1})-1}, \tag{11}$$

with $1 > (1-\alpha)(1-\tau)(1+\bar{r})$ and $(1-\alpha)(1-\tau)(1+i_{t+1}) > 1$.[5]

The steady state of the economy is determined by the initial distribution of wealth, Γ_0, and the policy set $\Phi = \{\tau, \nu\}$. Both in combination determine the long-run distribution of wealth Γ_* which in turn is determined by the distribution of the population between the two stable steady states x^u_{*,ϕ_*} and x^s_{*,ϕ_*}. The following proposition describes the steady state of our economy.

Proposition 2 *The steady state of the economy is characterized by a constant life expectancy,* $\phi_t = \phi_* \leq 1$, *which implies a constant interest rate*

$$i_* = \frac{\beta(1+\bar{r})}{(\beta-1)\phi_*} - 1,$$

a constant level of wealth for skilled and unskilled households

$$x^s_{*,\phi_*} == \frac{(1-\alpha)(1-\tau)(w^s\phi_* - h(1+\bar{r}))}{1-(1-\alpha)(1-\tau)(1+\bar{r})},$$

$$x^u_{*,\phi_*} == \frac{(1-\alpha)(1-\tau)w^u((1+\bar{r})(1-\tau)+\phi_*)}{1-(1-\alpha)(1-\tau)(1+\bar{r})},$$

$$x^{u,s}_{*,\phi_*} = \frac{(1-\alpha)(1-\tau)(h(1+i_*) - w^s\phi_*)}{(1-\alpha)(1-\tau)(1+i_*)-1},$$

with $1 > (1-\alpha)(1-\tau)(1+\bar{r})$ *and* $(1-\alpha)(1-\tau)(1+i_*) > 1$, *and a constant distribution of households between the two exterior steady states* x^s_{*,ϕ_*} *and* x^u_{*,ϕ_*}, *such that* $L^u_t = L^u_* \geq 0$ *and* $L^s_t = L^s_* \geq 0$ *with* $L^u_* + L^s_* = L$. *Therefore, the level of aggregate production* $Y_t = Y_*$ *is constant as well and reads*

$$Y_* = Y^u_* + Y^s_* = aL^u_* + b^{\frac{1}{1-\gamma}}\left(\frac{\gamma}{\bar{r}+\delta}\right)^{\frac{\gamma}{1-\gamma}} L^s_*,$$

such that the level of pollution, P, *tax revenues,* G, *public health expenditures,* H, *and abatement measures,* A, *are constant as well and given by*

[5] As can be verified easily: $\phi_{t+1} > (1-\alpha)(1-\tau)(1+\bar{r})$ implies that the exterior (conditional) steady states are stable while $(1-\alpha)(1-\tau)(1+i_{t+1}) > 1$ implies that the interior one is unstable.

$$P_* = \frac{\varepsilon_0 Y_* + \varepsilon_1(1-\nu)\tau G_*}{\eta},$$
$$G_* = \tau\big(\phi_*\big(y_t^s L_*^s + y_t^u L_*^u\big) + w_t^u L_*^u\big),$$
$$H_* = \nu G_*,$$
$$A_* = (1-\nu)G_*.$$

4 Policy

In this section we analyze changes in the policy set $\Phi = \{\tau, \nu\}$. Changes in Φ alter life expectancy and therefore, the skill composition of the population given the initial distribution of wealth Γ_0. Thus, economic policy has a direct impact on the evolution of inequality and the long-run performance of an economy. Changes in τ alter life expectancies, interest rates, disposable incomes of households, and resources available for abatement measures. In contrast, a change in ν leaves disposable incomes unaffected.

Before turning to changes in Φ, it will be instructive to analyze the effects of changes in life expectancies, ϕ, on the (conditional) steady states and x_t^{crit}. We thus analyze the dynamics of the conditional steady states (i.e. the shift of the respective intercepts with the 45°—line in Fig. 3), in response to the transition of ϕ towards its long-run value $\phi_* \le 1$. As we will see, the increase in ϕ gives rise to an endogenous take-off in terms of human capital investments. This take-off is essentially triggered by the dynamics of the conditional stable steady states, $x^u_{*,\phi_{t+1}}$, and $x^s_{*,\phi_{t+1}}$, while the dynamics of the wealth distribution and the composition of the population in terms of skilled and unskilled households is affected by x_t^{crit} and $x^{u,s}_{*,\phi_{t+1}}$.[6] The analytical results are summarized in Propositions 3 and 4. The following proposition summarizes the dependence of $x^u_{*,\phi_{t+1}}$ and $x^s_{*,\phi_{t+1}}$ in response to a change in ϕ.

Proposition 3 (Exterior Steady States, Changes in ϕ)

(i) *An increase in longevity increases the (conditional) long-run levels of wealth for skilled and unskilled households, i.e.* $\frac{\partial x^u_{*,\phi_{t+1}}}{\partial \phi_{t+1}} \ge 0$ *and* $\frac{\partial x^s_{*,\phi_{t+1}}}{\partial \phi_{t+1}} \ge 0$ *for* $\phi_{t+1} \in (0, \phi_*)$, *and* $\phi_* \le 1$, *where*

$$\frac{\partial x^u_{*,\phi_{t+1}}}{\partial \phi_{t+1}} = \frac{(1-\alpha)(1-\tau)w^u}{1-(1-\alpha)(1-\tau)(1+\bar{r})} \ge 0,$$
$$\frac{\partial x^s_{*,\phi_{t+1}}}{\partial \phi_{t+1}} = \frac{(1-\alpha)(1-\tau)w^s}{1-(1-\alpha)(1-\tau)(1+\bar{r})} \ge 0,$$

with equality, if $\phi_{t+1} = 1$. *Moreover,* $\frac{\partial x^u_{*,\phi_{t+1}}}{\partial \phi_{t+1}} < \frac{\partial x^s_{*,\phi_{t+1}}}{\partial \phi_{t+1}}$ *since* $w^u < w^s$.

[6] Remember that x_t^{crit} determines the necessary amount of wealth to become a skilled worker for a given life expectancy while $x^{u,s}_{*,\phi_{t+1}}$ determines the basin of attraction of the superior steady state for a given life expectancy.

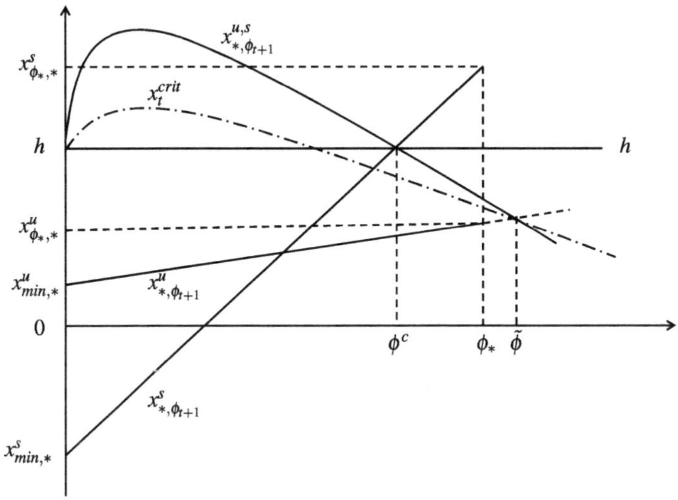

Fig. 4 The (conditional) steady states and x_t^{crit} as a function of life expectancy ϕ

(ii) *As life expectancy approaches zero, the (conditional) stable steady states of both population groups reach a minimum value*

$$x^s_{*,min} = -\frac{(1-\tau)(1-\alpha)(1+\bar{r})h}{1-(1-\alpha)(1-\tau)(1+\bar{r})} < 0,$$

$$x^u_{*,min} = \frac{(1-\tau)^2(1+\bar{r})(1-\alpha)w^u}{1-(1-\alpha)(1-\tau)(1+\bar{r})} > 0.$$

(iii) *Given that ϕ_* is such that $x^s_{*,\phi_*} > x^u_{*,\phi_*}$, it follows from* (i) *and* (ii) *that there exists a critical $\phi^c < \phi_*$ implying that $x^s_{*,\phi^c} = h$, such that agents with $x_t^j \geq h$ start to invest in skills for $\phi \geq \phi^c$.*[7]

In Fig. 4, we depict the conditional steady states $x^u_{*,\phi_{t+1}}$ and $x^s_{*,\phi_{t+1}}$ as linear functions of ϕ starting at $x^u_{*,min} > 0$ and $x^s_{*,min} < 0$, respectively. In light of item (i), both population groups benefit from increasing life expectancies, but the marginal effect is stronger for the skilled population group as $w^s > w^u$. Moreover, life expectancy must exceed a threshold ϕ^c in order to make investments in human capital profitable, in the sense that life expectancies above ϕ^c (items (ii) and (iii)) assure that lifetime utility of a skilled agent exceeds lifetime utility of remaining unskilled given that the amount of inherited wealth x_t^j is at least h. The threshold ϕ^c implies that the (conditional) steady state level of wealth for the skilled population group is as least as high as h. Since life expectancy is endogenous, our model is therefore able to generate an endogenous take-off in terms of human capital investment.

[7]Note that $x^u_{*,\tilde{\phi}} = x^s_{*,\tilde{\phi}}$ implies $w^s\tilde{\phi} - (1+\bar{r})h = (1-\tau)(1+\bar{r})w^u + \tilde{\phi}w^u$. This threshold, however, is irrelevant, since $h > x^s_*$.

The next feature of our framework is the dynamics of the wealth distribution as influenced by the evolution of life expectancies over time. For the long-run composition of the population in terms of skilled and unskilled households, the location of the minimum amount of wealth necessary to invest in skills, x_t^{crit}, and the location of the unstable interior (conditional) steady state, $x^{u,s}_{*,\phi_{t+1}}$, are crucial. More precisely, the distance between x_t^{crit} and $x^{u,s}_{*,\phi_{t+1}}$ is crucial for the long run distribution of the population. Though a decline in x_t^{crit} facilitates more unskilled households to invest in skills, it is the location of $x^{u,s}_{*,\phi_{t+1}}$ that demarcates the basin of attraction of the two exterior steady states. Hence, a reduction in x_t^{crit} given a distribution of wealth Γ_t is beneficial for unskilled households during the transition, but not necessarily in the long-run as long as the decline in x_t^{crit} is not accompanied by a decline in $x^{u,s}_{*,\phi_{t+1}}$ which assures that more unskilled dynasties transit towards the superior steady state. The following proposition shows the sensitivity of the interior (conditional) steady state and the minimum level of wealth necessary to invest in skills when life expectancy changes.

Proposition 4 (Interior Steady State, $x^{u,s}_{*,\phi_{t+1}}$, x_t^{crit}, and Changes in ϕ_{t+1}) *The minimum level of wealth, x_t^{crit}, necessary to become a skilled worker for agents with wealth $x_t^j < h$ and the interior unstable conditional steady state $x^{u,s}_{*,\phi_{t+1}}$ are hump-shaped in ϕ_{t+1}, whereby it holds that $x^{u,s}_{*,\phi_{t+1}} > x_t^{crit}$ if $0 < \phi_t < \tilde{\phi}$, where $\phi = \tilde{\phi}$ implies $x_t^{crit} = x^{u,s}_{*,\phi_{t+1}}$. Moreover, x_t^{crit} and $x^{u,s}_{*,\phi_{t+1}}$ are declining in ϕ for $\phi^c \leq \phi_{t+1} \leq \phi_*$. (Proof, see Appendix.)*

In light of Proposition 4, see also Fig. 4, it follows that an increase in life expectancy lowers the minimum level of wealth necessary to become a skilled worker, x_t^{crit}, and the unstable interior (conditional) steady state, $x^{u,s}_{*,\phi_{t+1}}$, in the relevant interval $\phi \in (\phi^c, \phi_*)$. Since, $x_t^{crit} = x^{u,s}_{*,\phi_{t+1}}$, if $\phi_t = \tilde{\phi}$, the distance between x_t^{crit} and $x^{u,s}_{*,\phi_{t+1}}$ declines with increasing life expectancy. Thus, the improvement in life expectancy is beneficial for descendants of unskilled households that wish to invest in skills. With the induced decline of $x^{u,s}_{*,\phi_{t+1}}$ the demarcation line of the two basins of attraction shrinks as well, such that more descendants of unskilled households that invested in skills may transit towards the superior long-run equilibrium x^s_{*,ϕ_*} which would have converged otherwise towards the inferior equilibrium x^u_{*,ϕ_*}. How sizable this effect is depends on the distribution of wealth, Γ_t, in the sense that a more equal distribution of wealth locates more unskilled households in the vicinity of $x^{u,s}_{*,\phi_{t+1}}$, such that more unskilled households investing in skills benefit from the improvement in ϕ and transit towards x^s_{*,ϕ_*}. In other words, the effect of improving health conditions on the long-run performance of the economy depends on Γ_t. Likewise, we will see that the effectiveness of economic policy in terms of Φ depends on Γ_t.

We next study the reaction of the (conditional) steady states in response to a change in the policy set $\Phi = \{\tau, \nu\}$. The following proposition summarizes the sensitivity of the two exterior stable (conditional) steady states in response to change in the tax rate τ:

Proposition 5 (Effects of τ on $x^s_{*,\phi_{t+1}}$ and $x^u_{*,\phi_{t+1}}$)

(i) *The (conditional) steady state of the skilled population group increases in response to an increase in* τ, *if*

$$(1-\tau)w^s\frac{\partial\phi_{t+1}}{\partial\tau} > \frac{w^s - h(1+\bar{r})}{1-(1-\tau)(1-\alpha)(1+\bar{r})} = \frac{x^s_{*,\phi_{t+1}}}{(1-\alpha)(1-\tau)} = y^s_{*,\phi_{t+1}}.$$

(ii) *The (conditional) steady state of the unskilled population group increases in response to an increase in* τ, *if*

$$(1-\tau)\left(\frac{\partial\phi_{t+1}}{\partial\tau} - (1+\bar{r})\right)w^u$$
$$> \frac{[(1+\bar{r})(1-\tau)+\phi]w^u}{1-(1-\tau)(1-\alpha)(1+\bar{r})} = \frac{x^u_{*,\phi_{t+1}}}{(1-\alpha)(1-\tau)} = y^u_{*,\phi_{t+1}}.$$

Proposition 5 states that an increase in τ benefits skilled and unskilled agents in terms of wealth as long as the marginal cost of taxes in terms of forgone lifetimes earnings (right-hand side of the equations) are smaller than the marginal increase in lifetime net-earnings generated through the improvement in life expectancy, $\frac{\partial\phi_{t+1}}{\partial\tau} > 0$ (left-hand side of the equations). Furthermore, the marginal benefit of the skilled population group exceeds the marginal benefit of the unskilled population group, as

$$\frac{w^s}{w^u} > \frac{\frac{\partial\phi_{t+1}}{\partial\tau} - (1+\bar{r})}{\frac{\partial\phi_{t+1}}{\partial\tau}},$$

since $w^s > w^u$ and $\frac{\frac{\partial\phi_{t+1}}{\partial\tau}-(1+\bar{r})}{\frac{\partial\phi_{t+1}}{\partial\iota}} < 1$. Moreover, the marginal benefit of the unskilled population group may even turn negative, if $\frac{\partial\phi_{t+1}}{\partial\tau} < (1+\bar{r})$. Interestingly, economic policy increases long-run inequality in terms of wealth through its asymmetric impact on the long-run levels of wealth for the skilled and the unskilled population group. However, this effect is at least partially compensated by a reduction in x^{crit}_t and $x^{u,s}_{*,\phi_{t+1}}$ in response to an increase in taxes.

Proposition 6 (Effects of τ on $x^{u,s}_{*,\phi_{t+1}}$ and x^{crit}_t)

(i) *The interior steady state declines in response to an increase in* τ, *i.e.* $\frac{\partial x^{u,s}_{*,\phi_{t+1}}}{\partial\tau} < 0$, *if*

$$\left[h(1+i_{t+1}) - w^s\phi_{t+1}\right] + (1-\tau)\frac{\partial\phi_{t+1}}{\partial\tau}\left[w^s - h\frac{\partial(1+i_{t+1})}{\partial\phi_{t+1}}\right]$$
$$< (1-\tau)\frac{\partial\phi_{t+1}}{\partial\tau}w^s\left[(1-\alpha)(1-\tau)\left[(1+i_{t+1}) - \frac{\partial(1+i_{t+1})}{\partial\phi_{t+1}}\phi_{t+1}\right]\right] \quad (12)$$

with $\frac{\partial(1+i_{t+1})}{\partial\phi_{t+1}} < 0$, *and* $h(1+i_{t+1}) - w^s\phi_{t+1}, (1-\alpha)(1-\tau)(1+i_{t+1}) - 1 > 0.$

(ii) *The minimum level of wealth necessary to invest in skills declines in response to an increase in taxes, since*

$$\frac{\partial x_t^{crit}}{\partial \tau} = \left[-w^u(1+\bar{r}) + h\frac{\partial(1+i)}{\partial \phi_t}\frac{\partial \phi_t}{\partial \tau} - \frac{\partial \phi_t}{\partial \tau}\left(w^s - w^u\right) \right]$$
$$- \frac{1}{i-\bar{r}}\left[(1-\tau)w^u(1+\bar{r}) + h(1+i) - \phi_t\left(w^s - w^u\right)\right] < 0,$$

as the first term in squared brackets is negative while the second term in squared brackets equals x_t^{crit} which is positive.

According to item (i) of Proposition 6, the change of $x^{u,s}_{*,\phi_{t+1}}$ in response to an increase in τ is ambiguous: Because of (11), an improvement in life expectancy financed by an increase in τ has an ambiguous effect on $x^{u,s}_{*,\phi_t}$ through the decline in i_{t+1}. On the one hand, it increases second period income $w^s\phi_t - h(1+i)$ (reduces debts), but diminishes the return on wealth. If condition (12) is met, the latter effect is lower than the former. Item (ii), in turn, states that x_t^{crit} is negatively associated to an increase in taxes. Both results in combination affect the dynamics of the wealth distribution. Given a favorable distribution of wealth, Γ_t, it is possible that the reduction in x_t^{crit} and $x^{u,s}_{*,\phi_{t+1}}$ affects all unskilled households, such that the long-run composition of households exhibits only skilled dynasties. Then, the induced asymmetric impact of τ on the exterior steady states doesn't play any role. The adverse impact on the inferior steady state gains in importance, however, the larger the initial amount of inequality is. Furthermore, we can not exclude the case that a decline in x_t^{crit} is accompanied by an increase in $x^{u,s}_{*,\phi_{t+1}}$, such that in an extreme case dynasties would start to invest in skills initially, but their descendants would cease to acquire human capital since the basin of attraction of the inferior steady state is at least temporarily larger.

We now turn our attention to the effects of a change in the public expenditure share for health care services, ν, given τ on the (conditional) steady states and on x_t^{crit}.

Proposition 7 (Effects of ν on $x^s_{*,\phi_{t+1}}$ and $x^u_{*,\phi_{t+1}}$) *The steady states of the skilled and the unskilled population group increase in response to an increase in ν*

$$\frac{\partial x^s_{*,\phi_{t+1}}}{\partial \nu} = \frac{(1-\tau)(1-\alpha)w^s}{1-(1-\tau)(1+\bar{r})(1-\alpha)}\frac{\partial \phi}{\partial \nu} > 0,$$
$$\frac{\partial x^u_{*,\phi_{t+1}}}{\partial \nu} = \frac{(1-\tau)(1-\alpha)w^u}{1-(1-\tau)(1+\bar{r})(1-\alpha)}\frac{\partial \phi}{\partial \nu} > 0,$$

such that $\frac{\partial x^s_{*,\phi_{t+1}}}{\partial \nu} > \frac{\partial x^u_{*,\phi_{t+1}}}{\partial \nu}$ *since $w^s > w^u$ and given $\phi_{t+1} < 1$.*

In light of the last proposition it becomes apparent that an increasing health expenditure share benefits both population groups while an increase in τ could benefit the skilled but harm the unskilled. Nevertheless, the skilled population group again benefits more in terms of long-run wealth, such that even this policy change

Table 1 Parameters

Technology	$\gamma = 0.3$; $\delta = 1$; $a = 0.2$; $b = 8$
Pollution	$\eta = 0.95$; $\varepsilon_0 = 0.1$; $\varepsilon_1 = 0.048$
Human capital	$\beta = 1.6$; $h = 0.515$; $\bar{r} = 4$
Preferences	$\alpha = 0.85$
$\Phi = \{\tau, \nu\}$	$\tau = 0.1$; $\nu = 0.8$

increases inequality in the long-run. But again, the long-run distribution of the population over the two exterior steady states is influenced by the dynamics of the conditional interior steady state, $x^{u,s}_{*,\phi_{t+1}}$, and x^{crit}_t. The reaction of $x^{u,s}_{*,\phi_{t+1}}$ and x^{crit}_t is summarized in the following proposition.[8]

Proposition 8 (Effects of ν on $x^{u,s}_{*,\phi_{t+1}}$ and x^{crit}_t)

(i) *The interior steady state declines in response to an increase in* ν, *if*

$$w^s - h\frac{\partial(1+i_{t+1})}{\partial\phi_{t+1}} < (1-\alpha)(1-\tau)\left((1+i_{t+1}) - \frac{\partial(1+i_{t+1})}{\partial\phi_{t+1}}\phi_{t+1}\right)w^s,$$

since $\frac{\partial(1+i_{t+1})}{\partial\phi_{t+1}} < 0$ *and* $\frac{\partial\phi_{t+1}}{\partial\nu} \geq 0$.[9]

(ii) *The minimum level of wealth necessary to invest in skills declines in response to an increase in taxes, as*

$$\frac{\partial x^{crit}_t}{\partial\nu} = \left[h\frac{\partial(1+i_{t+1})}{\partial\phi_{t+1}}\frac{\partial\phi_{t+1}}{\partial\nu} - \frac{\partial\phi_{t+1}}{\partial\nu}\left(w^s - w^u\right)\right]$$
$$- \frac{1}{i_{t+1}-\bar{r}}\left[(1-\tau)w^u(1+\bar{r}) + h(1+i_{t+1}) - \phi_{t+1}\left(w^s - w^u\right)\right] < 0,$$

since the first term in squared brackets is negative while the second term in squared brackets equals x^{crit}_t *which is positive.*

5 Numerical Experiments

In this section, we explore the dynamics of our economy numerically. The set of parameters is presented in Table 1. The capital income share, γ, is set to 0.3. Since one period encompasses approximately 30 years, we set the rate of capital depreciation, δ, equal to 1. $\bar{r} = 4$ implies an annual interest rate of 4.7 %. The weight of consumption in the utility function, α, is 0.85. The parameter $\beta = 1.6$ implies a borrowers' interest rate of 9 % p.a., if $\phi_t = 1$. As far as the evolution of the pollution stock is concerned, there are obviously several degrees of freedom. We therefore fix $\eta = 0.95$ which seems to be plausible over 30 years. The remaining parameters are

[8] The qualitative results are similar to Proposition 6.

[9] See the explanation following Proposition 6.

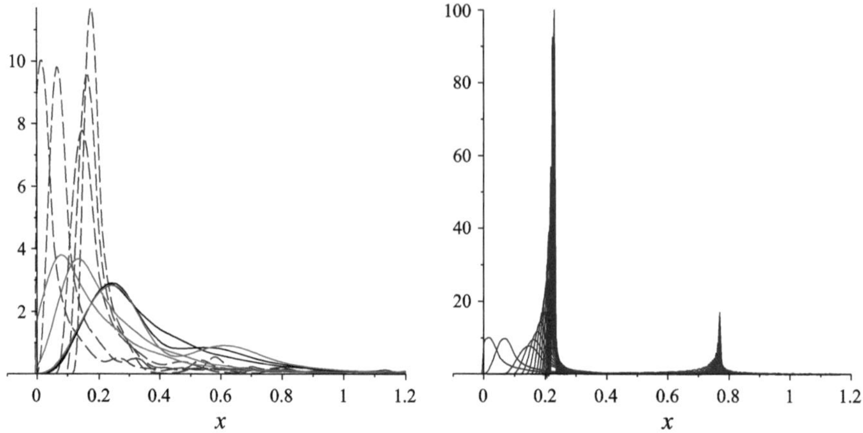

Fig. 5 Evolution of the wealth distribution (kernel density estimation)—left-hand panel: first five periods (*dashed*: increased inequality (mean preserving spread); *solid*: baseline scenario)—right-hand panel: complete transition of the baseline scenario

fixed in combination with the survival function, ϕ_t, (see Eq. (13)) in an iterative way assuring that life expectancy starts around 0.5 and the transition to the steady state is completed after 8 periods, i.e. 240 years.

We assume that life expectancy follows a logistic function in health expenditures and pollution

$$\phi_t = \frac{F + A * 0.0001}{0.0001 + \exp(-k * (\psi_H H_t - \psi_P P_t)) * (A/F - 1)}, \tag{13}$$

with $F = 0.5$; $k = 10$; $A = 200$; $\psi_H = 0.0038$; $\psi_P = 0.0005$.

In performing the numerical experiments, we generate an artificial sample of households ($N = 1000$) and draw the initial level of wealth x_0^j, $j \in N$ out of a log-normal distribution $\mathscr{F}_0 \sim (\mu_x; \sigma_x)$. Since our experiments will deal with different amounts of initial inequality, we increase the variance of initial wealth, but keep the mean of the distribution constant and assure therefore the comparability between the experiments.

Our first experiment deals with different amounts of initial inequality in terms of wealth. The evolution of the wealth distribution over time is depicted in Fig. 5: the left-hand panel shows the kernel density estimation for the first periods for different amounts of initial inequality and the right-hand panel depicts the overall transition of the wealth distribution. Since the population converges either to x^s_{*,ϕ_*} or x^u_{*,ϕ_*}, the wealth distribution collapses into two spikes located at x^u_{*,ϕ_*} and x^s_{*,ϕ_*}, with the height of the two spikes representing the amount of inequality.

In Fig. 6, we present the dynamics of the conditional steady states $x^u_{*,\phi_{t+1}}$, $x^{u,s}_{*,\phi_{t+1}}$, $x^s_{*,\phi_{t+1}}$ towards their respective stationary values indicated by subscript ϕ_*, and the critical level of wealth necessary to invest in skills, x_t^{crit}. The baseline scenario with low initial inequality is depicted in solid lines and the scenario characterized by

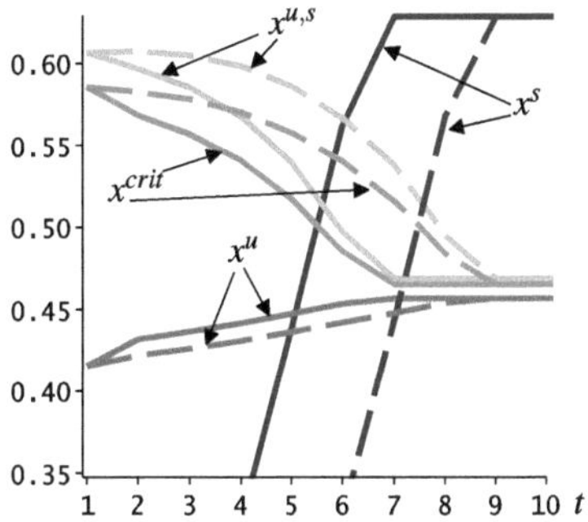

Fig. 6 Dynamics of the conditional steady states and x_t^{crit} with different amounts of initial inequality. High inequality = *dashed lines*; low inequality = *solid lines*

higher initial inequality is illustrated by dashed lines. As regards the response of the conditional steady states and x_t^{crit} to improvements in life expectancy, Fig. 6 reflects the insights of Fig. 4. Figure 7 shows the dynamics of aggregate output, Y, tax revenues, G, pollution, P, life expectancy, ϕ, the number of skilled and unskilled households, n^j, $j = s, u$, and the number of unskilled households investing in skills, $n^{u,s}$. The baseline scenario is again depicted in solid lines while dashed lines represent the scenario characterized by higher initial inequality.

During the initial stages of economic development, production is low which implies low levels of pollution. Tax revenues are low as well allowing only for low levels of public health care expenditures and hence, low levels of life expectancy. There are no incentives to invest in skills, since a low value of the life expectancy, ϕ, implies high interest rates of borrowers and a short amortization period of human capital investment. These dynamics are reflected by the fact that the conditional steady state of the unskilled, x^u_{*,ϕ_t}, exceeds the conditional steady state that results from skill investments, i.e. x^s_{*,ϕ_t} (see Fig. 6). Therefore, the entire population is composed of unskilled households, i.e. $n^u = N$. Since wealth evolves according to (6), the tax base of the government will however increase over time allowing for an increase in health expenditures and consequently an increase in life expectancy. Although the increase in life expectancy is ineffective with respect to human capital investment, as long as $\phi < \phi^c$, it is effective insofar as an increase in ϕ increases labor supply, and therefore, the level of aggregate production. Moreover, higher life expectancy increases the second period income of unskilled households, i.e. lifetime earnings and therefore, the accumulation of wealth. Both factors in combination increase tax revenues and life expectancy even further. In the simulation with low inequality, the threshold level of $\phi = \phi^c$ is reached in period $t = 5$. From period five onwards, the conditional steady state of unskilled households falls below the conditional steady state of skilled households ($x^u_{*,\phi_{t+1}} < x^s_{*,\phi_{t+1}}$), inducing households with $x_t^j > x_t^{crit}$ to invest in skills. Since further increases in ϕ reduce borrowers' interest rate i_{t+1}, and thereby x_t^{crit} and $x^{u,s}_{*,\phi_t}$, more and more unskilled households are not only willing to invest in skills but are also attracted by the superior steady state in

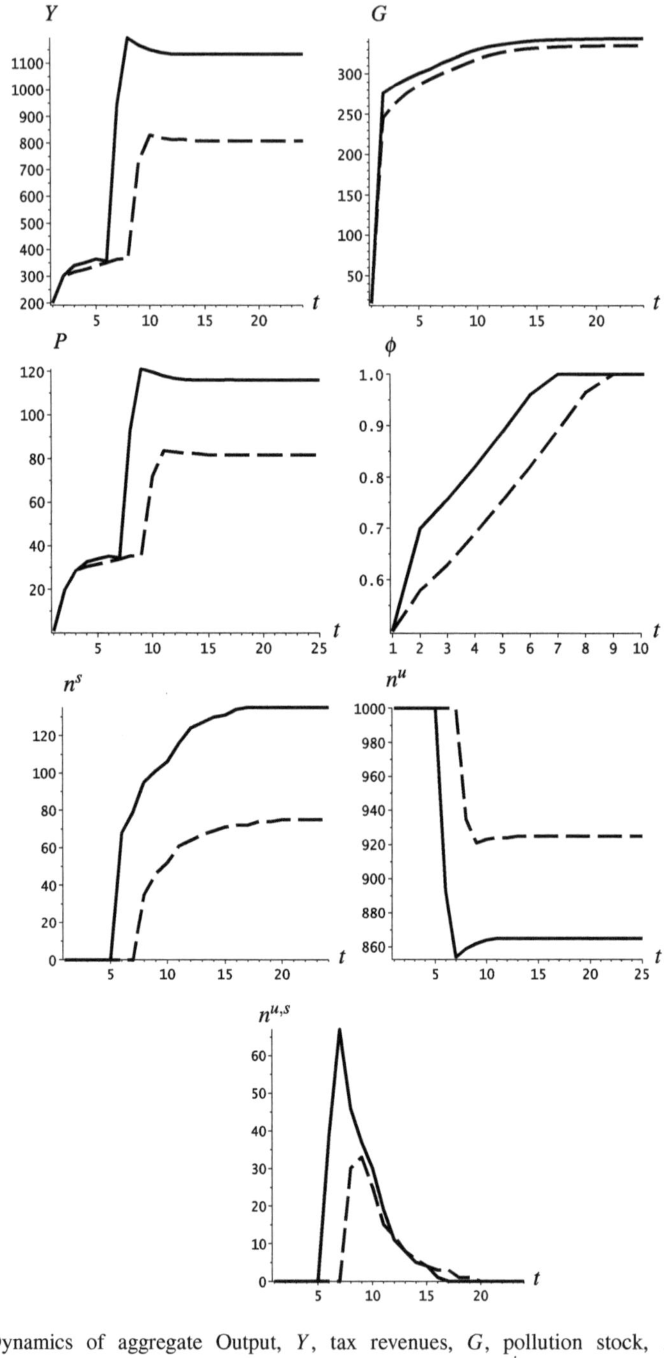

Fig. 7 Dynamics of aggregate Output, Y, tax revenues, G, pollution stock, P, life expectancy, ϕ, the number of skilled and unskilled households n^j, $j = u, s$ and the number of unskilled households investing in skills, $n^{u,s}$, with different amounts of initial inequality. High inequality = *dashed lines*; low inequality = *solid lines*

the long-run.[10] The endogenous switch to human capital accumulation increases the level of output since a more productive technology is used. As we assume a symmetric impact of the skilled and the unskilled sector on the environment, the pollution stock increases as a consequence of an increased level of aggregate production.[11]

An increase in initial inequality has no impact on the long-run values of wealth per household $x^u_{*\phi_*}, x^{u,s}_{*,\phi_*}$ and x^s_{*,ϕ_*} given that $\phi_* = 1$.[12] Nevertheless, a higher amount of initial inequality reduces the tax base since agents accumulate wealth at a slower pace. Consequently, life expectancy rises more slowly as compared to the low-inequality scenario. Accordingly, since lifetime earnings are lower, the levels of the conditional exterior steady states are reduced as well (see Fig. 6). Consequently, the threshold level ϕ^c is only reached in period $t = 7$ and the take-off in terms of human capital accumulation is delayed by two periods. Since the transition of life expectancy is delayed due to a higher initial inequality, the levels of $x^{u,s}_{*\phi_{t+1}}$ and x^{crit}_t are also higher during the transition.[13] The latter has far reaching consequences for the long-run performance of the economy. Although the conditional steady states and x^{crit}_t as well as the level of life expectancy converge to the same long-run values, the transitory higher value of x^{crit}_t hinders more unskilled households to invest in human capital while more unskilled households that invested in human capital are still in the basin of attraction of the inferior steady state caused by the increase in $x^{u,s}_{*,\phi_{t+1}}$. Therefore, the economy exhibits a lower number of skilled households as compared to the low-inequality scenario. Summing up, a larger amount of initial inequality reduces the tax base and therefore, the effectiveness of the policy set on life expectancy ϕ, for any exogenously fixed level of Φ and initial level of pollution. The lower production level (due to less skilled workers) reduces the pollution level.

Having explored the effects of different degrees of initial inequality on the long-run performance of the economy, as reflected by the level of aggregate output and pollution, we turn our focus now to changes in the policy set Φ. We start with a change in the income tax rate τ and compare the results (as presented in Fig. 8) to the baseline scenario. Thereafter, we analyze the impact of higher initial inequality in the high-tax regime with the low-tax regime (see Fig. 9). Finally, we conduct the same experiments with different shares of public health care, i.e. ν.

In Fig. 8, the high-tax scenario is depicted by dashed lines while the low-tax scenario is presented by solid lines. Higher taxes increase expenditures for public health care and abatement measures. Therefore, life expectancy ϕ increases faster as compared to the reference scenario. Accordingly the critical level of life expectancy,

---

[10]Remember the reduction in x^{crit}_t and $x^{u,s}_{*,\phi_t}$ reduces the basin of attraction of the inferior steady state.

[11]Note that the model is also able to account for an Environmental Kuznets Curve, if we would assume an asymmetric impact of both sectors on the environment in the sense that the s—sector would harm the environment less as compared to the u—sector. In this case, the evolution of the pollution stock would read as $P_{t+1} = (1-\eta)P_t + \varepsilon^u_0 Y^u_t + \varepsilon^s Y^s_t - \varepsilon_1 A_t$ with $\varepsilon^u_0 > \varepsilon^s_0$.

[12]Off course an extreme case of initial inequality could result in a long-run value of ϕ_* below 1 such that the long-run values of x would be reduced as well.

[13]This effect stems from lower life expectancies and higher interest rates.

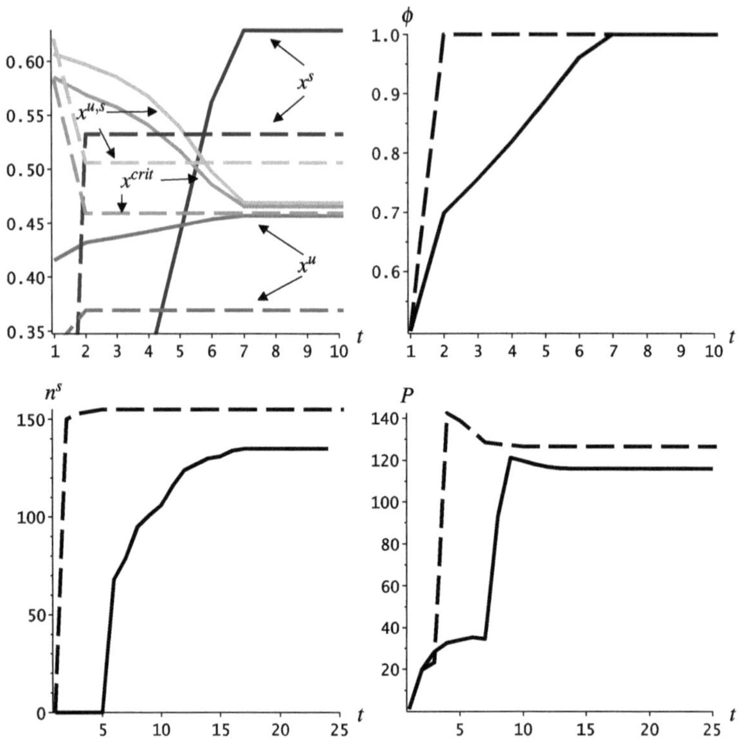

Fig. 8 Dynamics of the conditional steady states, x_t^{crit}, life expectancy, ϕ, the pollution stock, P, and the number of skilled households, n^s, with different tax rates ($\tau = 0.15$: *dashed lines*; $\tau = 0.1$: *solid lines*)

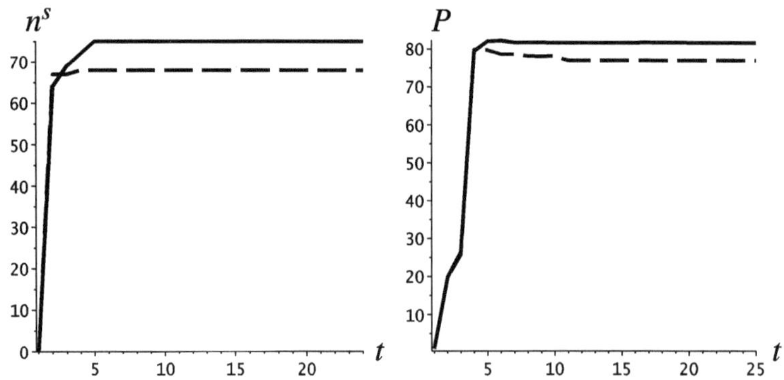

Fig. 9 Dynamics of the number of skilled households, n^s, and the pollution stock, P with different tax rates ($\tau = 0.15$: *dashed lines*; $\tau = 0.1$: *solid lines*) and higher initial inequality

ϕ^c, is reached already in period $t = 2$. From now on the conditional steady state that applies for households that wish to acquire human capital is above the conditional steady state of the unskilled population group. During the transition, the conditional steady states of the high-tax regime are above the ones of the low-tax regime. This fact is owed to the fast increase in life expectancy which compensates households for the lower post-tax income at a given level of life expectancy. In later stages of economic development, the effect of reduced post-tax incomes overcompensates further gains in life expectancy such that the long-run levels of wealth for both population groups are reduced compared to the reference scenario with lower taxes. Unskilled households with a relatively high level of inherited wealth may benefit from a faster increase in life expectancy inasmuch as the critical level of wealth necessary to acquire skills and the interior conditional steady state not only declines faster but converges also to a lower long-run steady state. Thus, the basin of attraction of the superior exterior steady state extends. Accordingly, the favorable evolution of life expectancy increases the number of skilled households, thereby aggregate production and pollution compared to the reference scenario.

Nevertheless, it is worth to notice that there may be a draw back of a higher tax rate depending on the initial distribution of wealth. As we have shown, the favorable evolution of life expectancy induces a fast decline in x_t^{crit} and $x_{*,\phi_{t+1}}^{u,s}$. This increases human capital investments and enhances the possibility of households with a level of wealth below h to transit in the long-run towards the superior steady state. However, this possibility depends on the initial amount of inequality. Given a more unfavorable initial distribution of wealth, the fast decline in $x_t^{u,s}$ may block the transition of unskilled households that acquired human capital towards the superior equilibrium, such that the number of skilled households shrinks. This effect is shown in Fig. 9. There, we consider two regimes that are characterized by high initial inequality but different tax rates. The high-tax regime is again depicted in dashed lines while the low-tax regime is presented in solid lines. Apparently, higher inequality reduces the favorable impact of higher taxes on human capital investments, but reduces the level of pollution.

In our next experiment we analyze a change in the expenditure share for public health care, ν, see Fig. 10. Contrary to a change in τ, a change in ν leaves (c.p.) disposable incomes of households unaffected. The long-run values of wealth for different population groups remain unaffected given that $\phi_* = 1$. Like an increase in τ, an increase in ν induces a faster increase in life expectancy which favors human capital investments.[14] Since the long-run values of wealth are unaffected by a change in ν, and the number of skilled households has increased, the economy's tax base will increase compared to the base line scenario. That is, although economic activity and its adverse impact on the environment have increased, there are also more resources available for abatement measures due to an increase in the tax base. Therefore, the level of pollution is similar compared to the one resulting from an

[14]Like in Fig. 8, the increase in ν induces also a fast decline in x_t^{crit} and $x_{\phi_t,*}^{u,s}$ such that the benefit of an increased ν depends also in this case on the initial distribution of wealth.

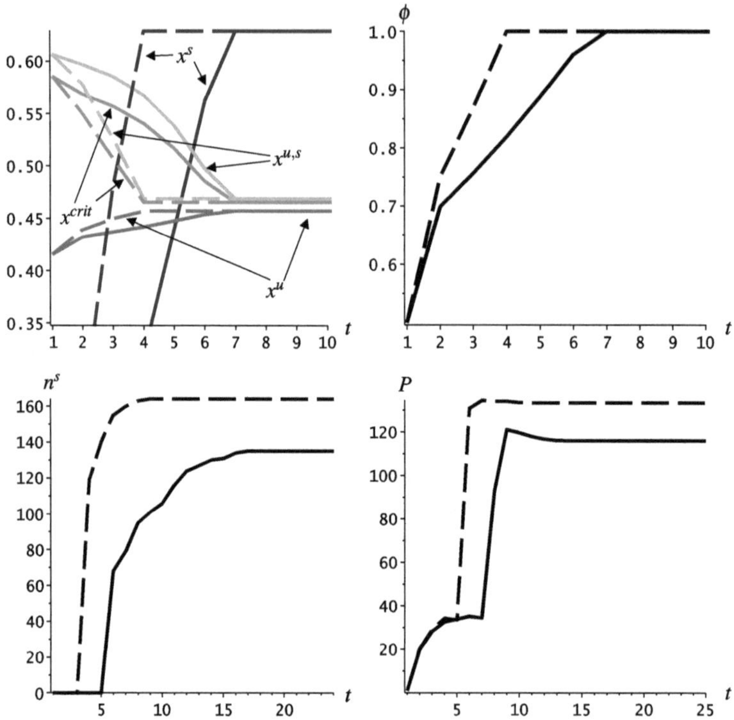

Fig. 10 Dynamics of the conditional steady states, x_t^{crit}, life expectancy, ϕ, the pollution stock, P, and the number of skilled households, with different expenditures shares ν for public health cares ($\nu = 0.85$: *dashed lines*; $\nu = 0.8$: *solid lines*)

increase in taxes although the number of skilled households and hence, the level of economic activity has increased even further (compare Figs. 8 and 10).

6 Discussion

By only focusing on public health care we aim to emphasize the role of the allocation of public expenditures between health care and pollution abatement on economic growth (see also Agenor and Neanidis 2011). On one hand health expenditures reduce income through taxes thereby also reducing spending on pollution abatement, on the other hand higher taxes allow for an increase in health expenditures thereby inducing a higher life expectancy and fostering human capital accumulation and economic growth.

Our analysis is valid for a given policy set, Φ, but we refrain from modeling explicitly the political process which may explain this policy set. That is, we analyze for a given policy set, the interplay between economic inequality, human capital investment and life expectancy while the latter is, in turn, influenced by the specific policy set, Φ, whose effectiveness depends on the amount of inequality.

Nevertheless, our framework opens an avenue for future research which takes account of the endogenous determination of Φ. Obviously, the determination of Φ requires a two-stage optimization with respect to ν and τ.

Assume that the government sets the tax rate given an optimal division of expenditures between health and abatement policy, ν. Then the government would seek to maximize the utility of a representative agent, j, for example the median voter, which could be either a skilled agent, an unskilled agent or a descendant of an unskilled household that invests in skills. Thus, the government would maximize agent j's lifetime utility (4)

$$\max_{0\leq\tau\leq 1}\left\{\phi_t\left[\bar{\alpha}+\ln\left[(1-\tau)y_t^j\right]\right]\right\}.$$

The associated first-order condition reads

$$\frac{\partial\phi_t}{\partial\tau}\left[\bar{\alpha}+\ln\left[(1-\tau)y_t^j\right]\right]+\frac{\phi_t}{(1-\tau)y_t^j}\left[-y_t^j+(1-\tau)\frac{\partial y_t^j}{\partial\phi_t}\frac{\partial\phi_t}{\partial\tau}\right]=0.$$

The term $-y_t^j$ in large squared brackets on the left hand side captures the marginal cost of taxes while the first term on the left hand side and $(1-\tau)\frac{\partial y_t^j}{\partial\phi_t}\frac{\partial\phi_t}{\partial\tau}$ captures the marginal benefit of an increase in taxes due to the associated increase in the survival probability. Obviously, the marginal benefit of raising taxes is zero, if $\phi_t=1$ and/or $\frac{\partial\phi_t}{\partial\tau}=0$, such that the first-order condition reduces to $-\frac{\phi_t}{(1-\tau)}$.[15] As has been discussed already, an income tax is welfare improving as long as the benefits that result from an increase in life expectancy and the associated increase in lifetime earnings overcompensate the income loss. Thus, the preferred tax rate depends on the sensitivity (curvature) of the survival function in response to changes in P and H and the income level of the representative agent. In addition it seems reasonable that wealthier agents are less exposed to pollutants as compared to poorer agents. Thus, the marginal benefit of wealthier agents may be lower which explains that the new entrepreneurial class saw no point in paying taxes for sanitation in cities. We come back to this point further below in this section.

As the life expectancy, ϕ_t, is increasing in aggregate health expenditures H_t, the welfare maximizing expenditure share would be $\nu=1$. This result can be explained by the fact that the current pollution stock cannot be reduced by current abatement measures, $A_t=(1-\nu)G_t$, and that the generation born in $t-1$ behaves entirely selfish in the sense that their impact on the future pollution stock is not internalized by their lifetime utility function. Thus, non-altruistic agents would opt for zero abatement measures. This result would change, if we assume at least imperfect altruism with respect to the future stock of pollution. An illustrative example may be represented by the following utility function:

$$u_t^j=\phi_t\left[\bar{\alpha}+\ln\left[(1-\tau)y_t^j\right]\right]-\pi\ln[P_{t+1}],$$

[15] Note that $\phi_t=1$ implies $\frac{\partial\phi_t}{\partial\tau}=0$ but $\frac{\partial\phi_t}{\partial\tau}=0$ may also result for $\phi_t<1$.

where π reflects the weight of environmental damages for the future generation captured by the pollution stock in the subsequent period. The first-order condition thus reads

$$\frac{\partial \phi_t}{\partial \nu}\left[\bar{\alpha} + \ln\left[(1-\tau)y_t^j\right]\right] + \frac{\phi_t}{(1-\tau)y_t^j}\left[(1-\tau)\frac{\partial y_t^j}{\partial \nu}\frac{\partial \phi_t}{\partial \nu}\right] - \frac{\pi}{P_{t+1}}\frac{\partial P_{t+1}}{\partial \nu} = 0.$$

In light of the last expression it becomes apparent that although ϕ_t is an increasing function in ν it is not necessarily welfare maximizing to set $\nu = 1$ given that the marginal benefit from abatement weighted by π is sufficiently high.

As we just clarified, we may expect that the marginal benefit of taxation for wealthier agents is low or even zero, a phenomenon which is very much in line with historical observations. The significance of water as an industrial raw material has been documented by Hassan (1985): fresh water was used for commercial purposes while the new entrepreneurial class saw no point in spending money for sanitation and sewage treatment plants. This finding points directly at the role of economic inequality especially as wealthier households moved to cleaner areas with significantly lower exposures to local pollutants. In this line of argumentation Szreter (1997) stresses

> … there is indeed something intrinsically dangerous and socially destabilizing in the wake of economic growth.…

He underlines his statement by the following two observations: (1) local authorities were failing the management of their environments, and, as regards the role of inequality, (2) as a consequence of it wealthier citizens moved to the periphery of the cities.

Our setting does not include spatial segregation and differential exposures to pollutants. Nevertheless, it opens an avenue to analyze the emerging social conflict between wealthier citizens living in cleaner areas and the majority of people living in overcrowded and polluted cities. This can be achieved by regionally differing survival functions capturing the local impact of pollutants on citizen's survival probability (see Schaefer 2013).

7 Summary and Conclusions

Based on the work by Galor and Zeira (1993), we analyzed the link between economic inequality in terms of wealth, life expectancy and pollution. The distribution of wealth is decisive for the number of households investing in human capital. Moreover, the willingness to invest in human capital is affected by agents' life expectancy since it triggers the length of the amortization period of human capital investments. In our framework, life expectancy is endogenous and positively affected by public health care expenditures but adversely affected by the pollution stock stemming from aggregate production.

Due to a low level of health expenditures owed to a low level of tax revenues, life expectancies are low in earlier stages of economic development such that there is no incentive to invest in human capital. Nevertheless, increasing accumulation of wealth increases step by step the tax base and agents' life expectancy. After a threshold level of life expectancy has been passed, households begin to acquire skills. Our model is therefore able to take account of an endogenous take-off in terms of human capital accumulation.

Higher initial inequality retards the take-off, since the tax base of the economy is reduced which implies lower expenditures on public health care and abatement measures. Consequently, the increase in life expectancy is slower. The latter in turn reduces human capital investments, such that the economy is characterized by a lower aggregate long-run level of production and pollution. In other words, higher inequality reduces the effectiveness of a given set of economic policies in terms of health and abatement expenditures.

An increase in taxes or the expenditure share for public health care increases life expectancy. The more favorable evolution of the life expectancy is beneficial for human capital investment given that the initial distribution of wealth is not too unequal. An increase in taxes, given a higher amount of initial inequality, may be harmful in the sense that less descendants of unskilled households invest in skills. In general, an increase in taxes reduces long-run wealth of households. Contrary to an increase in taxes, an increase in the public health expenditure share leaves (c.p.) disposable incomes of agents constant, such that the long-run levels of wealth per household remain unaffected or even increase. Since the more favorable evolution of life expectancy increases the tax base, higher health care expenditure shares increase in general (like an increase in the tax rate) the number of skilled households and the level of aggregate production in the long-run. The advantage compared to an increase in taxes however is the achieved increase in the tax base which induces more expenditures on health care and abatement measures, such that an even higher economic activity is compatible with a similar level of long-run pollution.

Acknowledgements We would like to thank Thomas Steger, Timo Trimborn, seminar participants at the University of Leipzig, and an anonymous referee for valuable comments and constructive suggestions. The usual disclaimer applies.

Appendix

Proof of Proposition 4

We first show that $x^{u,s}_{\phi_t,*}$ is hump-shaped. In the second step, we demonstrate that x^{crit}_t is hump-shaped as well and always below $x^{u,s}_{\phi_t,*}$.

1. Since $\lim_{\phi_t \to 0} x_{\phi_t,*} = h$ and $x^{s}_{\phi_t,*} = x^{u,s}_{\phi_t,*} = h$, if $\phi_t = \phi^c = \frac{h}{(1-\alpha)(1-\tau)w^s}$ it follows that $x^{u,s}_{\phi_t,*}$ is non-monotonous is ϕ_t. As moreover

$$\lim_{\phi_t \to 0} \frac{\partial x^{u,s}_{\phi_t,*}}{\partial \phi_t} = \frac{(\beta - 1)h}{\beta(1-\alpha)(1-\tau)(1+\bar{r})} > 0,$$

it follows outright that $x^{u,s}_{\phi_t,*}$ is hump-shaped.

2. Since $\lim_{\phi_t \to 0} x_t^{crit} = h$ and

$$\lim_{\phi_t \to 0} \frac{\partial x_t^{crit}}{\partial \phi_t} = \frac{\beta - 1}{\beta}\big((1-\tau)w^u + h\big) > 0,$$

it follows that $x_t^{crit} = x_{\phi_t,*}^{u,s} = h$ as ϕ_t approaches zero and that x_t^{crit} has a initially a positive slope as well. Moreover, x_t^{crit} has a unique intercept with $x_{\phi_t,*}^{u,s}$ and $x_{\phi_t,*}^{u}$, such that $x_t^{crit} = x_{\phi_t,*}^{u,s} = x_{\phi_t,*}^{u}$, if $\phi_t = \tilde{\phi}$. Since $x_{\phi_t,*}^{u,s}$ cuts $x_{\phi_t,*}^{s}$ from above at ϕ^c while $x_{\phi_t,*}^{u}$ is strictly below $x_{\phi_t,*}^{s}$ for $\phi_t > \phi^c$, it follows that $\tilde{\phi} > \phi^c$. Since the intercept between x_t^{crit} and $x_{\phi_t,*}^{u,s}$ is unique it follows that x_t^{crit} is hump-shaped in ϕ_t as well and always below $x_{\phi_t,*}^{u,s}$ for $\phi_t < \tilde{\phi}$.
3. From 1. and 2. it follows immediately that x_t^{crit} and $x_{\phi_t,*}^{u,s}$ are declining in ϕ_t for $\phi^c \leq \phi_t \leq \phi_*$.

References

Agenor, P.-R., & Neanidis, K. C. (2011). The allocation of public expenditure and economic growth. *Manchester School*, *79*, 899–931.

Aisa, R., & Pueyo, F. (2006). Government health spending and growth in a model of endogenous longevity. *Economics Letters*, *90*, 249–253.

Chay, K. Y., & Greenstone, M. (2003). The impact of air pollution on infant mortality: evidence from geographic variation in pollution shocks induced by a recession. *The Quarterly Journal of Economics*, *118*, 1121–1167.

Galor, O., & Zeira, J. (1993). Income distribution and macroeconomics. *Review of Economic Studies*, *60*, 35–52.

Haines, M. R. (2004). Growing incomes, shrinking people—can economic development be hazardous to your health? *Social Science History*, *28*, 249–270.

Hassan, J. A. (1985). The growth and impact of the British water industry. *The Economic History Review*, *38*, 531–547.

John, A., & Pecchenino, R. (1994). An overlapping generations model of growth and the environment. *The Economic Journal*, *104*, 1393–1410.

John, A., Pecchenino, R., Schimmelpfennig, D., & Schreft, S. (1995). Short-lived agents and the long-lived environment. *Journal of Public Economics*, *58*, 127–141.

Komlos, J. (1998). Shrinking in a growing economy? The mystery of physical stature during industrial revolution. *The Journal of Economic History*, *58*, 779–802.

Mariani, F., Perez-Barahona, A., & Raffin, N. (2010). Life expectancy and the environment. *Journal of Economic Dynamics & Control*, *34*, 798–815.

Raffin, N., & Seegmuller, T. (2012). *Longevity, pollution and growth*. University of Paris West—Nanterre la Défense, EconomiX. EconomiX Working Paper, No. 2012-47.

Schaefer, A. (2013). *The growth drag of pollution*. University of Leipzig. Working Paper.

Szreter, S. (1997). Economic growth, disruption, disease and death: on the importance of the politics of public health for development. *Population and Development Review*, *23*, 693–728.

World Bank (2007). http://www.worldbank.org/en/news/press-release/2007/07/11/statement-world-bank-china-country-director-cost-pollution-china-report.

Uncertain Climate Policy and the Green Paradox

Sjak Smulders, Yacov Tsur, and Amos Zemel

Abstract Unintended consequences of announcing a climate policy well in advance of its implementation have been studied in a variety of situations. We show that a phenomenon akin to the so-called "Green-Paradox" holds also when the policy implementation date is uncertain. Governments are compelled, by international and domestic pressure, to demonstrate an intention to reduce greenhouse gas emissions. Taking actual steps, such as imposing a carbon tax on fossil energy, is a different matter altogether and depends on a host of political considerations. As a result, economic agents often consider the policy implementation date to be uncertain. We show that in the interim period between the policy announcement and its actual implementation the emission of green-house gases increases vis-à-vis business-as-usual.

1 Introduction

An increasing body of economic literature suggests that the very large potential damage due to emissions-induced climate change calls for effective regulation measures to limit the accumulation of atmospheric pollution. The costly measures would be justified only if the response they entail actually advances the desired goal of reduced emissions. Recent studies reveal, however, that this is not always the case,

S. Smulders
Department of Economics and CentER, Tilburg University, P.O. Box 90153, Tilburg 5000, The Netherlands
e-mail: J.A.Smulders@uvt.nl

Y. Tsur
Department of Agricultural Economics and Management, The Hebrew University of Jerusalem, P.O. Box 12, Rehovot 76100, Isràel
e-mail: tsur@agri.huji.ac.il

A. Zemel (✉)
Department of Solar Energy and Environmental Physics, The Jacob Blaustein Institutes for Desert Research, Ben Gurion University of the Negev, Sede Boker Campus 84990, Israel
e-mail: amos@bgu.ac.il

E. Moser et al. (eds.), *Dynamic Optimization in Environmental Economics*, Dynamic Modeling and Econometrics in Economics and Finance 15,
DOI 10.1007/978-3-642-54086-8_7,

and climate policies may paradoxically give rise to more emissions relative to the laissez-faire scenario. For example, partial participation in an international emission reduction program may introduce a leakage effect, whereby the response of the non-participating parties more than offsets the reduction activities of the participants. The resulting "Green Paradox" is analyzed, for example, in Sinn (2008) and in Eichner and Pethig (2011). A similar paradoxical outcome may stem from the regulator's wish to allow the parties prepare in advance to the proposed policy measures and spread their adjustment efforts over time. A model based on this mechanism has been developed in Di Maria et al. (2012) where a study of the response of coal or oil fields owners to an advance announcement of an anticipated climate policy finds that the inelastic supply of the non-renewable resources might induce them to lower prices prior to the policy implementation, encouraging enhanced emissions. The robustness of the paradoxical outcome to various assumptions concerning the cost of backstop substitutes to the polluting resource is studied in van der Ploeg and Withagen (2012).

At the core of the mechanisms driving these results lies a finite resource stock that owners wish to exploit before the announced policy interrupts their supply activities. A recent contribution, Smulders et al. (2012), shows that scarcity is not the sole driver of such effects and obtain the paradoxical outcome in a model with an unlimited supply of fossil energy. Introducing regulation via a carbon tax, which effectively raises the price of fossil energy, and assuming that the regulator announces the plan to levy the tax well in advance, they show that the early announcement distorts resource allocation processes in a number of ways. In particular, it reduces consumption and increases saving, thus giving rise to a larger capital stock. The larger capital stock, in turn, enhances the demand for fossil energy by firms that use capital, energy and labor as factors of production. Thus, announcing a policy aimed at reducing the use of fossil energy well in advance gives rise to the opposite effect until the policy is actually realized. The result holds both when the regulation policy involves a mild tax rate which reduces fossil use but does not induce the use of alternative, clean (solar) energy as well as when the tax rate is high enough to trigger a transition to solar energy.

In this work we extend the results of Smulders et al. (2012) by considering uncertainty as yet another driver of paradoxical effects. We incorporate uncertainty into the model by assuming that the government announces the intention to levy the carbon tax, but the date of implementation depends on political conditions and is therefore uncertain. The distinction appears to be important as it affects the underlying mechanism that drives the paradox. In particular, the continuity of the consumption process plays a key role in deriving the early announcement effect when the implementation date is known in advance. In contrast, under uncertain implementation date, the consumption path undergoes a discontinuous jump at the (random) time when the policy is implemented. Nevertheless, we establish the "green paradox" also under uncertainty, and show that it is driven by the same economic forces: anticipating that the tax will reduce energy use in the future induces households to enhance saving today in order to accumulate more capital that can substitute for the lower energy input. Prior to implementation of the tax policy, the increased capital stock is associated with increased energy input, hence the paradoxical outcome. Indeed, since uncertainty regarding implementation appears to be a common feature characterizing climate policies, the negative effect of the paradox may be significant.

Of course, the saving efforts must come at the expense of consumption, and the realization of the effect depends on a condition relating the production elasticity of capital to the elasticity of marginal utility of consumption. As explained in Smulders et al. (2012), this condition would be satisfied in any empirically relevant calibration, and the paradoxical nature of the uncertainty effect appears to be robust.

2 Setup

We begin with a brief summary of the unregulated case on which the early announcement analysis is based.

2.1 The Unregulated Economy

Early responses to expectations regarding the future introduction of a climate policy are studied in the framework of Tsur and Zemel (2011) where the penetration of solar technologies into competitive energy markets is analyzed. We outline briefly the main components of this model and the results that drive the present analysis. The economy consists of a final good sector, an energy sector, and capital owning households. The final goods are produced using energy x and capital k as inputs. We employ the Cobb-Douglas (CD) production technology

$$y(k,x) = Fk^{\alpha}x^{\gamma} \tag{1}$$

with $\alpha + \gamma < 1$ and $F > 0$.[1] The energy sector consists of fossil energy firms that supply energy at the price ζ and of solar energy firms that invest in solar infrastructure (capital) s. Once the latter has been installed, the generation of solar energy entails no additional cost but is limited by the available stock s of solar capital. The two sources of energy are perfect substitutes, hence

$$x = x^f + bs, \tag{2}$$

where x^f is fossil energy and $b > 0$ is an efficiency parameter measuring how much solar power can be delivered from one unit of solar capital.[2] Solar energy is supplied

[1] All quantities are given in per capita terms, hence the labor input is omitted. The CD specification is not essential for our analysis, but it allows for a simple and transparent derivation.

[2] This formulation evidently abstracts from many features characterizing the fossil-solar competition such as the stochastic nature of the solar input, the significant ongoing improvements in the efficiency of solar energy generation or new discoveries of fossil resources. This simplification is consistent with our general approach of focusing only on those features that are directly relevant to the green paradox effect considered here.

at the going market price and the forward-looking solar firms base their investment decisions on their forecast regarding the evolution of future energy demand. The solar stock, then, evolves according to

$$\dot{s} = \iota - \delta s, \tag{3}$$

where ι is the investment rate and $\delta > 0$ is the capital depreciation rate.

Households have a concave utility function $u(\cdot)$ over consumption c of final goods and seek to maximize the present-value utility stream over an infinite horizon

$$\int_0^\infty u\big(c(t)\big)e^{-\rho t}dt \tag{4}$$

subject to the budget constraint[3]

$$\dot{k} = y\big(k, x^f + bs\big) - \zeta x^f - \iota - \delta k - c, \tag{5}$$

where ρ is the pure (utility) rate of discount.

Absent market failures, the competitive equilibrium processes are determined by finding nonnegative $\{c(t), x^f(t), \iota(t)\}$ that maximize (4) subject to (3), (5), $k(0) = k_0 > 0$ and $s(0) = 0$.

The competitive allocation is characterized in Tsur and Zemel (2011) in terms of the critical price

$$\zeta^c = (\rho + \delta)/b, \tag{6}$$

and three conditions:

1. The condition for fossil energy use, equating its price to the marginal product of energy

$$y_x = F\gamma k^\alpha x^{\gamma-1} = \zeta \tag{7}$$

yielding

$$x = \left(\frac{F\gamma}{\zeta}\right)^{1/(1-\gamma)} k^{\alpha/(1-\gamma)}. \tag{8}$$

2. A steady state (Ramsey) condition,

$$y_k = F\alpha k^{\alpha-1} x^\gamma = \rho + \delta, \tag{9}$$

yielding

$$x = \left(\frac{\rho + \delta}{F\alpha}\right)^{1/\gamma} k^{(1-\alpha)/\gamma}. \tag{10}$$

---

[3]Observe that the same depreciation rate δ has been assumed for both types of capital. In fact, both k and s consist of numerous types of equipment, each with its particular depreciation rate. For simplicity of presentation we take δ to represent an average rate holding for both capital stocks.

3. A simultaneous growth condition, equating the marginal product for both types of capital

$$y_k = by_x \tag{11}$$

yielding

$$x = (b\gamma/\alpha)k. \tag{12}$$

The following characterization is established in Tsur and Zemel (2011):

Proposition 1 (i) *When the fossil energy price ζ falls short of ζ^c, no investment in solar ever takes place, $s(\cdot) \equiv 0$, and the competitive processes converge to a steady state $(\hat{k}, \hat{x})$ determined by conditions* (7) *and* (9). (ii) *When the fossil energy price ζ exceeds ζ^c the competitive processes converge to an exclusively solar steady state with $(\hat{k}, \hat{x})$ determined by conditions* (9) *and* (11), *where $\hat{x}^f = 0$ and $\hat{s} = \hat{x}/b$.*

Economies satisfying condition (i) are referred to as fossil-based economies, while those satisfying condition (ii) are called solar-based. These terms describe long term behavior. In the interim, when the initial capital stock k_0 is small, energy is derived exclusively from fossil sources and investment in solar capital is delayed (or avoided if the economy is fossil-based), while fossil energy use is determined by (8).

2.2 Regulation

The discussion so far has focused on the economic and technological aspects of the distinction between fossil and solar technologies, ignoring the externalities associated with the use of the former, due, e.g. to the polluting emissions it entails. A common policy addressing such externalities entails imposing Pigouvian taxes on emissions. In our setting, such a policy is equivalent to increasing the fossil price ζ. If the "carbon tax" τ is imposed abruptly, the parties will respond promptly by switching from the competitive processes corresponding to the initial (low) price ζ^l to the higher price $\zeta^h = \zeta^l + \tau$. Imposing such a policy by surprise entails discontinuities in the consumption and saving processes, which may raise political opposition. Support-seeking regulators, thus, may choose to announce the tax policy well ahead of its actual implementation in order to allow agents to adjust gradually to the forthcoming changes. The early announcement effects of this policy were shown in Smulders et al. (2012) to give rise to a 'green paradox', whereby the use of fossil energy will actually increase, rather than decrease, during the intermediate period between the announcement of the tax policy and its actual implementation. This result holds both when the tax rate leaves the originally fossil-based economy at the same type classification (albeit less energy intensive) and when τ is large enough to bring ζ^h well above the critical price ζ^c of (6), turning the economy into a solar-based type. In both cases, agents know the implementation date precisely and adjust their behavior so as to ensure a smooth consumption process, even though this entails results that diametrically oppose the regulator's original aim.

Here we extend the analysis to situations where the regulator announces the intention to levy the tax, but is unable or unwilling to commit to a specific date of implementing it. When the policy actually takes place, it implies a prompt adjustment to the higher fossil energy price and discontinuous disruptions cannot be avoided. The agents' response, therefore, differs from that following a pre-specified (known) implementation date. We refer to this scenario as 'uncertain announcement' and investigate whether it can also give rise to paradoxical outcomes. We restrict attention to the case of a mild tax rate which leaves the economy as a fossil-based type also after the tax is imposed. Higher tax rates implying a transition to solar-type economies entail a more tedious analysis, but the paradoxical effects are expected to be driven by the same mechanism, as in Smulders et al. (2012).

2.3 Allocation Dynamics

The analysis is based on a comparison of the competitive processes following an uncertain announcement to those corresponding to a fixed low price ζ^l free of regulation. Here we characterize the latter processes. Employing the energy input at its demand (cf. (8)) gives the output

$$y = F\left(\frac{F\gamma}{\zeta}\right)^{\gamma/(1-\gamma)} k^{\alpha/(1-\gamma)}, \tag{13}$$

and implies

$$\zeta x = y_x x = F\gamma k^{\alpha} x^{\gamma} = \gamma y.$$

Net production, then, can be expressed as a function of capital only:

$$y - \zeta x = (1-\gamma) y = F(1-\gamma)\left(\frac{F\gamma}{\zeta}\right)^{\gamma/(1-\gamma)} k^{\alpha/(1-\gamma)} \equiv A(\zeta) k^{\beta}, \tag{14}$$

where

$$\beta \equiv \alpha/(1-\gamma) < 1 \tag{15}$$

is the effective capital share and

$$A(\zeta) \equiv F(1-\gamma)\left(\frac{F\gamma}{\zeta}\right)^{\gamma/(1-\gamma)} \tag{16}$$

decreases in the fossil price ζ. Fossil based economies with different fossil prices follow the same dynamics, differing only in the parameter $A(\zeta)$. The optimization problem (4), thus, reduces to a single state (k) and single control (c) problem whose solution is governed by the pair of dynamic equations

$$\dot{k} = A(\zeta) k^{\beta} - \delta k - c \tag{17}$$

and

$$\dot{c} = c\sigma(c)\big[A(\zeta)\beta k^{\beta-1} - (\rho+\delta)\big], \tag{18}$$

where

$$\sigma(c) = -u'(c)/\big[u''(c)c\big] \tag{19}$$

is the intertemporal elasticity of substitution.

The steady state $(\hat{k}, \hat{c})$ of this system is given by the relations

$$A(\zeta)\beta\hat{k}^{\beta-1} = \rho+\delta \tag{20}$$

and

$$\hat{c} = A(\zeta)\hat{k}^{\beta} - \delta\hat{k} = \hat{k}\big[(\rho+\delta)/\beta - \delta\big] \equiv r_{\infty}\hat{k}, \tag{21}$$

where

$$r_{\infty} = (\rho+\delta)/\beta - \delta \tag{22}$$

is independent of ζ. The steady state consumption-capital relation coincides with the straight line $\hat{c} = r_{\infty}\hat{k}$ for all values of the fossil price below the critical price ζ^{c}.

For the autonomous system at hand we can write $c = c(k)$, hence $\dot{c} = c'(k)\dot{k}$ and Eqs. (17)–(18) imply

$$c'(k) = \frac{\sigma(c(k))c(k)}{k}\frac{A(\zeta)\beta k^{\beta} - (\rho+\delta)k}{A(\zeta)k^{\beta} - \delta k - c(k)}. \tag{23}$$

Combined with the boundary condition $c(\hat{k}) = \hat{c}$, Eq. (23) determines consumption for every positive capital stock:[4]

Proposition 2 *If $\beta\sigma(c) < 1$ for all c then the $c(\cdot)$ curve lies* ***above*** *the straight line $c = r_{\infty}k$ for all $k \in (0, \hat{k})$ and it lies* ***below*** *this straight line for all $k > \hat{k}$.*[5]

Proof At $k = \hat{k}$, $c(\hat{k}) = r_{\infty}\hat{k}$ and Eq. (23) cannot be used directly to determine c' because both numerator and denominator vanish. However, $c'(\hat{k})$ can be obtained by applying l'Hôpital's rule, yielding the quadratic equation

$$\Theta\big(c'\big) \equiv c'^{2} - \rho c' - r_{\infty}\sigma(\hat{c})[\rho+\delta](1-\beta) = 0 \tag{24}$$

with $\Theta(0) < 0$, while $\Theta(r_{\infty}) = r_{\infty}(r_{\infty}+\delta)(1-\beta)(1-\beta\sigma(\hat{c})) > 0$ hence the positive root $c'(\hat{k})$ of (24) is smaller than r_{∞}. Just below $\hat{k}$, then, the $c(\cdot)$ curve lies above the straight line $c = r_{\infty}k$. Suppose that the two curves cross at some state

[4]Strictly speaking, (23) corresponds to the market solution only for $k \le \hat{k}$. For our purpose, however, it turns out expedient to characterize the properties of its formal solutions also at larger capital stocks.

[5]Symmetric considerations show that if $\beta\sigma(c) > 1$ for all c then the relation between $c(\cdot)$ and the straight line $c = r_{\infty}k$ is reversed. In this work we maintain the condition $\beta\sigma(c) < 1$ cited in the Proposition, because it corresponds to any empirically relevant calibration.

$0 < \tilde{k} < \hat{k}$ where $c(\tilde{k}) = r_\infty \tilde{k}$. Then $c'(\tilde{k}) \geq r_\infty$. However, at $\tilde{k}$ we can use (22) and (23) to obtain

$$c'(\tilde{k}) = \sigma(r_\infty \tilde{k}) r_\infty \frac{A(\zeta)\beta\tilde{k}^\beta - (\rho+\delta)\tilde{k}}{A(\zeta)\tilde{k}^\beta - (\delta + r_\infty)\tilde{k}} = \beta\sigma(r_\infty\tilde{k})r_\infty < r_\infty, \tag{25}$$

and the curves cannot cross. The relation at $k > \hat{k}$ is established in a symmetric manner. □

2.4 Different Fossil Energy Prices

Next we compare two unregulated $c(\cdot)$ curves corresponding to different fossil prices. We consider the prices $\zeta^h > \zeta^l$ and use the superscripts h and l to denote all quantities associated with the high and low price, respectively. We assume that even the higher price ζ^h is insufficient to induce the economy to use solar energy, hence the dynamics of the previous subsection hold for both processes. Observe that r_∞ is independent of ζ and the steady-states corresponding to both fuel prices lie on the straight line $c = r_\infty k$. According to (20), $\hat{k}^l > \hat{k}^h$ and therefore $\hat{c}^l$ is proportionately larger than $\hat{c}^h$.

According to Proposition 2, $c^l(\hat{k}^h) > r_\infty \hat{k}^h = c^h(\hat{k}^h)$, hence the low-price consumption curve lies above its high-price counterpart at $k = \hat{k}^h$. We establish now that this property holds for all capital stocks.

Proposition 3 *If $\beta\sigma(c) < 1$ for all c then the $c^l(\cdot)$ curve lies above the $c^h(\cdot)$ curve for all $k > 0$.*

Proof The Proposition holds for $k = \hat{k}^h$. Suppose that the two curves cross at some point $(\tilde{k}, \tilde{c})$ with $\tilde{k} \in (0, \hat{k}^h)$. It follows that $dc^l(\tilde{k})/dk \geq dc^h(\tilde{k})/dk$. Using (23) we find

$$\frac{A(\zeta^l)\beta\tilde{k}^\beta - (\rho+\delta)\tilde{k}}{A(\zeta^l)\tilde{k}^\beta - \delta\tilde{k} - \tilde{c}} \geq \frac{A(\zeta^h)\beta\tilde{k}^\beta - (\rho+\delta)\tilde{k}}{A(\zeta^h)\tilde{k}^\beta - \delta\tilde{k} - \tilde{c}}. \tag{26}$$

All terms of (26) are positive, because both k and c increase below their corresponding steady states. Thus,

$$(\rho+\delta)\tilde{k}A(\zeta^h) + \beta(\delta\tilde{k}+\tilde{c})A(\zeta^l) \leq (\rho+\delta)\tilde{k}A(\zeta^l) + \beta(\delta\tilde{k}+\tilde{c})A(\zeta^h),$$

or

$$\beta(\delta\tilde{k}+\tilde{c})\left[A(\zeta^l) - A(\zeta^h)\right] \leq (\rho+\delta)\tilde{k}\left[A(\zeta^l) - A(\zeta^h)\right].$$

Now, $A(\zeta^l) > A(\zeta^h)$, yielding

$$\beta(\delta\tilde{k}+\tilde{c}) \leq (\rho+\delta)\tilde{k}$$

or, using (22)

$$\tilde{c} \leq r_\infty \tilde{k},$$

violating Proposition 2. It follows that the two consumption curves do not meet in the interval $(0, \hat{k}^h]$.

At $k > \hat{k}^h$ the inequality (26) and the signs of its terms are reversed, but a crossing of the consumption curves can be ruled out via the same considerations, recalling that the curves lie below the straight line $c = r_\infty k$ when the capital stock k exceeds their respective steady states. □

Proposition 3 lies at the core of the early announcement effects studied in Smulders et al. (2012). We proceed now to investigate how the analysis can be extended to study uncertain announcements.

3 Uncertain Implementation Date

Suppose that implementation of the carbon tax τ, under which the price of fossil energy increases from ζ^l to $\zeta^h = \zeta^l + \tau$, is considered to take place at some unknown future date T. The realization of T may depend on the successful ratification and implementation of some international treaty, or on other developments in the global arena, and is taken as exogenous to the economy under consideration. Thus, from the vantage point of the economy, the hazard rate π corresponding to the random T is constant. The payoff, conditional on T, is

$$\int_0^T u\big(c(t)\big)e^{-\rho t}dt + e^{-\rho T}v\big(k(T)|\zeta^h\big), \tag{27}$$

where $v(k|\zeta)$ represents the value given a constant fossil price ζ:

$$v(k|\zeta) = \max_{\{c(t)\}} \int_0^\infty u\big(c(t)\big)e^{-\rho t}dt \tag{28}$$

subject to (17), given $k(0) = k$. Note that $dv(k|\zeta^h)/dk = \lambda^h(k) = u'(c^h(k))$, where λ^h is the current-value shadow price of capital under the optimal policy corresponding to $v(k|\zeta^h)$.

A hazard process $\pi(\cdot)$ is related to the survival function and density of the event occurrence time, $S(t) = Pr\{T > t\}$ and $f(t) = -S'(t)$, according to

$$\pi(t)\Delta \equiv Pr\big\{T \in (t, t+\Delta) \mid T > t\big\} = \frac{f(t)}{S(t)}\Delta = -\frac{S'(t)}{S(t)}\Delta.$$

Thus, $\pi(t) = -d\ln(S(t))/dt$. Integrating, using the initial value $S(0) = 1$, gives

$$S(t) = \exp\bigg(-\int_0^t \pi(\theta)d\theta\bigg)$$

and

$$f(t) = \pi(t)S(t).$$

The first term of (27) is written as $\int_0^\infty u(c(t))I(T > t)e^{-\rho t}dt$ where the indicator $I(\cdot)$ obtains the value of unity when its argument holds true and zero otherwise, so that its expected value with respect to the T-distribution is $E_T\{I(T > t)\} = S(t)$. The expectation of the second term is written as $E_T\{v(k(T)|\zeta^h)e^{-\rho T}\} = \int_0^\infty f(t)v(k(t)|\zeta^h)e^{-\rho t}dt$. A constant hazard π implies that T is exponentially distributed (with $f(t) = \pi e^{-\pi t}$ and $S(t) = e^{-\pi t}$) and the expected payoff is

$$E_T\left\{\int_0^T u\big(c(t)\big)e^{-\rho t}dt + e^{-\rho T}v\big(k(T)|\zeta^h\big)\right\}$$
$$= \int_0^\infty \big[u\big(c(t)\big) + \pi v\big(k(t)|\zeta^h\big)\big]e^{-(\rho+\pi)t}dt.$$

The allocation problem with uncertain carbon tax date T becomes

$$v^\pi\left(k_0|\zeta^l, \zeta^h\right) = \max_{\{c(t)\}} \int_0^\infty \big[u\big(c(t)\big) + \pi v\big(k(t)|\zeta^h\big)\big]e^{-(\rho+\pi)t}dt \tag{29}$$

subject to (17) with $\zeta = \zeta^l$, given $k(0) = k_0$. We compare the emission path corresponding to $v(k_0|\zeta^l)$, under which no carbon tax is contemplated, with that corresponding to $v^\pi(k_0|\zeta^l, \zeta^h)$, under which a carbon tax τ will be imposed at an uncertain time T.

The capital process $k^\pi(\cdot)$ corresponding to $v^\pi(k_0|\zeta^l, \zeta^h)$ follows (17) with $\zeta = \zeta^l$ (the prevailing price until the tax is imposed) while Eq. (18) becomes

$$\dot{c}^\pi(t) = \sigma\big(c^\pi(t)\big)c^\pi(t)\big[A\big(\zeta^l\big)\beta k^\pi(t)^{\beta-1} - (\rho+\delta) + P\big(k^\pi(t)\big)\big], \tag{30}$$

where

$$P(k) \equiv \pi\left(\frac{u'(c^h(k))}{u'(c^\pi(k))} - 1\right). \tag{31}$$

Comparing (30) with (18), we see that the uncertainty in T, with $\pi > 0$, is represented by the $P(k)$ term, the sign of which depends on the relative magnitudes of $c^h(k)$ and $c^\pi(k)$. We turn now to study the effects of this term.

3.1 *The Consumption-Capital Trajectory*

We consider the capital dependence of consumption under the π regime. Equation (23) becomes

$$\frac{dc^\pi(k)}{dk} = \sigma\left(c^\pi(k)\right)c^\pi(k)\frac{A(\zeta^l)\beta k^{\beta-1} - (\rho+\delta) + P(k)}{A(\zeta^l)k^\beta - \delta k - c^\pi(k)}, \tag{32}$$

with the steady state values $\hat{k}^\pi$ and $\hat{c}^\pi$, given by

$$A\big(\zeta^l\big)\big(\hat{k}^\pi\big)^\beta - \delta\hat{k}^\pi - \hat{c}^\pi = 0 \tag{33}$$

and

$$\beta A\big(\zeta^l\big)\big(\hat{k}^\pi\big)^{\beta-1} - (\rho+\delta) + P\big(\hat{k}^\pi\big) = 0. \tag{34}$$

We compare these steady state values with their regulation-free counterparts.

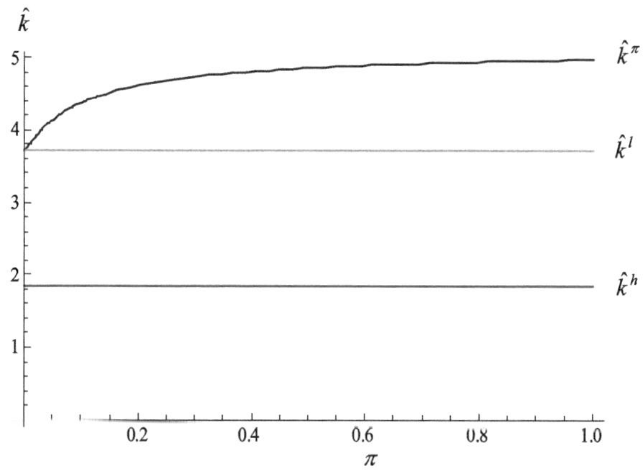

Fig. 1 The steady state capital $\hat{k}^{\pi}$ as a function of the hazard rate π. The upper and lower horizontal lines indicate $\hat{k}^{l}$ and $\hat{k}^{h}$, respectively. The curves in all figures were derived under the above function specifications and the parameter values: $\alpha = \gamma = 1/3$, $F = 1$, $\sigma = 1$, $\rho = \delta = 5$ % annually, $\zeta^{l} = 1$ and $\zeta^{h} = 2$

From (20) and (34) we obtain

$$A(\zeta^{l})\beta\left[(\hat{k}^{l})^{\beta-1} - (\hat{k}^{\pi})^{\beta-1}\right] = P(\hat{k}^{\pi}). \tag{35}$$

According to (31), $P(\hat{k}^{\pi})$ is small when π is small, hence $\hat{k}^{\pi}$ is close to $\hat{k}^{l}$ and (33) implies that

$$c^{\pi}(\hat{k}^{\pi}) = \hat{c}^{\pi} \approx \hat{c}^{l} = c^{l}(\hat{k}^{l}) \approx c^{l}(\hat{k}^{\pi}) > c^{h}(\hat{k}^{\pi}).$$

With $u''(\cdot) < 0$, it follows that $u'(c^{\pi}(\hat{k}^{\pi})) < u'(c^{h}(\hat{k}^{\pi}))$ and $P(\hat{k}^{\pi}) > 0$. Turning again to (35) and recalling that $\beta - 1 < 0$, we find that $\hat{k}^{\pi} > \hat{k}^{l}$ when the hazard rate π is small. We show that this relation between the steady states extends to arbitrary positive values of π. Consider the steady state $\hat{k}^{\pi}$ as a function of π and assume that at some π value this function crosses the constant $\hat{k}^{l}$ so that the left hand side of (35) vanishes. However, (33) holds for both $c^{\pi}(\cdot)$ and $c^{l}(\cdot)$ hence $c^{\pi}(\hat{k}^{\pi}) = c^{l}(\hat{k}^{\pi}) > c^{h}(\hat{k}^{\pi})$. According to (31) $P(\hat{k}^{\pi}) > 0$ hence the right hand side of (35) is positive, while the left hand side vanishes. Thus, the crossing cannot occur. We conclude, therefore that

$$\hat{k}^{\pi} > \hat{k}^{l} \quad \forall \pi > 0, \tag{36}$$

as Fig. 1 illustrates.

Next we compare the complete consumption curves by relating $c^{\pi}(k)$ to $c^{l}(k)$. Since $\hat{k}^{l}$ represents the steady state for the $k^{l}(\cdot)$ process, it follows that $\dot{k}^{l}(t) = 0$ at this state. However, the steady state $\hat{k}^{\pi}$ of $k^{\pi}(\cdot)$ exceeds $\hat{k}^{l}$, hence $\dot{k}^{\pi}(t) > 0$ when $k^{\pi}(t) = \hat{k}^{l}$. Thus, (17) implies $c^{l}(\hat{k}^{l}) > c^{\pi}(\hat{k}^{l})$. We show that this relation cannot reverse at other capital states. Suppose otherwise, that $c^{l}(k^{*}) = c^{\pi}(k^{*})$ (hence $P(k^{*}) > 0$) at some capital state $k^{*} < \hat{k}^{l}$ but $c^{l}(k) > c^{\pi}(k) \forall k \in (k^{*}, \hat{k}^{l}]$. It follows that $dc^{l}(k^{*})/dk \geq dc^{\pi}(k^{*})/dk$. However, we can write (32) as

$$\frac{dc^{\pi}(k^{*})}{dk} = \frac{dc^{l}(k^{*})}{dk} + \frac{\sigma(c^{l}(k^{*}))c^{l}(k^{*})P(k^{*})}{A(\zeta^{l})k^{*\beta} - \delta k^{*} - c^{l}(k^{*})} > \frac{dc^{l}(k^{*})}{dk},$$

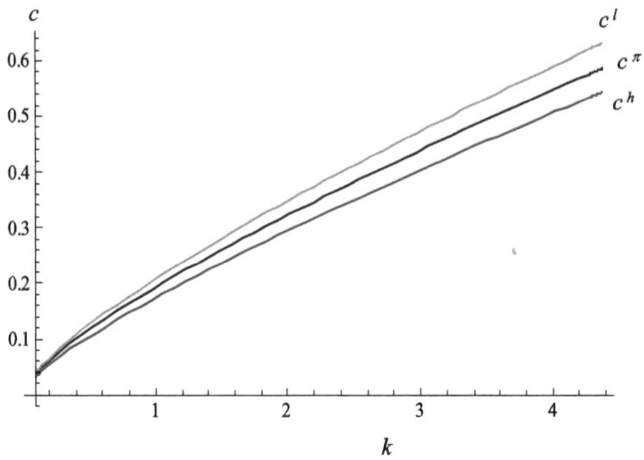

Fig. 2 Consumption curves as functions of capital under uncertain T (c^π), low fossil energy price (c^l) and high fossil energy price (c^h). In this and the following figures we use the value $\pi = 0.1$ corresponding to $E\{T\} = 10$ years

because the denominator of the second term is also positive at k^*. A crossing of the consumption curves (with $dc^l(k^*)/dk \leq dc^\pi(k^*)/dk$) can be ruled out also for $k^* > \hat{k}^l$ using the same argument, since the denominator is negative above $\hat{k}^l$. Thus,

$$c^\pi(k) < c^l(k) \quad \forall k > 0.$$

We wish to compare the uncertain consumption curve also to its high price counterpart, $c^h(\cdot)$. We use (36) to deduce from (35) that $P(\hat{k}^\pi) > 0$ hence $c^h(\hat{k}^\pi) < c^\pi(\hat{k}^\pi)$. To establish the same relation for smaller capital stocks, we assume otherwise, that $c^h(\tilde{k}) = c^\pi(\tilde{k})$ at some stock $\tilde{k} < \hat{k}^\pi$, where $dc^h(\tilde{k})/dk \leq dc^\pi(\tilde{k})/dk$ but $P(\tilde{k}) = 0$. This, however, implies (26) which can be ruled out via the same arguments used to establish Proposition 3. We summarize these considerations in Fig. 2 and in

Proposition 4 *If $\beta\sigma(c) < 1\ \forall c$, then $c^h(k) < c^\pi(k) < c^l(k)\ \forall k \in (0, \hat{k}^\pi]$.*

Uncertainty, then, reduces consumption but not by as much as would be implied by a prompt implementation of the tax.

3.2 *The "Green Paradox"*

The time trajectories of k^l and k^π are given, respectively, by the implicit solutions of (17):

$$t = \int_{k_0}^{k^l(t)} \frac{dk}{A(\zeta^l)k^\beta - \delta k - c^l(k)} dk,$$

and

$$t = \int_{k_0}^{k^\pi(t)} \frac{dk}{A(\zeta^l)k^\beta - \delta k - c^\pi(k)} dk.$$

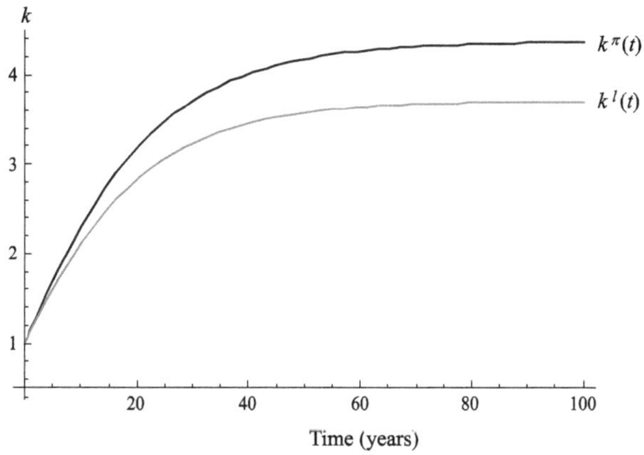

Fig. 3 Time trajectories of capital stocks under uncertain T (k^{π}) and low fossil energy price (k^{l})

Thus, the relation $c^{l}(k) > c^{\pi}(k)$ implies that

$$k^{l}(t) < k^{\pi}(t) \quad \forall t > 0,$$

as indicated in Fig. 3. Indeed, this result provides the manifestation of the "Green Paradox" effect in the case of uncertain T. Since both $k^{l}(\cdot)$ and $k^{\pi}(\cdot)$ proceed under the same price of fossil energy and with the same production technology, the larger $k^{\pi}(\cdot)$ process entails enhanced energy use at each point of time (until implementation), in contrast to the original purpose of the announcement. As in the case of a certain early announcement, preparing for the anticipated tax consists of accumulating a larger capital stock so that when the tax is eventually levied, the larger capital stock will partly compensate for the reduced energy use implied by the tax.

Interestingly, a comparison of the corresponding consumption time trajectories does not display the same simple pattern in time: With a higher steady state consumption, $c^{\pi}(t)$ must exceed $c^{l}(t)$ at large time (but prior to actual implementation). This relation between the consumption processes, however, cannot extend all the way back to $t = 0$ (when the capital stock equals k_0 under both regimes) because if it did, the relation between the capital processes displayed in Fig. 3 would be reversed. The two consumption processes, therefore, must cross at some finite time, as shown in Fig. 4. Efforts to prepare for the tax (in terms of reduced consumption) are concentrated at the early stages of the growth process, while at later times, parts of the fruits of the oversized capital (relative to the prevailing low fossil energy price) are used again to finance enhanced consumption.

4 Concluding Comments

The model presented in this work suggests yet another mechanism to produce "paradoxical" outcomes of climate policies without resorting to the scarcity of the fossil resource. Here, the effects are due to uncertainty regarding the timing of introducing the carbon tax. While the economic forces at work are similar to those driving

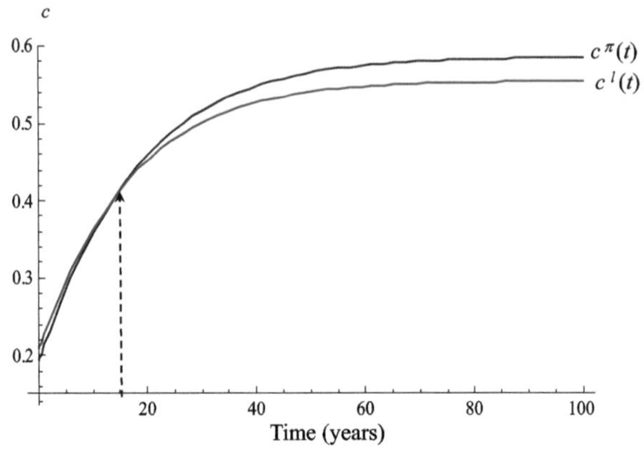

Fig. 4 Time trajectories of consumption processes under uncertain T (c^{π}) and low fossil energy price (c^{l}). The arrow indicates the time when the trajectories cross

the early announcement model of Smulders et al. (2012), the two mechanisms operate differently because in the present model economic agents cannot predict the tax implementation date at which they must ensure a smooth transition of the consumption process. In fact, consumption will undergo a discontinuous jump on this date and the adopted processes are tuned so as to minimize the expected utility loss associated with the jump. The solution involves delicate tradeoffs as manifested by the crossing of the time profiles of the consumption processes displayed in Fig. 4. Nevertheless, the "paradoxical" effect of increased fossil energy use persists at all times until the tax policy is realized.

For brevity and simplicity of exposition, the results are presented in terms of the simplest specification of a Cobb-Douglas technology, constant hazard rate and a mild tax rate which does not imply a transition to a solar-based economy. As indicated in Smulders et al. (2012), none of these assumptions is essential and the "paradoxical" effect can be obtained in a more general setting, albeit at the cost of more tedious derivations.

Acknowledgements This research has been carried out while Amos Zemel spent a sabbatical year in Tilburg University. The support of NWO, the Netherlands, is gratefully acknowledged.

References

Di Maria, C., Smulders, S., & van der Werf, E. (2012). Absolute abundance and relative scarcity: environmental policy with implementation lags. *Ecological Economics*, *74*, 104–119.

Eichner, T., & Pethig, R. (2011). Carbon leakage, the green paradox and perfect future markets. *International Economic Review*, *52*, 767–805.

Sinn, H.-W. (2008). Public policies against global warming. *International Tax and Public Finance*, *15*, 360–394.

Smulders, S., Tsur, Y., & Zemel, A. (2012). Announcing climate policy: can a green paradox arise without scarcity? *Journal of Environmental Economics and Management*, *64*, 364–376.

Tsur, Y., & Zemel, A. (2011). On the dynamics of competing energy sources. *Automatica*, *47*, 1357–1365.

van der Ploeg, F., & Withagen, C. (2012). Is there really a green paradox? *Journal of Environmental Economics and Management*, *64*, 342–363.

Uniqueness Versus Indeterminacy in the Tragedy of the Commons: A 'Geometric' Approach

Franz Wirl

Abstract This paper characterizes continuous Markov perfect equilibria as smooth connections between an 'initial', i.e., at the origin of the state space, and an 'end' manifold that result from patching with the boundary solution. The major result is that multiple equilibria require a non-monotonic initial manifold. This necessary condition for multiple equilibria can be tested without (or prior to) solving the Hamilton-Jacobi-Bellman equation. Application to a familiar dynamic tragedy of the commons with nonlinear instead of linear-quadratic utilities shows that the elasticity of marginal utility is the crucial property: If this elasticity is (everywhere) greater than $\frac{n-1}{n}$, $n =$ number of polluters, then the Nash equilibrium is unique. Assuming the opposite inequality (globally) implies that no saddle-point equilibrium exists. Therefore, the 'focal' point equilibrium is gone and all conceivable boundary conditions determine a corresponding equilibrium, e.g. 'anything goes' for power utility functions.

1 Introduction

The much discussed possibility of a family of multiple equilibria in dynamic games and the search for underlying economic and arithmetical reasons motivates this investigation. This interest is in particular due to the much quoted and discussed paper of Dockner and van Long (1993) with its claim that nonlinear Markov strategies instead of the familiar linear ones in a linear-quadratic differential game 'solve' the tragedy of the commons. However, nonlinear strategies have only local support. This requires an ad hoc and ex post (i.e., after having obtained the solutions) restriction of the state space to ensure subgame perfection. Rubio and Casino (2002) is a first attempt to reduce the set of equilibria, Wirl (2008a) investigates a stochastic version of a stock pollution game, and Jorgensen et al. (2010) is a comprehensive survey of dynamic pollution games. Harstad (2012) is a recent application of this dynamic tragedy of the commons to global warming extended for technology (renewable energy) and (incomplete) contracts. Rowat (2000, 2006) gives a rigorous analysis of

F. Wirl (✉)
Department of Business Studies, University of Vienna, Brünnerstr. 72, 1210 Vienna, Austria
e-mail: franz.wirl@univie.ac.at

E. Moser et al. (eds.), *Dynamic Optimization in Environmental Economics*, Dynamic Modeling and Econometrics in Economics and Finance 15,
DOI 10.1007/978-3-642-54086-8_8,

the Dockner-Long game and shows that the local nonlinear strategies can be extended by patching with the boundary strategy ('zero' = no emissions) in order to obtain globally defined strategies. This raises the vague hope that imposing Inada conditions, which rule out a priori a 'zero' strategy, lead to a unique equilibrium due to the implicit requirement of globally interior strategies. The common denominator of all these variations is that nonlinear and multiple equilibrium strategies mitigate the tragedy of the commons; Rowat (2006) conjectures that some kind of folk theorem is behind this optimistic result. However, Wirl (2007) rejects the generality of this conclusion—the multiple equilibria increase conservation—by showing that multiple equilibria aggravate the tragedy of the commons for utility functions with hyperbolic risk aversion. The following paper differs from this rather narrow objective in Wirl (2007) by establishing general criteria for multiple equilibria applicable at the outset (i.e., ideally without solving the differential game) and by ruling out hyperbolic risk aversion.

More precisely, this paper proposes an alternative view and uses geometric means in order to characterize Markov perfect equilibria in differential games but restricted to one-state. Other non-standard characterizations of such equilibria are: Dockner and Wagener (2008) introduce a system of auxiliary ordinary differential equations (thus also restricted to a single state) and Rincon-Zapatero et al. (1998) apply the Hamiltonian approach, derive a system of quasi-linear partial differential equations, and this approach is then applicable also to higher dimensional state spaces. Speaking in formal terms, the paper exploits a particular result from Rowat (2006) and the optimality conditions of value matching and smooth pasting at the level(s) of patching. This allows to describe global, continuous and stable Markov strategies as a continuous connection between two manifolds, 'initial' and 'end'. The major result is the link between unique or multiple equilibria and the shape of the initial manifold. This in turn allows to solve the question of potential multiple equilibria without solving the Hamilton-Jacobi-Bellman equation. The application to the above quoted dynamic tragedy of the commons shows that it is the magnitude of the elasticity of marginal utility (or the degree of relative risk aversion) that determines whether the equilibrium is (i) unique (and then given by a singular or 'saddle-point' path, which is the linear strategy in linear-quadratic games) or whether (ii) an entire family of equilibria exists without the saddle-point equilibrium, or (iii) the standard case in linear-quadratic differential games of a unique singular combined with a family of uncountably many strategies. In other words, preferences, more precisely the implied flexibility and thus the elasticity of demand determine in this application whether multiple equilibria are possible or not. Even given this possibility of multiple equilibria, further conditions can render uniqueness, e.g., low external costs in the application below.

The paper is organized as follows. Section 2 introduces the framework. The optimality conditions are derived and analyzed in Sect. 3. Section 4 studies initial and end manifolds of (continuous, stationary) Markov-strategies and derives conditions for uniqueness as well as for 'maximal' indeterminacy (i.e. each conceivable boundary condition induces a Nash equilibrium). Section 5 presents some examples and explains the striking differences in outcomes—uniqueness and maximal indeterminacy—which occur even within the same class of utility functions (CRRA).

2 Framework

In spite of the general applicability of the following analysis, this paper focuses on the much discussed dynamic tragedy of the commons. Following Dockner and van Long (1993), each player $i = 1, \dots, n$ chooses a non-negative x_i ('emission') such that the individual net present value (using the discount rate $r > 0$) is maximized

$$V_i\big(X(0)\big) = \max_{\{x_i(t)\geq 0\}} \int_0^\infty e^{-rt}\big[u\big(x_i(t)\big) - C\big(X(t)\big)\big]dt. \tag{1}$$

The instantaneous payoff consists of the individual benefit u minus the external costs

$$C(X) = \frac{c}{2}X^2. \tag{2}$$

The stock of pollution X accumulates over time,

$$\dot{X}(t) = x_i(t) + \sum_{j\neq i} x_j(t) - \delta X(t), \quad X(0) = X_0 \text{ given}, \ \delta \geq 0, \tag{3}$$

and provides the constraint for the optimization in (1). Using a linear-quadratic utility u and quadratic damages (as in (2)), Dockner and van Long (1993) claim that nonlinear strategies can mitigate the tragedy of the commons. Martin-Herran and Rincon-Zapatero (2005) use the method developed in Rincon-Zapatero et al. (1998) to characterize Pareto-efficient Nash equilibrium outcomes. The non-negativity constraint in (1) accounts for irreversibility—it is impossible to reduce actively the stock of pollution ($x < 0$ is infeasible)—and may require patching of interior strategies with the boundary strategy $x = 0$ in order to obtain globally defined strategies. Although local strategies are considered in many papers, this contradicts the demand for a subgame perfect equilibrium, because the strategies must be defined ex ante over the entire state space.

The following assumptions are made: first the standard assumptions about the strategy space (A1) and benefits (A2), and then some more specific assumptions about the preferences, (B1)–(B3):

- (A1) the analysis is restricted to symmetric Nash equilibria in continuous, globally defined, stationary Markov strategies,

  $$x = \varphi(X) \quad \forall X \in [0, \infty), \ \varphi \in C^0,$$

  which are stable ($\dot{X} = n\varphi(X) - \delta X = 0$ has a stable steady state). The value function $V(X)$ is twice differentiable in the interior, $x > 0$. These assumptions are standard except for demanding a global support for the strategies. Furthermore, the value of the cooperative solution, denoted by $W(X)$, is finite for all $X \geq 0$; this provides an upper bound on $V(X)$.

- (A2) benefit $u = u(x)$ need not be linear-quadratic as in Dockner and van Long (1993) and its follow ups in Rubio and Casino (2002) and in Rowat (2006) but satisfies the usual properties $u'(0) > 0$, $u'' < 0$; if existing,

$$\widehat{x} := \arg\max u(x),$$

denotes saturation and $u'(0) =$ the choke price (see discussion below).

- (B1) $u(0) = 0$,
- (B2) the elasticity of marginal utility,

$$\sigma(x) \equiv -\frac{u''(x)x}{u'(x)}, \quad (4)$$

is non-decreasing, i.e., $\sigma' \geq 0$, and

- (B3) the familiar Inada conditions, $\lim_{x\to 0} u'(x) \to \infty$ and $\lim_{x\to\infty} u'(x) \to 0$.

The standard Assumptions (A1) and (A2) are assumed throughout the paper without further mentioning. Assumption (B1) simplifies the boundary solution, but is ignored when imposing Inada conditions, because (B3) rules out $x = 0$ (hence, the value of $u(0)$ is irrelevant anyway) and because familiar examples do not satisfy this normalization. The marginal willingness to pay (u') defines the inverse demand function, $u'(x) = p$ in a partial equilibrium framework for the (hypothetical) price p per unit of emission x. Since the price elasticity is then $\varepsilon = 1/\sigma$, (B2) stipulates that (individual) demand is getting not less elastic as prices increase. This assumption of a non-decreasing elasticity σ generalizes a (crucial) property of the much used linear-quadratic utility. It also holds for the often used specifications implying either constant relative risk aversion (short CRRA since $u''x/u'$ is constant) or constant absolute risk aversion (short CARA since u''/u' is constant). Wirl (2007) considers utilities with hyperbolic risk aversion (known as HARA) and shows that multiple equilibria are then worse than the counterpart of the linear strategy (= singular or saddle-point strategy) if $\sigma' < 0$.

Remark 1 The following results can be easily generalized into two directions. Firstly, by allowing general external cost functions $C(X)$. This changes little as long as $C(0) = 0$ is assumed but eliminates the analytical solution along the boundary $x = 0$ (derived below). Secondly and similarly, nonlinear depreciation $D(X)$ with $D(0) = 0$, $D'(0) > 0$, $D'' \leq 0$ instead of δX has little effect (as becomes clear below) yet at the cost of eliminating again the explicit determination of the value function along the boundary.

3 Optimality Conditions

The value functions must satisfy the Hamilton-Jacobi-Bellman equation (see e.g. Dockner et al. 2000):

$$rV_i(X) = \max_{x_i \geq 0}\Big\{u(x_i) - C(X) + V_i'(X)\Big(x_i + \sum_{j\neq i} x_j - \delta X\Big)\Big\}. \tag{5}$$

Maximization on the right hand side yields the optimal strategy x_i either as a function of the derivative of the value function (by the implicit function theorem since $u'' < 0$) or as boundary solution due to the non-negativity constraint:

$$x_i^* = \varphi(X) \equiv \begin{Bmatrix} f(V_i'(X)),\ f' = -\frac{1}{u''} \\ 0 \end{Bmatrix} \quad \text{if } V_i'(X) \begin{Bmatrix} > \\ \leq \end{Bmatrix} - u'(0). \tag{6}$$

Substituting (6) into (5) not only for player i but by assumed symmetry also for $j \neq i$ yields along the interior ($x > 0$):

$$rV = u\big(f(V')\big) - C(X) + nV' f\big(V'\big) - \delta X V' \quad \text{if } V_i'(X) > -u'(0), \tag{7}$$

and along the boundary, $x = 0$, using Assumption (B1), $u(0) = 0$,

$$rV = -C(X) - \delta X V' \quad \text{if } V_i'(X) \leq -u'(0). \tag{8}$$

Any value function $V(X)$ of this game is bounded from above by the value of the cooperative solution (social optimum). Let y denote the socially optimal level of pollution by each player and W the corresponding value function, then this value function must satisfy the following Hamilton Jacobi-Bellman equation,

$$\begin{aligned} rW(X) &= \max_{y\geq 0}\big\{u(y) - C(X) + W'(X)(ny - \delta X)\big\} \\ &= \begin{Bmatrix} u(g(W')) - C + W'(ng(W') - \delta X W') \\ -C - \delta X W' \end{Bmatrix} \quad \text{if } W' \begin{Bmatrix} > \\ \leq \end{Bmatrix} - \frac{u'(0)}{n}. \end{aligned} \tag{9}$$

Therefore, $V(X) \leq W(X)$ for all $X \geq 0$ and $y^* = g(W') \equiv u'^{-1}(-nW')$ is the socially optimal interior pollution policy obtained from the maximization of the right hand side in the first line in (9) with the derivative $g' = -n/u'' > 0$ determined by the implicit function theorem.

3.1 Boundary Solution ($x = 0$)

The differential equation in (8) has an analytical solution (at least for the quadratic specification of costs which explains their assumed specification),

$$V_b(X) = AX^{-r/\delta} - \frac{c}{2}\frac{X^2}{(r + 2\delta)}. \tag{10}$$

The subscript b indicates that this value function applies only in the boundary domain ($x = 0$) and A is the integration constant. $V_b(X)$ describes the net present value of benefits minus costs including those from future emissions, because X declines in this stopping domain due to depreciation.

3.2 Patching

Whenever an interior strategy hits the boundary $x = 0$ at the level $\overline{X}$ (such threshold levels are identified by the upper bar), patching of interior and boundary strategies is necessary, a point stressed in Rowat (2006) in order to obtain global strategies. This requires that the corresponding value functions, i.e. the solutions of (7) and (8), can be connected in a continuous (value matching) and differentiable (smooth pasting) manner. These conditions are mostly ignored including Rowat (2006) and if not as, e.g., in Fershtman and Kamien (1987) and in Benchekroun (2003), then they are relegated to appendices and their economic consequences are not fully used. One explanation is that these conditions seem irrelevant for the linear equilibrium strategy (but are of course satisfied). However off the equilibrium and/or for particular equilibrium strategies, decisions have to be made about optimal 'stopping'. For instance if $\delta = 0$, then the nonlinear strategies advocated in the linear-quadratic model of Dockner and van Long (1993) will hit the constraint $x \geq 0$ in finite time. Or consider a case where the players are in a situation of $X > \overline{X}$ (of course, due to erroneous moves in the past). Thus they emit nothing but each must decide when to 're-start' emitting. Accounting for the individual rationality of these decisions implies smooth pasting in analogy to the real options literature, see Dixit and Pindyck (1994, p. 109). Furthermore, applying this smooth pasting condition can render uniqueness, e.g., it eliminates the nonlinear strategies in Wirl (1994) and in Wirl and Dockner (1995) as equilibria of the games. Unfortunately, it cannot render uniqueness in general.

Proposition 1 *Assume* (B1) *and that interior and boundary solutions are joined at* $\overline{X}$*, then*

$$\lim_{X \to \overline{X}^-} V(X) = \lim_{X \to \overline{X}^+} V_b(X) = A\overline{X}^{-r/\delta} - \frac{c}{2}\frac{\overline{X}^2}{(r+2\delta)}, \tag{11}$$

$$\lim_{X \to \overline{X}^-} V'(X) = -u'(0) = \lim_{X \to \overline{X}^+} V_b'(X) = -\frac{r}{\delta}A\overline{X}^{(-r/\delta-1)} - \frac{c\overline{X}}{(r+2\delta)}. \tag{12}$$

Proof See Appendix. □

Proposition 2 $x > 0$ *at* $X = 0$.

A proof of Proposition 2 can be found in Rowat (2000). Proposition 2 states an economically rather obvious property: why should one refrain from emitting into a pristine environment given that polluting is advisable for positive stocks and the public nature of the externality? It is a consequence of the explicit solution (10) along the boundary and of the continuity of the derivative. Combined they imply that equilibrium emissions must be positive at (and by continuity close to) the origin of the state space. This allows to exclude at least some of the (local) strategies that are often considered as (local) equilibria, e.g., in Dockner and van Long (1993).

4 Equilibria

Differentiating (7), solving for the second derivative, and arranging as a ratio yields

$$V'' = \frac{P(X, V')}{Q(X, V')} = \frac{(r+\delta)V' + cX}{nf - \delta X - \frac{n-1}{u''}V'}, \quad X < \overline{X}. \tag{13}$$

Note that V is not twice continuously differentiable at $\overline{X}$ (although social value W defined in (9) must be, see Wirl 2008b), since

$$\lim_{X\to\overline{X}^+} V'' = \frac{(r+\delta)u'(0) - c\overline{X}}{\delta\overline{X}} \neq \lim_{X\to\overline{X}^-} V'' = \frac{(r+\delta)u'(0) - c\overline{X}}{\delta\overline{X} - (n-1)\frac{u'(0)}{u''(0)}}.$$

The optimal strategy is according to (6) a monotonic transformation of V'. Therefore the solution curves V' of the differential equation (13) are candidates for Nash equilibrium strategies. In order to characterize these solution curves the following three crucial loci in the (X, V') phase plane are analyzed. The second order differential equation (13) is given by a ratio and has thus the following two crucial loci in the (X, V') phase plane: First, the set of local extrema, $\varphi' = -\frac{V''}{u''} = 0$ implies that the numerator in (13) vanishes,

$$P\big(X, V'\big) = 0 \iff V' = -\frac{cX}{r+\delta}.$$

This condition is as in the linear-quadratic case because numerator P is independent of the specification of the benefits u. Second, the set of singularities ($|V''| \to \infty$) is given by the roots of the denominator,

$$Q\big(X, V'\big) = 0 \iff (n-1)V' - \big(nf\big(V'\big) - \delta X\big)u''\big(f\big(V'\big)\big) = 0. \tag{14}$$

Hence, no strategy can cross $Q = 0$ except through the intersection with $P = 0$.[1] It is this passing through a uniquely determined point (coupled with the requirement of stable strategies) that renders uniqueness for the optimization problem (i.e. for W)

[1] In order to simplify notation $Q = 0$ refers to the set $\{(X, V') \mid Q(X, V') = 0\}$ or to its corresponding manifold, $\Omega : Q = 0$; analogous for $P = 0$.

but not necessarily for the game. Third, the set of steady states is implicitly defined by:

$$\dot{X} = nf(V') - \delta X = 0.$$

Since the derivative of $\dot{X}$ with respect to V' exists and since $nf' = -n/u'' > 0$, the implicit function theorem implies the existence of a monotonically increasing function

$$V' = s(X), \qquad s' = -\frac{\delta u''}{n} > 0, \qquad s(0) = -u'(0) \quad \text{along } \dot{X} = 0,$$

which starts at the origin of the state space at the level $V' = -u'(0)$. This determines simultaneously the level of V' at which the interior solution joins the boundary solution $x = 0$. If u satisfies the first Inada condition, $\lim_{x \to 0} u'(x) = \infty$, then $V' \to -\infty$ for $X \to 0$ and patching with $x = 0$ is infeasible.

Lemma 1 *Assume that u satisfies* (B2), *then $Q(V', X) = 0$ defines a unique $V' = q(X) < 0$ for each $X < \frac{n\widehat{x}}{\delta}$, iff the elasticity of marginal utility (σ defined in* (4)) *satisfies*

$$\sigma(\widehat{x}) > \frac{n-1}{n - \frac{\delta X}{\widehat{x}}}. \tag{15}$$

Hence, finite saturation ($\widehat{x} < \infty$) is sufficient for the existence of $q(X)$. If u satisfies in addition Assumption (B3) *(thus $\widehat{x} \to \infty$), then q exists iff*

$$\lim_{x \to \infty} \sigma(x) = \lim_{V' \to 0} \sigma(f(V')) > \frac{n-1}{n}. \tag{16}$$

If q exists, then $q > s$ (for $X < \frac{n\widehat{x}}{\delta}$) and $q' > 0$ (due to Assumption (B2)).

Proof See Appendix. □

The characteristic of an interior solution at the origin of the state space (Proposition 2) and the conditions of value matching (11) and smooth pasting (12) at the level of joining interior and boundary solutions (Proposition 1) imply the following characterization: Nash equilibria (in stationary, continuous and stable Markov strategies as assumed in (A1)) are those solutions of the Hamilton-Jacobi-Bellman equation (7) that allow to connect the associated *initial* and *end* manifolds, which are derived and investigated below. Given that Nash equilibria are smooth connections between these two manifolds the question is: What are the implications of the properties of these manifolds on equilibria and in particular whether the equilibrium is unique or not if a family of such connections can exist. The analysis starts with the initial manifold, followed by the end manifold, and the conditions for unique or indeterminate outcomes.

4.1 Initial Manifold, $X = 0$

Proposition 2 ensures an interior solution at the origin of the state space. Therefore evaluating (7) at $X = 0$ determines the initial manifold from which any equilibrium strategy must originate:

$$I \equiv \{(V'(0), V(0))\}$$
$$\text{s.t.} \quad V(0) \in (-u'(0), 0], V(0) = \frac{u(f(V'(0))) + nV'(0) f(V'(0))}{r}. \tag{17}$$

Note that the tuple $(V'(0), V(0))$ does not refer to the value function (which still needs to be determined) but to the relation between V' and V that any value function candidate must satisfy at $X = 0$. This notational convenience—treating $V'(0)$ as something like an independent variable—is also applied in the following characterizations of the initial manifold I and in the examples.

Proposition 3 *The initial manifold I is decreasing in that domain of $V'(0)$ where*

$$\sigma\big(u\big(f\big(V'(0)\big)\big)\big) < \frac{n-1}{n},$$

and increasing where

$$\sigma\big(u\big(f\big(V'(0)\big)\big)\big) > \frac{n-1}{n}.$$

Assuming that (B2) *holds, the local minimum of I is at*

$$\sigma\big(u\big(f\big(V'(0)\big)\big)\big) = \frac{n-1}{n}; \tag{18}$$

if existing, it is unique, coincides with $Q = 0$ evaluated at $X = 0$ and separates I into an increasing (I_1) and a decreasing (I_2) part.

Proof Differentiating the expression for $V(0)$ in (17)

$$\frac{dV(0)}{dV'(0)} = \frac{1}{r}\left(nf - \frac{u'}{u''} - \frac{nV'}{u''}\right) = \frac{nf}{r}\left(1 - \frac{n-1}{n\sigma}\right)$$

proves the claimed slopes of the initial manifold and it remains to show that $Q = 0$ simplifies at $X = 0$ to (18). Setting $X = 0$ in (14) implies:

$$V'(0) = \frac{n}{n-1} u''\big(f\big(V'(0)\big)\big) f\big(V'(0)\big). \tag{19}$$

A solution of this implicit relation (19) requires that

$$\frac{u''(f(V'(0))) f(V'(0))}{V'(0)} = \underbrace{-\frac{u''(f(V'(0))) f(V'(0))}{u'(f(V'(0)))}}_{\equiv \sigma} = \frac{n-1}{n}.$$

Therefore, the value of $Q = 0$ at $X = 0$ is determined by the level of $V'(0)$ that would induce emissions $x(0)$ such that the associated elasticity of marginal utility (σ) is equal to the ratio $\frac{n-1}{n}$.

The uniqueness of the minimum (under the proviso of its existence) and its consequence of separating I follow from Assumption (B2), $\sigma' \geq 0$. □

Remark 2 Consider the **increasing part** of I denoted by I_1: $V_1(0) > V_2(0) \Longleftrightarrow V_1'(0) > V_2'(0) \Longleftrightarrow x_1(0) > x_2(0)$, i.e., a higher value (at $X = 0$) is associated with larger emissions. Furthermore the condition for the existence of an increasing part of the initial manifold ensures in turn the existence of $V' = q(X)$ for $X > 0$ due to Lemma 1.

Remark 3 A larger payoff requires less pollution along the subset I_2 where $V(0)$ **decreases** with respect to $V'(0)$.

Remark 4 The initial manifold of the cooperative solution (setting $X = 0$ in (9)) consists only of the increasing part, $dW(0)/dW'(0) = ny > 0$, which is the reason for the uniqueness of the socially optimal policy, and consequently, it will be the decreasing part that allows for multiple equilibria, as shown below.

Remark 5 The initial manifold I as well as the characterization of the manifold $Q = 0$ does not depend on the specification of the costs and depreciation except for the assumptions $C(0) = 0 = D(0)$.

In order to demonstrate the applicability of the above, the following simple example of a non-saturating logarithmic utility function is introduced,

$$u(x) = \ln(1 + x), \tag{20}$$

in order to go beyond the usual linear-quadratic case although it has nevertheless similar implications. It leads to the following Hamilton-Jacobi-Bellman equation in the interior ($x^* = -1 - \frac{1}{V'} > 0$):

$$rV = -n - \frac{c}{2}X^2 + \ln\left(-\frac{1}{V'}\right) - (n + \delta X)V',$$

which reduces at $X = 0$ to

$$rV = \ln\left(-\frac{1}{V'}\right) - n\left(1 + V'\right). \tag{21}$$

This initial manifold (21) is shown in Fig. 1. It consists of the two parts, $I = I_1 \cup I_2$, increasing (I_1) for $V' > -\frac{1}{n}$ and decreasing (I_2) for $V' \leq -\frac{1}{n}$ since $\sigma = \frac{x}{1+x} \in [0, 1)$ covers both domains addressed in Proposition 3.

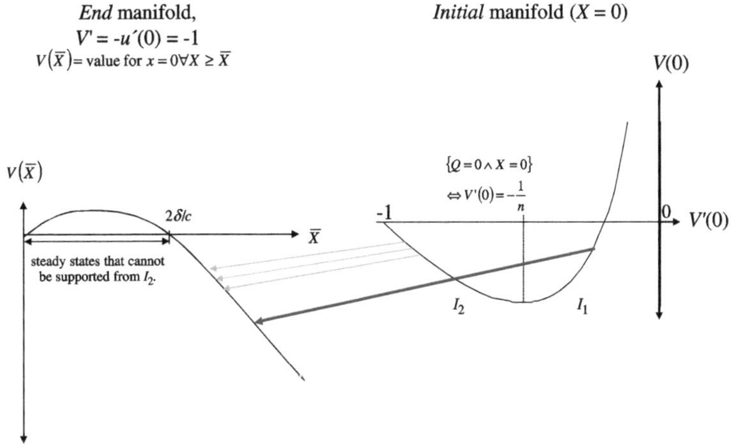

Fig. 1 *Initial* and *end* manifold for the example $u(x) = \ln(1+x)$; starting at I_2 avoids the crossing of the singularity $Q = 0$ and thus allows for multiple equilibria, while starting at I_1 requires to pass through $Q = 0$ which renders uniqueness if departing from I_1

4.2 End Manifold, $X = \overline{X}$

This manifold characterizes the value of any value function at any level $X = \overline{X}$ where an interior strategy $x > 0$ is joined with the boundary strategy, $x = 0$. Feasibility of $x = 0$ rules out utility functions satisfying the first Inada condition in Assumption (B3) and therefore, $u'(0) < \infty$ and normalization (B1) are assumed in the following. Solving (12) for A and substituting into (10) yields a quadratic polynomial:

$$V(\overline{X}) = \frac{u'(0)\delta}{r}\overline{X} - \frac{c}{2r}\overline{X}^2. \tag{22}$$

This *end* manifold is shown on the left hand side in Fig. 1.

4.3 Conditions for Uniqueness and Indeterminacy

Summarizing the analysis so far: *a stationary, stable and continuous Markov perfect equilibrium strategy must satisfy the differential equation (13) and must connect 'initial' (17) and 'end' (22) manifolds in a continuous manner*. Although Fig. 1 is based on the example (20) it highlights the general reasons and geometry for multiple equilibria: 'starting' from the decreasing part of the initial manifold (17), $V'(0) \in I_2$, means to start at $X = 0$ below the critical set $Q = 0$. This allows the corresponding 'strategy'[2] $V'(X)$ to bypass the singularity $Q = 0$ on its way to

[2]Of course, $x = f(V'(X))$ is the Markovian strategy, yet completely determined as a monotonic transformation of V', which is thus for short also called strategy.

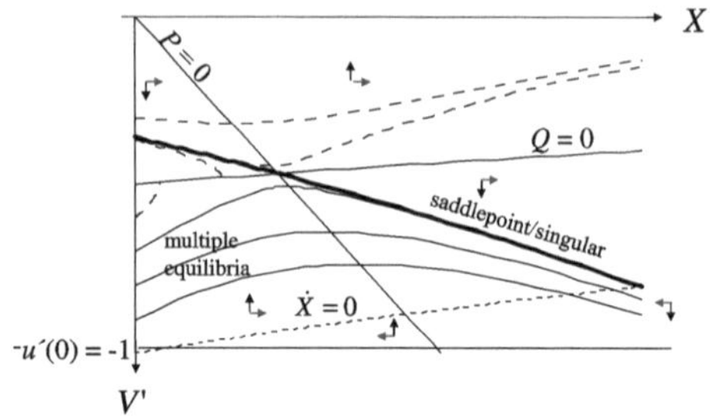

Fig. 2 Phase diagram (X, V'), $u(x) = \ln(1+x)$; *bold* = saddle-point (or) singular solution, the *dashed solution curves* of (23) are not equilibria

the stopping (22) and patching level $V'(\overline{X}) = -u'(0)$. Or conversely, each solution curve associated with a patching level $\overline{X}$, $V(\overline{X})$ from (22) and $V'(\overline{X}) = -u'(0)$ that results from integrating (13) backwards and that ends up at the initial manifold I characterizes an equilibrium.

The example in Fig. 1 implies negative values already at the origin, $V(0) < 0$, along the decreasing part of the initial manifold and thus for all multiple equilibria (if existing). This constrains at the outset the set of thresholds $\overline{X}$ that can be supported by strategies originating from this part (I_2), because $V(\overline{X}) < V(0)$ due to $V' < 0$ along emitting.

Starting at the increasing part of the manifold, $V'(0) \in I_1$, means to begin with $V'(0) > V' \in \{Q = 0\} \cap \{X = 0\} > V'(\overline{X}) = -u'(0)$ (both if existing otherwise the limit) and this allows, at least in principle for $V(0) > 0$ although Fig. 1 shows a negative one. However, any solution $V'(X)$ starting from the increasing part I_1 must pass through the singularity $Q = 0$. This is only possible through the intersection $\{P = 0\} \cap \{Q = 0\}$, which is unique given the declining locus $P = 0$ and the increasing manifold $Q = 0$ if existing as addressed in Lemma 1. This in turn renders conditional uniqueness to solutions that originate from I_1.

Therefore, two different kinds of smooth connections and thus of equilibria are conceivable in general and in particular for (20): the unique saddle-point (or "singular" as Rowat 2006 calls it) equilibrium originating from I_1 and a family of multiple equilibria from I_2. A phase diagram based on (20),

$$V'' = \frac{P(X, V')}{Q(X, V')} = \frac{(-V')[cX + (r + \delta)V']}{1 + (n + \delta X)V'},$$

$$Q = 0 \Leftrightarrow V' = q(X) = -\frac{1}{n + \delta X},$$

$$\dot{X} = 0 : V' = s(X) = -\frac{n}{n + \delta X}, \tag{23}$$

looks similar to the linear-quadratic model, see the phase plane analysis in Fig. 2. Therefore it has the same qualitative implications: a family of multiple equilibria can exist (three corresponding examples are shown in Fig. 2), which is more conservationist than the saddle-point strategy (the point Dockner and van Long 1993 stress). Of course this phase diagram highlights also the properties addressed in Lemma 1 such as the existence of $q(X)$ and that $q > s$.

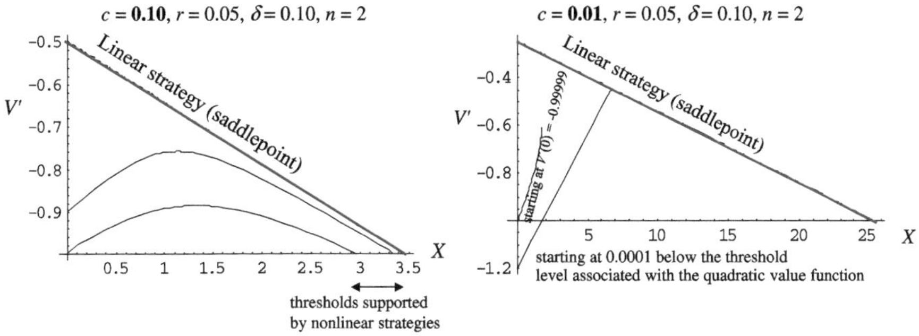

Fig. 3 Two examples for $u = x - 1/2x^2$ and $n = 2$: one with ($c = 0.10$ and thus higher external cost) and one without ($c = 0.01$) multiple equilibria in nonlinear strategies due to $x(0) > 0$

However, the absence of such smooth connections between the decreasing part of the initial manifold (i.e., of I_2) and the end manifold establishes uniqueness due to Proposition 1. A particular example is Fig. 2 in Rubio and Casino (2002) in which global nonlinear strategies would have to start with $x = 0$. Indeed it is easy to construct corresponding numerical examples even within the Dockner and van Long (1993) framework of (normalized) linear-quadratic utility, $u(x) = x - \frac{1}{2}x^2$, see Fig. 3. In these two examples, the one on the left hand side allows for smooth connections between initial and end manifold staying in the feasible domain, but the one on the right hand side does not. Therefore, the parameters shown on the right hand side of Fig. 3, which are characterized by low external costs, allow only for the saddle-point outcome, which is then the unique equilibrium. Therefore,

Proposition 4 *The existence of the decreasing part of the initial manifold is necessary but not sufficient for multiple equilibria. In particular, the requirement $x > 0$ at $X = 0$ from Proposition 2 can render uniqueness in particular cases.*

Utility functions satisfying the Inada condition $\lim_{x\to 0} u'(x) \to \infty$ rule out $x = 0$. This implicit requirement of a global interior strategy suggests uniqueness of the Nash equilibrium. This would render the multiplicity of equilibria an artifact of particular utility functions, which may be implausible anyway (such as the linear-quadratic description that implies saturation).

Proposition 5 *Assume that* (B2) *and* (B3) *hold and that the $Q(X, V') = 0$ implicitly defined function $V' = q(X)$ exists, then the equilibrium is unique.*

Proof $P = 0$ crosses linearly through the entire relevant quadrant, $X \geq 0$, $V' \leq 0$. The set of steady states, s, is below $Q = 0$ and both functions are increasing and have a pole at the origin of the state space, $V' \to -\infty$ for $X \to 0$. Therefore, any strategy must begin (i.e. at $X = 0$) above $Q = 0$ ($\to -\infty$ for $X \to 0$) and must thus cross $Q = 0$ on its way to s (and beyond). This is only possible through the intersection of $Q = 0$ with $P = 0$. Hence, only a single solution curve allows for a global, continuous, stationary and stable strategy; see the phase diagram in Fig. 4 (for a corresponding example discussed in Sect. 5). □

Remark 6 Actually, monotonicity of the $Q = 0$ (and thus Assumption (B2)) is not necessary for uniqueness given (B3). All that matters is that $Q = 0$ and $P = 0$ intersect only once.[3]

The above proposition might suggest that Inada conditions ensure unique equilibria. This turns out to be wrong according to the major result classifying either uniqueness or indeterminate outcomes:

Theorem 1 *Consider the nonlinear differential game* (1)–(3) *and suppose that utility* $u(x)$ *satisfies the Assumptions* (B2) *and* (B3).

1. *If* u *such that* $\sigma \geq \frac{n-1}{n}$ $\forall x > 0$, *then the Nash equilibrium is unique.*
2. *If* u *such that* $\sigma < \frac{n-1}{n}$ $\forall x > 0$, *then* $\{Q = 0\} = \emptyset$. *As a consequence, each solution of* (13) *determines an equilibrium for any of the boundary conditions*

$$V'(0) \in (0, -\infty), \qquad V(0) = \frac{u(f(V'(0))) + nV^{'}(0)f(V'(0))}{r},$$

and the saddle-point equilibrium is missing.

Proof See Appendix. □

Remark 7 The existence of $q(X)$ and of multiple equilibria is linked to the elasticity of marginal utility in the same way as the properties of increasing and decreasing of the initial manifold (17). Therefore, already the shape of the initial manifold (17)—increasing or decreasing or both—determines the nature of equilibria saving a much more involved phase diagram analysis.

Although this theorem is expressed in terms of global properties of the elasticity of marginal utility this can be reduced to a single check due to the assumption that σ is non-decreasing.

Corollary 1 *Given Assumptions* (B2) *and* (B3), *then*

$$\lim_{x \to 0} \sigma > \frac{n-1}{n}$$

[3]Multiple intersections and associated saddle-point equilibria would still induce a substantial reduction from the uncountably many. Indeed Wirl (2007) shows an example of two intersections of $Q = 0$ and $P = 0$ through which the nevertheless unique singular equilibrium strategy must pass.

is sufficient for the uniqueness of the Nash equilibrium in Markov strategies. In contrast,

$$\lim_{x\to\infty} \sigma < \frac{n-1}{n}$$

is sufficient for multiple equilibria and in fact for 'maximal indeterminacy', because a saddle-point equilibrium is then ruled out.

Corollary 2 *Given Assumption* (B2) *then*

$$\lim_{x\to 0} \sigma < \frac{n-1}{n}$$

is necessary for multiple equilibria.

Remark 8 $\sigma > \frac{n-1}{n}$ holds for all satiating utilities for $x \to \widehat{x}$ (= saturation level). Hence, multiple equilibria are located in the domain where x is small—strictly speaking, x must be only small at $X = 0$—yielding the implicit conservation observed in Dockner and van Long (1993). Hence, this conservation is a consequence of $\sigma' > 0$ and might be turned upside down for utilities with $\sigma' < 0$; and so it is, see Wirl (2007).

Summarizing, the properties of uniqueness versus multiple equilibria (including the case of maximal indeterminacy) are linked to the value of the elasticity of marginal utility (or the degree of relative 'risk aversion', or the intertemporal substitution elasticity) such that a relatively 'large' value of this elasticity (and at least greater than 1) ensures uniqueness (high risk aversion), while a low value of this elasticity (little risk aversion) can provide for a considerable freedom of equilibrium choice. An economic explanation of this result is given in the following section.

5 Examples

This section presents examples with unique or indeterminate outcome and gives an economic reason for this stark difference. The linear-quadratic utility and the above example (20) blend these two different outcomes. This blending extends to all utilities for which the elasticity of marginal utility varies sufficiently to cover both cases of Theorem 1 and CARA utilities provide another example since a constant absolute risk aversion of ρ implies $\sigma = \rho x$ and thus a variation in σ that allows for both, a saddle-point equilibrium and a family of equilibria.

Proposition 5 ensures uniqueness for utility functions satisfying the Inada conditions and what seems a rather technical condition, the existence of the function q. However, the set $Q = 0$ can be empty even for standard specifications such as power functions:

$$u(x) = \frac{1}{a}x^a, \quad 0 < a < 1. \tag{24}$$

Fig. 4 Phase diagram (X, V'), $u(x) = xa$, $a < 1/n$, unique equilibrium

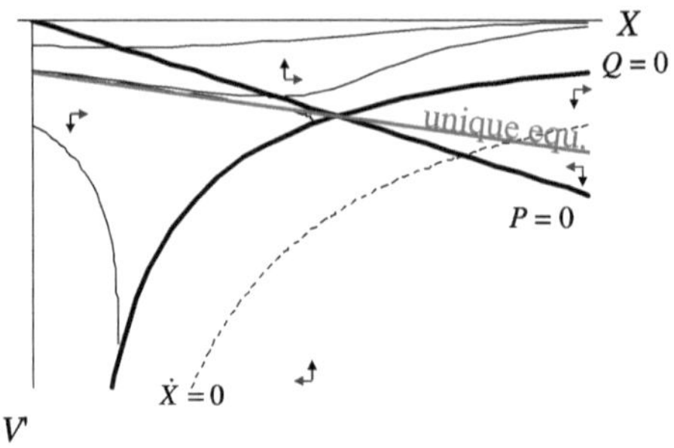

This specification implies a constant elasticity, $\sigma = 1 - a$, $x^* = (-V')^{\frac{1}{a-1}}$, and

$$V'' = \frac{P(X, V')}{Q(X, V')} = \frac{(a-1)[cX + (r+\delta)V']}{(an-1)(-V')^{\frac{1}{a-1}} + (1-a)\delta X}. \tag{25}$$

As we know already, the set $P = 0$ is unaffected by the choice of u and

$$s = -\left(\frac{\delta X}{n}\right)^{a-1} \tag{26}$$

has the familiar shape (starting at $-\infty$, increasing through the entire relevant quadrant, $s \to 0$ for $X \to \infty$). The initial manifold in this example is

$$rV(0) = \frac{1-an}{a}(-V'(0))^{\frac{a}{a-1}}, \tag{27}$$

so that its slope (either positive or negative) depends on the sign of $(1 - an)$. Therefore one must differentiate between the two cases addressed in Proposition 3 and Theorem 1.

If $a < \frac{1}{n} \Leftrightarrow \sigma < \frac{n-1}{n}$, the initial manifold is increasing, $Q = 0$ exists,

$$V' = -\left(\frac{(1-a)\delta X}{1-an}\right)^{a-1},$$

and the equilibrium is unique. Figure 4 highlights the reasons for the uniqueness of the Nash equilibrium: since only the increasing and 'unstable' part of the initial manifold exists, one must start at $X = 0$ above $Q = 0$ and a later crossing of $Q = 0$, which is necessary to reach a steady state (i.e., s), is only possible through the intersection of $P = 0$ with $Q = 0$.

If $\sigma > \frac{n-1}{n} \Longleftrightarrow 1 - an < 0$, the initial manifold is decreasing (and negative, $V(0) < 0$) and the set $Q = 0$ is empty according to Theorem 1 and the singularity is gone. Hence, the right hand side of the differential equation (25) is defined over the entire and relevant quadrant, $X \geq 0$ and $V' < 0$. Therefore, all solution curves starting at $X = 0$ at an arbitrary $V'(0) \in (-\infty, 0)$ and $V(0)$ from (27), and satisfying the Hamilton-Jacobi-Bellman equation are globally defined, continuous

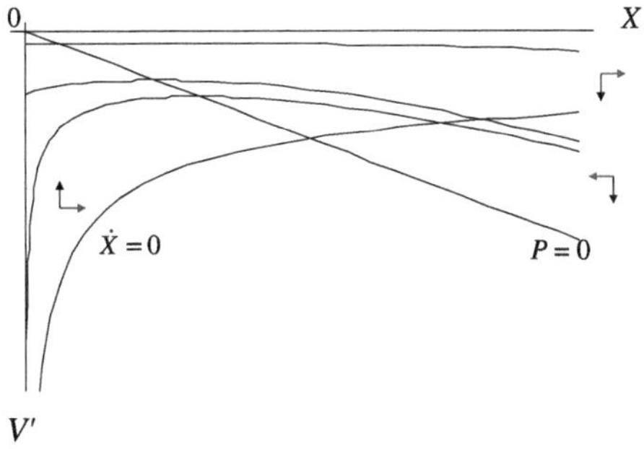

Fig. 5 Phase diagram (X, V'), $u(x)=xa$, $a>1/n$, "maximal indeterminacy" (three examples)

(actually everywhere differentiable), and stable. Hence, they determine corresponding Nash equilibria $x=(-V'(X))^{\frac{1}{a-1}}$; Fig. 5 highlights this case and shows three examples from the uncountably many equilibria. As a consequence, not only the dimension of indeterminacy is increased, but a potential focal point, the saddle-point path equilibrium, is gone if $\{Q=0\}=\emptyset$.

This feature—(almost) anything goes (to quote Paul Feyerabend) for $a>\frac{1}{n}$—is puzzling and applies whenever the number of polluters is sufficiently large. The good news is that this indeterminacy allows for more conservationist strategies by choosing a smaller $V'(0)$. In spite of this high degree of flexibility, however, stationary pollution remains excessive compared with the social optimum:

Proposition 6 *Given* (24), *stationary pollution must exceed*

$$\underline{X} \equiv \left(\frac{nc}{nr+\delta}\right)^{\frac{1}{a-2}}\left(\frac{n}{\delta}\right)^{\frac{a-1}{a-2}} > X^* = \left(\frac{nc}{r+\delta}\right)^{\frac{1}{a-2}}\left(\frac{n}{\delta}\right)^{\frac{a-1}{a-2}}$$

and thus in particular X^*, which is the efficient stationary outcome.

Proof See Appendix. □

The extreme difference in outcomes—uniqueness on the one hand, maximal indeterminacy on the other hand—and all this within the same class of utility functions is astonishing. The crucial economic magnitude is the elasticity of marginal utility (σ), which is the reciprocal of the price elasticity, $\varepsilon=\frac{1}{\sigma}$, for the individual players' emission demands. Sufficiently elastic demand with a price elasticity greater $\frac{n}{n-1}$ suggests substantial flexibility. And given such flexibility globally as in the case of $\frac{1}{n}<a<1$ for the above power function eliminates the saddle-point equilibrium with the consequence of maximal indeterminacy. In contrast, less elastic, $\varepsilon<\frac{n}{n-1}$, and thus in particular inelastic (e.g. for the familiar CRRA utilities $u(x)=-x^{-\alpha}$, $\alpha>1$) demand stands for too little flexibility. If this limited flexibility holds globally, then it is impossible to implement more cooperative, non-saddle-point equilibria such that the equilibrium is unique.

6 Summary

This paper reconsidered the much discussed multiplicity of Nash equilibria in Markov strategies in dynamic games; in line with this literature, the analysis was restricted to stationary, continuous and stable Markov strategies. Using the Dockner and van Long (1993) framework of a stock externality the following was introduced: (1) Global strategies (instead of local ones), which may require patching with the boundary solution (here: no emissions) in the case of multiple equilibria and in particular for nonlinear strategies in a linear-quadratic model. (2) This demand for global strategies implies mostly neglected boundary conditions at the origin of the state space and at the point of patching with the boundary solution (value matching and smooth pasting). (3) These conditions allow to characterize Nash equilibria as smooth connections of an 'initial' with an 'end' manifold. (4) Extension to general nonlinear instead of linear-quadratic utilities. These extensions and different characterizations have the following implications. The different kinds of Nash equilibria—on the one hand a unique saddle-point equilibrium (a likely focal point of the game that implies a linear strategy in the linear-quadratic model) and on the other hand the possibility of a continuous family of Nash equilibria—are related to two different parts of an initial manifold (increasing and decreasing). This holds for the familiar linear-quadratic framework as well as for nonlinear extensions. In the latter case, Inada conditions do not ensure uniqueness despite their implicit demand for global interior strategies.

The Nash equilibrium is unique if the elasticity of marginal utility (σ) is everywhere greater than $\frac{n-1}{n}$, because this ensures that only the increasing part of the initial manifold exists; uniqueness can also follow from the property of positive emissions at the origin of the state space despite the existence of the decreasing part of the initial manifold. In contrast, maximal indeterminacy—each initial condition $V'(0) < 0$ determines a corresponding global Nash equilibrium—results if the elasticity of marginal utility is everywhere less than $\frac{n-1}{n}$. The reason is that only the decreasing part of the initial manifold exists, which eliminates the saddle-point equilibrium and thus a potential focal point among the many equilibria. This is not a theoretical artifact since it arises for the familiar power function, $u(x) = x^a$, $\frac{1}{n} < a < 1$, where the inequality is always satisfied for a sufficient number of players. Despite this enormous flexibility, feasible conservation is restricted such that the implementation of the socially optimal stationary stock of pollution is impossible, let alone of the efficient policy.

An economic explanation of this puzzling difference—uniqueness versus maximal indeterminacy—for similar utility functions (power functions) with similar properties (constant elasticities of marginal utility) is the following: The (absolute) price elasticity of demand equals the reciprocal of the elasticity of marginal utility, here from emissions (within the caveat of no income effects or of a partial equilibrium framework). Sufficient flexibility, i.e., a price elasticity greater $\frac{n}{n-1}$, allows for multiple equilibria (and no saddle-point equilibrium and maximal indeterminacy if this property holds globally). In contrast, too less elastic (and in particular inelastic) demand implies little flexibility, which makes it impossible to implement more cooperative, non-saddle-point equilibria such that a unique equilibrium results if the elasticity is globally less than $\frac{n}{n-1}$. Unfortunately, this seems to apply to global warming due to the inelastic demand for energy. A blend of the above extreme cases—a saddle-point equilibrium as well as a family of (non-saddle-point) equilibria—applies to utility functions that include both domains of the elasticity, e.g. the familiar linear-quadratic utility, and CARA utilities.

These results can be applied to other differential games in which utilities depend nonlinearly on the strategies; a nonlinear evaluation of the stock only will not change the qualitative properties from the linear-quadratic game (e.g., in the public good game of Fershtman and Nitzan 1991). Finally, as in the related literature, the analysis was restricted to symmetric equilibria, yet little is known about the possibility and properties of asymmetric ones (Rowat 2000 makes some attempts in this direction).

Acknowledgement I acknowledge helpful comments from an anonymous referee.

Appendix

7.1 Proof of Proposition 1

First, value matching is obvious to rule out arbitrage from increasing or reducing X marginally so that the proof is confined to establish smooth pasting.

After moving for some time along the boundary (thus assuming the normalization (B1) and also symmetry), the optimal strategy of an individual player to start emitting again at T is

$$\max_T \int_0^T e^{-rt}\big(-C\big(X(t)\big)\big) + e^{-rT} V\big(X(T)\big),$$
$$\dot{X} = -\delta X,$$

in which the interior value function V describes the scrap value. Defining the Hamiltonian, $H = -C - \lambda\delta X$, the condition for the optimal choice of T is

$$H(T) = -C\big(X(T)\big) - \lambda(T)\delta X(T) = rV\big(X(T)\big).$$

Using $V_b' = \lambda$ from continuous dynamic programming on the left hand side, (7) and continuity of the strategies ($x(T) = 0$) on the right hand side, yields:

$$-C\big(X(T)\big) - V_b'\big(X(T)\big)\delta X(T) = -C\big(X(T)\big) - \delta X(T)V'\big(X(T)\big),$$

which requires $V_b' = V'$. □

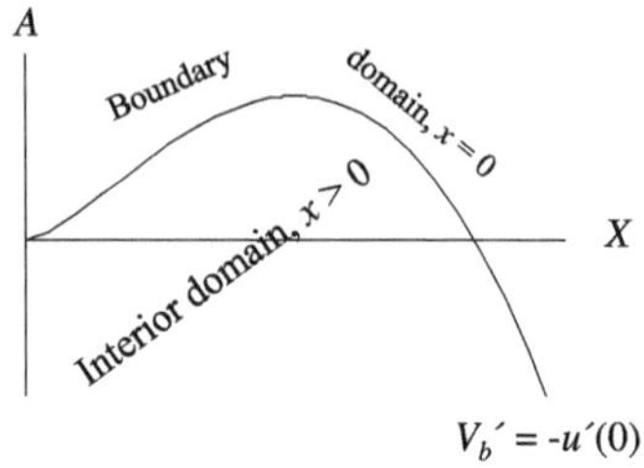

Fig. 6 Interior and boundary domain in terms of the state X and integration constant A, similar to Rowat (2006) for the linear-quadratic case

7.2 *Proof of Proposition 2*

The following sketch of a proof follows Rowat (2000, 2006). Due to (12),

$$V_b' = -\frac{Ar}{\delta}X^{-\frac{r+\delta}{\delta}} - \frac{cX}{(r+2\delta)} = -u'(0)$$

characterizes the border between interior and boundary solutions, which establishes a relation between the state X and the constant of integration A as shown in Fig. 6. If $A > 0$ then for $x = 0$: $V = V_b \to \infty$ for $X \to 0$, contradicting the in (A1) assumed upper bound. The boundary strategy is impossible for $A < 0$ in a surrounding of $X = 0$ (see Fig. 1) and the knife-edge case, $A = 0$, can be excluded by showing that a deviating strategy, $x > 0$ at least for X close to 0, is profitable (see Rowat 2006). □

7.3 *Proof of Lemma 1*

We show below that the inequality (15) is necessary and sufficient for the existence of $Q = 0$ for any $X < \frac{n\widehat{x}}{\delta}$. From this follows immediately the first part for satiating utilities, $\widehat{x} < \infty$, since then $\sigma \to \infty$ for $x \to \widehat{x}$ and this inequality (15) is trivially fulfilled. Assuming instead non-satiating utilities, inequality (16) follows readily from the second Inada condition, $\widehat{x} = \infty$.

Sufficiency: We have to show that inequality (15) ensures the existence of $Q = 0$. Rearranging (14) and holding $X < \frac{n\widehat{x}}{\delta}$ fixed yields

$$\sigma\left(f\left(V'\right)\right) = \frac{n-1}{n - \frac{\delta X}{f(V')}}, \tag{28}$$

i.e., two functions—one on the left hand side and the other on the right hand side—defined for $V' \leq 0$. The elasticity of marginal utility, σ, is positive and non-decreasing in x (due to Assumption (B2)) and thus also in V' (due to (6)). Hence, the left hand side is positive, non-decreasing, $\sigma' f' \geq 0$, and reaches its maximum $\sigma(f(0)) = \sigma(\widehat{x})$ at $V' = 0$, since $V' \to 0$ implies $u' \to 0$ and thus $x \to \widehat{x}$. The right hand side declines monotonically for all $V' \in (s(X), 0]$ and assumes all positive real numbers from $[\frac{n-1}{n-\frac{\delta X}{\widehat{x}}}, \infty)$. The upper bound results from $V' \to s(X)$, the lower from the above implication of $V' = 0$. Summarizing: (a) the right hand side of (28) is declining and covering $[\frac{n-1}{n-\frac{\delta X}{\hat{x}}}, \infty)$; (b) the left hand side is positive, increasing (or constant) and surpassing the minimal level of the right hand side if inequality (15) holds. Therefore, there exists a unique intersection that determines the value of V' uniquely for each value of $X < \frac{n\hat{x}}{\delta}$ along $Q = 0$ and thus the function, $V' = q(X)$.

Necessity follows because the opposite of the inequality of (15) rules out an intersection of the left hand and right hand sides in (28) so that $\{Q = 0\} = \emptyset\ \forall X \geq 0$ in contradiction to the assumed existence.

The claim that $q > s$ holds in fact in general and without Assumption (B2). For this purpose, rewrite (14) moving the term defining s to the right hand side

$$Q = 0 \quad \Longleftrightarrow \quad \frac{(n-1)V'}{u''} = (nf - \delta X) = \dot{X}.$$

This implies that V' from $Q = 0$ is in the domain of $\dot{X} > 0$ (for any $X < \frac{n\hat{x}}{\delta}$ since $V' = q < 0$). Hence $q(X) > s(X)$ since it takes a larger negative value of V' to induce $\dot{X} = 0$.

Finally, the last claim, $q' > 0$, is proven. The choice of X has no effect on the left hand side of (28), but a larger value of X reduces the support on the right hand side $(s(X), 0]$, because $s(X)$ is increasing. This implies in turn a larger value of V' (i.e. a smaller negative number), shifts the curve defined by the right hand side of (28) upwards and, as a consequence, the point of intersection at V' must increase (from applying Assumption (B2) to the left hand side). Hence the set $\{Q = 0\}$ defines (if existing) an increasing function in the (X, V') plane. □

7.4 *Proof of Theorem 1*

The first part of the Theorem for $\sigma > \frac{n-1}{n}\ \forall x > 0$ follows directly from Lemma 1 and Proposition 5 since only passing through the then unique intersection between $P = 0$ and $Q = 0$ allows for a globally defined strategy.

If in contrast $\sigma < \frac{n-1}{n}\ \forall x > 0$ and thus in particular for all $V' \in (-\infty, 0]$, then the left hand side of (28) is always below the right hand side for $V' \in (s(X), 0]$, and always above for $V' \in (-\infty, s(X))$ (irrespective of the sign of σ'). Therefore, the set $Q = 0$ is empty for all X. $P = 0$ is linearly declining and thus crossing the entire relevant quadrant $X \geq 0$ and $V' < 0$. The set s is monotonically increasing and also covering the entire relevant domain if u satisfies the two Inada conditions: $s(X) \to -\infty$ for $X \to 0$, $s(X) \to 0$ for $X \to \infty$. As a consequence, any solution curve $V'(X)$ starting at an arbitrary initial value $V'(0) < 0$ (and $V(0)$ determined from the initial manifold (17)) must stay within the quadrant $X \geq 0$ and $V' < 0$ since it must start above s, must increase until it intersects $P = 0$ at a feasible value of $V' < 0$ and must decline after this intersection. Therefore this solution has global, interior support.

It remains to show that each strategy originating from an arbitrary $V'(0) < 0$ ensures a unique and stable steady state. Stability of the associated Markov strategy requires that such a solution curve V' crosses s from above (and only from above). And clearly given the shapes of $P=0$ and s any of these strategies must cross s, which ensures stability of this steady state. Therefore further and in particular unstable steady states must be excluded. Indirectly, assume such an unstable steady state at X_2, (i.e., intersections from below yet clearly in the domain $V''>0 \Longleftrightarrow P<0$):

$$V'' = \frac{-u''[cX+(r+\delta)V']}{(n-1)V'} > s' = -\frac{\delta u''}{n} \quad \Leftrightarrow \quad V' < -\frac{ncX}{nr+\delta} \text{ at } X = X_2.$$

Since $V'(0) > \lim_{X\to 0} s = -\infty$ it must follow an intersection from above at $X_1 < X_2$: $V'' < s' \Leftrightarrow V'(X_1) > -\frac{ncX_1}{nr+\delta}$. Therefore, $V'(X_1) > V'(X_2)$, which is impossible in the domain $P<0$ and thus $V''>0$. Contradiction. □

7.5 *Proof of Proposition 6*

The proof falls into three parts. First, the socially optimal policy is derived. Second a lower bound on steady states attainable by Nash behavior is derived. Finally, this lower bound is compared with the efficient stationary pollution.

7.5.1 Derivation of the Social Optimum for Utility (24)

The Hamilton-Jacobi-Bellman equation for the corresponding value function W is

$$rW(X) = \max_{x\geq 0}\left\{n\left(\frac{x^a}{a} - \frac{c}{2}X^2\right) + W'(X)(nx-\delta X)\right\}. \tag{29}$$

Maximization on the right hand side yields for interior policies:

$$x^* = (-W')^{\frac{1}{a-1}},$$

which implies for (29):

$$rW = n\left(\frac{(-W')^{\frac{a}{a-1}}}{a} - \frac{c}{2}X^2\right) + W'\left(n(-W')^{\frac{1}{a-1}} - \delta X\right). \tag{30}$$

Differentiating (30), simplifying and representing as a ratio yields,

$$W'' = \frac{P(X,W')}{Q(X,W')} = \frac{cnX+(r+\delta)W'}{n(-W')^{\frac{1}{a-1}} - \delta X}.$$

Therefore

$$P(X,W') = 0 \quad \Longleftrightarrow \quad W' = -\frac{ncX}{r+\delta},$$

$$Q(X, W') = 0 \quad \Longleftrightarrow \quad W' = -\left(\frac{\delta X}{n}\right)^{a-1}.$$

Moreover, $\{Q = 0\}$ coincides with s. Therefore, the intersection of $P = 0$ with $Q = 0$ determines simultaneously a point of the saddle-point path and the steady state. Hence, solving the equation

$$-\frac{ncX}{r+\delta} = -\left(\frac{\delta X}{n}\right)^{a-1}$$

yields

$$X^* = \left(\frac{nc}{r+\delta}\right)^{\frac{1}{a-2}} \left(\frac{n}{\delta}\right)^{\frac{a-1}{a-2}}.$$

□

7.5.2 Lower Bound for $\bar{X}$

Figure 5 suggests that the solution curves cut the steady state line s only after the intersection of $P = 0$ with s. This intersection between P and s occurs at

$$\widetilde{X} = \left(\frac{c}{r+\delta}\right)^{\frac{1}{a-2}} \left(\frac{n}{\delta}\right)^{\frac{a-1}{a-2}}.$$

This bound clearly exceeds the social optimum, $\widetilde{X} > X^*$, but one cannot exclude in principle the possibility of an intersection at a lower level, i.e. when $V'' > 0$. A stable intersection (i.e. one from above) requires

$$V' > -\frac{ncX}{nr+\delta},$$

which is independent from the specification of u. Intersecting now this general (lower) bound of V' with the specific s from (26) yields

$$-\frac{ncX}{nr+\delta} = -\left(\frac{\delta X}{n}\right)^{a-1}.$$

The solution of this equation determines the lower bound on stable steady states

$$\underline{X} = \left(\frac{cn}{nr+\delta}\right)^{\frac{1}{a-2}} \left(\frac{n}{\delta}\right)^{\frac{a-1}{a-2}}.$$

Therefore it remains to compare the social optimum with the above lower bound,

$$\frac{X^*}{\underline{X}} = \frac{(\frac{nc}{r+\delta})^{\frac{1}{a-2}} (\frac{n}{\delta})^{\frac{a-1}{a-2}}}{(\frac{cn}{nr+\delta})^{\frac{1}{a-2}} (\frac{n}{\delta})^{\frac{a-1}{a-2}}} = \left(\frac{nr+\delta}{r+\delta}\right)^{\frac{1}{a-2}} = \left(\frac{r+\delta}{nr+\delta}\right)^{\frac{1}{2-a}} < 1. \tag{31}$$

□

References

Benchekroun, H. (2003). Unilateral production restrictions in a dynamic duopoly. *Journal of Economic Theory*, *111*, 214–239.

Dixit, A. K., & Pindyck, R. S. (1994). *Investment under uncertainty*. Princeton: Princeton University Press.

Dockner, E. J., & van Long, N. (1993). International pollution control: cooperative versus non-cooperative strategies. *Journal of Environmental Economics and Management*, *25*, 13–29.

Dockner, E. J., & Wagener, F. O. O. (2008). Markov-perfect Nash equilibria in models with a single capital stock. Mimeo. CeNDEF, Universiteit van Amsterdam.

Dockner, E. J., Jorgensen, S., van Long, N., & Sorger, G. (2000). *Differential games in economics and management science*. Cambridge: Cambridge University Press.

Fershtman, C., & Kamien, M. I. (1987). Dynamic duopolistic competition with sticky prices. *Econometrica*, *55*, 1151–1164.

Fershtman, C., & Nitzan, S. (1991). Dynamic voluntary provision of public goods. *European Economic Review*, *35*, 1057–1067.

Harstad, B. (2012). Climate contracts: a game of emissions, investments, negotiations, and renegotiations. *Review of Economic Studies*, *79*, 1527–1557.

Jorgensen, S., Martin-Herran, G., & Zaccour, G. (2010). Dynamic games in the economics and management of pollution. *Environmental Modeling & Assessment*, *15*, 433–467.

Martin-Herran, G., & Rincon-Zapatero, J. P. (2005). Efficient Markov perfect Nash equilibria: theory and application to dynamic fishery games. *Journal of Economic Dynamics & Control*, *29*, 1073–1096.

Rincon-Zapatero, J. P., Martinez, J., & Martin-Herran, G. (1998). New method to characterize subgame perfect Nash equilibria in differential games. *Journal of Optimization Theory and Applications*, *96*, 377–395.

Rowat, C. (2000). Additive externality games. King's College, Cambridge University. http://socscistaff.bham.ac.uk/rowat/research/master4.pdf. Accessed 17 Jun. 2013.

Rowat, C. (2006). Nonlinear strategies in a linear quadratic differential game. *Journal of Economic Dynamics & Control*, *31*, 3179–3202.

Rubio, S. J., & Casino, B. (2002). A note on cooperative versus non-cooperative strategies in international pollution control. *Resource and Energy Economics*, *24*, 251–261.

Wirl, F. (1994). Pigouvian taxation of energy for stock and flow externalities and strategic, non-competitive pricing. *Journal of Environmental Economics and Management*, *26*, 1–18.

Wirl, F. (2007). Do multiple Nash equilibria in Markov strategies mitigate the tragedy of the commons? *Journal of Economic Dynamics & Control*, *31*, 3723–3740.

Wirl, F. (2008a). Tragedy of the commons in a stochastic dynamic game of a stock externality. *Journal of Public Economic Theory*, *10*, 99–124.

Wirl, F. (2008b). Reversible stopping ("switching") implies super contact. *Computational Management Science*, *5*, 393–401.

Wirl, F., & Dockner, E. J. (1995). Leviathan governments and carbon taxes: costs and potential benefits. *European Economic Review*, *39*, 1215–1236.

Part II
Optimal Extraction of Resources

Dynamic Behavior of Oil Importers and Exporters Under Uncertainty

Lucas Bretschger and Alexandra Vinogradova

Abstract We consider long-run incentives for oil-importing and -exporting countries when the arrival of a backstop technology is uncertain. Oil importers invest in research and development to avoid their dependence on foreign oil; the arrival of the oil substitute is modeled with a Poisson process. The optimum resource extraction path is determined by the optimization of oil exporters, which take the Poisson rate as exogenously given. We provide clear-cut solutions of the optimization plans for both types of countries. We find that the optimal consumption rate of oil importers may be either constant or falling during substitute development; it is a decreasing function of the oil price. When the substitute arrives, the rate of resource depletion jumps down and the depletion pace during the post R&D phase is slower than during the R&D phase, provided that the Poisson arrival rate is sufficiently large. We also show under which conditions there is never strategic behavior and when strategic behavior may take place.

1 Introduction

Extraction and the use of the world's most important natural resource, oil, divides the world into two types of countries: oil-importing and oil-exporting. The costs for oil imports are high, and have even increased in the new millennium. Shares of GDP spent for these imports vary between 2.1 percent for Germany and 2.5 percent for the United States to 2.8 percent for China and 2.9 percent for Japan up to 7.6 percent for India.[1] Besides current and future costs, oil-importers worry about the market power of the suppliers organized in a cartel, the OPEC, which controls more than 75 % of proven world reserves. Another issue is the political volatility in the oil extracting

[1]Data are for the year 2008 and taken from the BP Statistical Review of World Energy 2010 and WDI online data of the World Bank.

L. Bretschger (✉) · A. Vinogradova
Center of Economic Research, CER-ETH, Zürichbergstrasse 18, 8092 Zürich, Switzerland
e-mail: lbretschger@ethz.ch

A. Vinogradova
e-mail: avinogradova@ethz.ch

E. Moser et al. (eds.), *Dynamic Optimization in Environmental Economics*, Dynamic Modeling and Econometrics in Economics and Finance 15,
DOI 10.1007/978-3-642-54086-8_9,

regions and the associated possibility of supply shortages. Moreover, there are major concerns about the environmental impacts, particularly with climate change.

Taken together, the significantly increased interest in the development of a substitute for fossil fuels can well be explained. Especially in the US, there has been a notable increase of Research, Development and Demonstration (RD&D) spending on renewables and on energy sources just after the 2008 peak of oil costs, while in the European countries the same can be observed with a minor delay; the sharp rise in research spending was in the years 2010 and 2011.[2] Importing countries coordinate on energy policy and energy security issues through various international organizations such as the international energy agency (IEA), OECD and EU, and cooperate in the development of renewable alternatives. As in all the sectors in the economy, technical development cannot be predicted precisely; it involves risks and uncertainties, affecting the behavior of both importers and exporters.

The model developed in this chapter reflects the division of the world into the two asymmetric types of countries and analyzes the long-run incentives for both oil-importing and oil-exporting countries. Our main focus is on the impact of uncertainty in the development of a substitute for oil on innovative behavior and resource extraction. The arrival of the substitute is modeled with a Poisson process. We derive two possible cases and show that the optimal R&D expenditure on substitute development may be completely independent of the oil price and thus of the actions of oil-exporting countries. We also show the analytics of resource extraction under the chosen type of uncertainty. We establish that the extraction rate declines over time, like in the standard theory of exhaustible resource extraction. Finally, a comparison to extraction without uncertainty is drawn.

The paper is related to different strands of literature. The original model of a monopolist extraction firm facing a backstop resource is given in Hoel (1978). A series of earlier papers focuses on optimal timing of when to adopt an alternative, fixed technology at a given exogenous cost, see e.g. Heal (1976), Dasgupta et al. (1983), Gallini et al. (1983), Olsen (1993). The seminal paper by Kamien and Schwartz (1978) endogenizes the uncertain arrival date of the substitute through investment in R&D. Hung and Quyen (1993) go further to determine the optimal time to initiate the R&D project. Their R&D investment policy is simplified to a single-date expenditure, after which a backstop may arrive with a constant Poisson rate. We adopt a similar assumption in the present paper. As regards the strategic aspects of the decisions, these papers use various assumptions on the ability to commit and the timing of moves. In a more recent contribution, Gerlagh and Liski (2011) study a (deterministic) game in which the importer can trigger a process which ends with the introduction of the substitute. The delay between the decision to develop the substitute and the arrival of the technology acts as a commitment device. An incremental process of backstop development has been considered by Tsur and Zemel (2003), where the socially optimal case is derived. Because the marginal cost of research activities is constant, the planner steers the economy to the steady-state process as

[2]Data extracted from OECD iLibrary in March 2013.

quickly as possible. In Tsur and Zemel (2005), R&D activities gradually reduce the backstop cost. Depending on the initial conditions the paper derives a wide variety of optimum investment solutions. Van der Ploeg and Withagen (2012) show that, with monopolistic supply of a polluting resource, an (exogenous) decrease of the cost of the substitute may lead to a permanent reduction of resource use. Valente (2011) analyzes a two-phase endogenous growth model in which the optimal timing of switching to a backstop resource is determined by welfare maximization. The major contribution on dynamic interactions under uncertainty is Harris and Vickers (1995) who model a probabilistic R&D process of an oil importer which affects the extraction path of the exporter. The optimal path then follows a modified Hotelling rule and may exhibit non-monotonicity due to strategic considerations.

To add to the literature, our model takes up the important issue of uncertainty in research and combines it with other crucial features, specifically with an endogenous R&D process, increasing marginal costs of R&D, and a gradual introduction of the backstop technology. Although we do not explicitly analyze strategic behavior of oil importers and exporters, we provide a clear-cut solution for their respective optimization programs and discuss conditions under which strategic behavior may take place. Non-monotonicity of the extraction path, as in Harris and Vickers (1995), may arise in our model even without strategic considerations on behalf of the exporter. The remainder of the chapter is organized as follows. Section 2 presents the optimization problem and the results for the oil importers. Section 3 derives the optimum for resource exporters. Section 4 concludes.

2 Oil-Importing Country

Consider a resource-importing country (RIC) which has no resource reserves of its own. We shall refer to this resource as oil. The country must therefore satisfy the entire demand by imports from abroad at a price P per unit. It seeks to develop a perfect substitute for oil, although the discovery of the substitute is uncertain. We assume that the arrival of the substitute is governed by the Poisson process with the mean arrival rate $\lambda(K)$, where K is the initial investment in the R&D project and $\lambda'(K) > 0$, $\lambda''(K) \leq 0$. By an initial investment we mean a one-time expenditure which determines the probability of a technological breakthrough. One may think of this expenditure as of a fixed cost of setting up an R&D facility. A bigger facility requires a bigger investment but also brings larger returns, in the sense that the chances of making a discovery are higher.

RIC produces a composite consumption good, with oil and labor as two essential inputs, according to a Cobb-Douglas technology $Y_t = L^\alpha R_t^{1-\alpha}$, where L denotes a constant labor input and R_t stands for oil. When the substitute arrives, a constant quantity B becomes available every period at negligible cost. Then RIC discontinues its imports of oil and uses only the substitute for final goods production.[3]

[3]We implicitly assume that the cost of producing a unit of the substitute is below the price of a unit of oil.

RIC's objective is to maximize the present discounted value of lifetime welfare given an infinite planning horizon and a constant discount rate, δ. Utility is derived from consuming the composite good and the utility function is such that $u'(c_t) > 0$, $u''(c_t) < 0$, where c_t denotes the consumption rate at time t. RIC's optimization problem reads

$$\max_{c_t, c_t^b, R_t, K} \int_0^\infty \left(\int_0^\tau u(c_t)e^{-\delta t}dt + \int_\tau^\infty u\left(c_t^b\right)e^{-\delta t}dt \right) f_\tau d\tau - u(K)$$

subject to[4]

$$\begin{aligned} c_t &= Y(R_t, L) - PR_t, \quad t \in [0, \tau], \\ c_t^b &= Y(B, L) \equiv \bar{Y}, \quad t > \tau, \end{aligned} \tag{1}$$

where $f_\tau = \lambda(K)e^{-\lambda(K)\tau}$ is the density of an exponentially distributed random variable.[5] RIC's optimization program therefore includes two phases: the first while the substitute has not yet arrived and the second after its arrival. Note that once the substitute is online, RIC's consumption rate in the second phase, c_t^b, becomes constant and equal to the output, $\bar{Y}$. The value of the second-phase program can therefore be written as

$$\int_\tau^\infty u\left(c_t^b\right)e^{-\delta t}dt = u(\bar{Y})\frac{e^{-\delta\tau}}{\delta}.$$

During the first phase, RIC must optimally choose the size of the R&D project, K, and the oil imports so as to maximize utility from consumption until the substitute arrives (if ever). Thus oil will be imported to the point where its marginal productivity equals its price, $\frac{\partial Y}{\partial R} = P$, which yields the oil demand

$$R = \left(\frac{1-\alpha}{P}\right)^{1/\alpha} L. \tag{2}$$

Substituting (2) into (1) and assuming that the resource price is constant, we obtain a constant optimal consumption rate in Phase 1

$$c = L^\alpha R^{1-\alpha} - (1-\alpha)L^\alpha R^{-\alpha} R = \alpha L^\alpha R^{1-\alpha} = \alpha L \left(\frac{1-\alpha}{P}\right)^{\frac{1-\alpha}{\alpha}}.$$

---

[4]The term $u(K)$ can be viewed as an approximation to the present value of all future costs if they are incurred at each point in time. It would be, perhaps, more appropriate to subtract investment expenditure evaluated by the marginal utility of consumption at time zero. This would, however, make the calculation more complicated. We therefore opted in favor of a simpler specification by ignoring the secondary effect of the investment expenditure on the marginal utility of consumption, while taking into account only the direct effect. In Hung and Quyen (1993), for instance, a lump-sum investment cost incurred at a single date τ is simply discounted to time zero at the rate of time preference and subtracted from the value function.

[5]Since the discovery of the substitute is governed by the Poisson process with arrival rate λ, the waiting time until the first arrival is an exponentially distributed random variable.

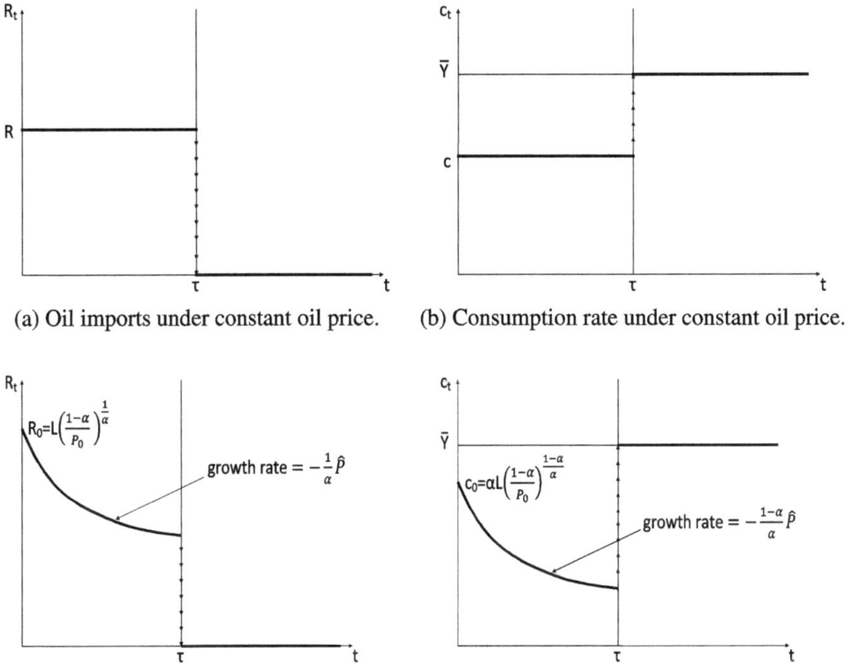

(a) Oil imports under constant oil price. (b) Consumption rate under constant oil price.

(c) Oil imports under increasing oil price. (d) Consumption rate under increasing oil price.

Fig. 1 Optimal consumption rate and oil imports

This implies that

$$\int_0^\tau u(c_t)e^{-\delta t}dt = u(c)\frac{1-e^{-\delta\tau}}{\delta}.$$

Proposition 1 *The optimal consumption rate of an oil importer is constant during the phase of substitute development and is a decreasing function of oil price. At the time of the invention, consumption jumps to a new, higher level and remains at this level forever.*

Note that if the oil price were to change over time, say increase, then the optimal consumption rate of an oil importer would decrease over time at the rate equal to $-\frac{1-\alpha}{\alpha}$ times the rate of increase of the oil price. The time paths of oil imports and the consumption rate are shown in Fig. 1. In Figs. 1a and 1b it is assumed that the oil price is constant. The oil imports are then constant over time until $t = \tau$, when they drop to zero since the substitute has become available. The consumption rate also remains constant until $t = \tau$ and jumps upward to $\bar{Y}$ at time τ. In Figs. 1c and 1d we assumed an increasing oil price over time. The oil imports are then decreasing at the rate $-\hat{P}/\alpha$ until $t = \tau$ and then drop to zero. The consumption rate also decreases but at a slower rate $-\frac{1-\alpha}{\alpha}\hat{P}$ and jumps upwards at $t = \tau$.

We turn next to the determination of the optimal size of the R&D project. It is such that

$$\frac{\partial}{\partial K}\left[\int_0^\infty \left(u(c)\frac{1-e^{-\delta\tau}}{\delta} + u(\bar{Y})\frac{e^{-\delta\tau}}{\delta}\right)\lambda(K)e^{-\lambda(K)\tau}d\tau\right] = u'(K).$$

Let us assume for simplicity that $\lambda(K) = \sigma K$, with $\sigma \in (0, 1)$ being the efficiency of R&D investment. The optimality condition with respect to K becomes

$$\int_0^\infty \left(u(c)\frac{1-e^{-\delta\tau}}{\delta} + u(\bar{Y})\frac{e^{-\delta\tau}}{\delta}\right)(\sigma - \tau\sigma K)e^{-\sigma K\tau}d\tau = u'(K).$$

Assuming log utility, i.e., $u(c) = \ln c$, it can be further simplified to yield a quadratic equation in K

$$\begin{aligned}
&\sigma\int_0^\infty \left(u(c)\frac{1-e^{-\delta\tau}}{\delta} + u(\bar{Y})\frac{e^{-\delta\tau}}{\delta}\right)e^{-\sigma K\tau}d\tau \\
&\quad - \sigma K\int_0^\infty \left(u(c)\frac{1-e^{-\delta\tau}}{\delta} + u(\bar{Y})\frac{e^{-\delta\tau}}{\delta}\right)\tau e^{-\sigma K\tau}d\tau = 1/K, \\
&\sigma u(c)\int_0^\infty \frac{e^{-\sigma K\tau} - e^{-(\sigma K+\delta)\tau}}{\delta}d\tau + \sigma u(\bar{Y})\int_0^\infty \frac{e^{-(\sigma K+\delta)\tau}}{\delta}d\tau \\
&\quad - \sigma u(c)K\int_0^\infty \tau\frac{e^{-\sigma K\tau} - e^{-(\sigma K+\delta)\tau}}{\delta}d\tau \\
&\quad - \sigma u(\bar{Y})K\int_0^\infty \tau\frac{e^{-(\sigma K+\delta)\tau}}{\delta}d\tau = 1/K, \\
&\frac{\sigma u(c)}{\delta}\left(\frac{1}{\sigma K} - \frac{1}{\delta+\sigma K}\right) + \frac{\sigma u(\bar{Y})}{\delta}\frac{1}{\delta+\sigma K} - \frac{\sigma u(c)K}{\delta}\left[\frac{1}{(\sigma K)^2} - \frac{1}{(\delta+\sigma K)^2}\right] \\
&\quad - \frac{\sigma u(\bar{Y})K}{\delta}\frac{1}{(\delta+\sigma K)^2} = 1/K.
\end{aligned}$$

Multiplying both sides by K and defining $\frac{K}{\delta+\sigma K} = x$, we get a quadratic equation in x

$$x^2\big(u(\bar{Y}) - u(c)\big) - x\big(u(\bar{Y}) - u(c)\big) + \frac{\delta\sigma + u(c)(1-\sigma)}{\sigma^2} = 0.$$

Or, setting $b \equiv u(\bar{Y}) - u(c) > 0$ and $d \equiv \frac{\delta\sigma+u(c)(1-\sigma)}{\sigma^2} > 0$, we have

$$bx^2 - bx + d = 0. \tag{3}$$

The roots are given by

$$x_{1,2} = \frac{b \pm \sqrt{b^2 - 4bd}}{2b}.$$

Three cases are then possible: (i) a unique root when $b^2 - 4bd = 0$; (ii) two real positive roots when $b^2 - 4bd > 0$; (iii) two complex roots when $b^2 - 4bd < 0$. Only the first two cases have economic sense and we shall therefore ignore the last one. The condition for real positive roots is $b - 4d \geq 0$ or, in terms of the relationship between the oil price, P, and the quantity of the substitute, B

$$B \geq e^{\omega} P^{-\nu}, \tag{4}$$

where $\omega \equiv \frac{4[(1-\sigma)\ln \alpha L + \delta\sigma] + [4(1-\sigma)+\sigma^2]\frac{1-\alpha}{\alpha}\ln(1-\alpha)}{\sigma^2(1-\alpha)} > 0$ and $\nu \equiv \frac{4(1-\sigma)+\sigma^2}{\alpha\sigma^2} > 0$.

Multiple solutions for the optimal R&D investment expenditure exist. Case (i): A unique root occurs only if (4) holds with equality. Then we have $x = 1/2$ and thus $K^* = \frac{\delta}{2-\sigma}$. The optimal R&D investment depends only on the discount rate δ and the efficiency of the R&D lab, σ. Case (ii): If (4) holds with strict inequality, then two roots exist and both of them are strictly positive: $x_1 = \frac{1+\sqrt{1-4d/b}}{2} > 0$ and $x_2 = \frac{1-\sqrt{1-4d/b}}{2} > 0$. Then $K_1^* = \frac{x_1\delta}{1+\sigma x_1}$ and $K_2^* = \frac{x_2\delta}{1+\sigma x_2} < K_1^*$.

The second-order condition for the optimum is such that the derivative of (3) with respect to K, evaluated at the extremum, must be negative. That is,

$$z \equiv (2xb - b)\frac{dx}{dK}\bigg|_{K=K^*} < 0.$$

The unique root $x = \frac{1}{2}$ clearly does not satisfy this condition since $dx/dK = 0$. Evaluating z at K_1^* yields

$$z|_{K=K_1^*} = b(2x_1 - 1)\frac{(1+\sigma x_1)^2}{\delta}.$$

The sign of this expression hinges on the sign of $2x_1 - 1$. This is clearly positive since $2x_1 - 1 = \sqrt{1 - 4d/b} > 0$ and thus K_1^* does not satisfy the second-order condition. Finally, evaluating z at K_2^* yields

$$z|_{K=K_2^*} = b(2x_2 - 1)\frac{(1+\sigma x_2)^2}{\delta}.$$

Again, the sign depends on $2x_2 - 1$, which is negative since $2x_2 - 1 = -\sqrt{1-4d/b} < 0$ and therefore K_2^* is the only root that satisfies the second-order condition.

Differentiating K_2^* with respect to P, we find that

$$\frac{\partial K_2^*}{\partial P} = -\frac{2du'(c)L(1-\alpha)^{1/\alpha}P^{-1/\alpha}}{b^2(1+\sigma x_2)^2\sqrt{1-4d/b}} < 0,$$

implying that an increase in the price of oil will lead to a smaller investment in the R&D activity. If an oil exporter has an objective of delaying the technological breakthrough, it would then restrain its exports thereby pushing the oil price up, so that RIC optimally chooses a smaller K.

The fact that the optimal investment in R&D decreases in oil price is somewhat counterintuitive. If the non-renewable resource becomes more and more expensive, an argument can be made that investment in a renewable substitute should increase, so that the arrival date of the substitute is shifted to the present. But, on the other hand, an increase in investment requires a larger sacrifice of current consumption and, moreover, with a higher resource price the budget constraint of the economy becomes tighter (since resource imports are now more expensive). Therefore, the sacrifice of current consumption is more difficult to implement. Overall, there are two conflicting forces at work. On the one hand, the urge to develop a backstop calling for a higher investment rate and, on the other hand, a tighter budget constraint calling for a lower investment rate. Under specific conditions one or the other effect dominates. In particular, the elasticity of intertemporal consumption substitution (EICS) plays a crucial role as it determines how willing the economy is to shift its consumption intertemporally. When EICS is relatively low, the economy cares relatively more about the time profile of consumption as opposed to the total discounted consumption. This is the case in our setting since we have assumed a logarithmic utility function, so that EICS equals unity.[6]

3 Oil-Exporting Country

The resource-exporting country (REC for short) believes that a substitute arrives at the constant Poisson rate λ, which it takes to be exogenous. It also knows that once the substitute is developed, the demand for oil drops to zero. Thus, REC's time horizon may be split in two phases: One where the substitute has not yet been invented, and the other where the substitute is online and oil exports are zero. REC's objective is to maximize its lifetime welfare with respect to consumption in phase 1, c_t^*, consumption in phase 2, c_t^{*b}, the extraction rate in both phases, R_t^*, and R_t^{*b} and the exports of oil in phase 1, R_t. The objective function can be written as[7]

$$\max_{c_t^*, c_t^{*b}, R_t^*, R_t^{*b}, R_t} \int_0^\infty \left(\int_0^\tau u\left(c_t^*\right) e^{-\delta t} dt + \int_\tau^\infty u\left(c_t^{*b}\right) e^{-\delta t} dt \right) f_\tau d\tau$$

subject to

$$c_t^* = Y\left(R_t^* - R_t, L^*\right) + P(R_t) R_t, \quad t \in [0, \tau], \tag{5}$$

$$c_t^{*b} = Y\left(R_t^{*b}, L^*\right), \quad t > \tau, \tag{6}$$

[6]More details about the relationship between oil price and investment in renewables R&D are provided in Vinogradova (2012). It is shown, in particular, that in the empirically relevant range of EICS the optimal response of the investment rate to an increase in oil price is negative, even when the economy can borrow in the international capital market at a constant interest rate.

[7]An alternative way of solving REC's optimization problem is to use the method presented in Boukas et al. (1990).

$$\begin{aligned} \dot{S}_t &= -R_t^*, \quad \text{if substitute is not online,} \quad S_0 \text{ given,} \\ \dot{S}_t &= -R_t^{*b}, \quad \text{if substitute is online,} \end{aligned} \tag{7}$$

$$P_t = (1-\alpha)L^{\alpha}R_t^{-\alpha}, \quad t \in [0, \tau], \tag{8}$$

where S_t denotes the oil stock at time t and $\dot{S}_t \equiv dS_t/dt$. (5) states that consumption in phase 1 is equal to total output plus the proceeds from oil sales. The output in phase 1 is produced with constant labor, L^*, and a quantity of oil, which is equal to total extraction, R_t^*, minus exports, R_t. (6) states that there are no oil exports in phase 2 and thus consumption is just equal to total output. Note that REC takes into account the demand for oil from RIC, (8). The solution to this program is found by splitting it into two subprograms, namely the one pertaining to phase 1 and the other pertaining to phase 2. The subprogram of phase 2 is standard since it does not involve any uncertainty: once the substitute has arrived, the demand for oil from RIC drops to zero, so that the oil extracted in a given period is used entirely domestically.

Phase 2 REC's optimization program in the second phase reads

$$\max_{c_t^{*b}, R_t^{*b}} \int_{\tau}^{\infty} u\left(c_t^{*b}\right)e^{-\delta t}dt$$

subject to (6), (7) and the initial resource stock S_τ. We shall assume that RIC and REC share the same production technology and differ only with respect to their labor endowment and there is no population growth, so that $Y(R_t^{*b}, L^*) = L^{*\alpha}R_t^{*b1-\alpha}$. The current-value Hamiltonian may be written as

$$H = u\left(c_t^{*b}\right) + \nu\left(Y\left(R_t^{*b}, L^*\right) - c_t^{*b}\right) - \mu R_t^{*b}$$

and the first-order conditions

$$\begin{aligned} c_t^{*b}&: \quad u'\left(c_t^{*b}\right) - \nu = 0, \\ R_t^{*b}&: \quad \nu\left(\partial Y/\partial R_t^{*b}\right) - \mu = 0, \\ S&: \quad 0 = \delta\mu - \dot{\mu}. \end{aligned}$$

Combining these three conditions and recalling our assumption of logarithmic utility function, we get $\delta = -\hat{c}_t^{*b} - \alpha\hat{R}_t^{*b}$. By (6), $\hat{c}_t^{*b} = (1-\alpha)\hat{R}_t^{*b}$ and thus $\hat{R}_t^{*b} = -\delta$, implying that $R_t^{*b} = R_\tau^{*b}e^{-\delta(t-\tau)}$. Inserting this extraction path in (7) and integrating yields $R_\tau^{*b} = \delta S_\tau$. The optimal consumption path is then equal to

$$c_t^{*b} = L^{*\alpha}(\delta S_\tau)^{1-\alpha}e^{-\delta(1-\alpha)(t-\tau)}$$

with the initial consumption rate in Phase 2 being $c_\tau^* = L^{*\alpha}(\delta S_\tau)^{1-\alpha}$.

Phase 1 We now turn to phase 1 and write the Hamilton-Jacobi-Bellman equation for REC's optimization under uncertainty

$$\max_{R_t^*, R_t} \left\{ u\big(Y\big(R_t^* - R_t, L^*\big) + P(R_t)R_t\big) - R_t^* \frac{\partial V_t}{\partial S_t} \right\} + \lambda\big(V_t^b - V_t\big) - \delta V_t = 0,$$

where we inserted (5) directly in the utility function, V_t is the value function in phase 1 (while the substitute has not yet arrived) and V_t^b is the value function in phase 2 (after the substitute has arrived). The first-order conditions are given by

$$R_t^*\colon \quad u'\big(c_t^*\big) \frac{\partial Y}{\partial (R_t^* - R_t)} - \frac{\partial V_t}{\partial S_t} = 0, \tag{9}$$

$$R_t\colon \quad u'\big(c_t^*\big)\left[-\frac{\partial Y}{\partial (R_t^* - R_t)} + R_t \frac{\partial P(R_t)}{\partial R_t} + P(R_t) \right] = 0, \tag{10}$$

$$S\colon \quad -R_t^* \frac{\partial^2 V_t}{\partial S_t^2} + \lambda\left(\frac{\partial V_t^b}{\partial S_t} - \frac{\partial V_t}{\partial S_t} \right) - \delta \frac{\partial V_t}{\partial S_t} = 0. \tag{11}$$

Equation (10) determines the optimal split of per period oil production between exports and domestic use. It equates the marginal revenue from oil sales to the marginal productivity of oil

$$R_t \frac{\partial P(R_t)}{\partial R_t} + P(R_t) = \frac{\partial Y}{\partial (R_t^* - R_t)}.$$

By (8), we obtain

$$(1-\alpha)^2 L^\alpha R^{-\alpha} = (1-\alpha) L^{*\alpha} \big(R^* - R\big)^{-\alpha}$$

and thus oil exports represent a constant fraction of per period oil extraction

$$R = R^* \left[1 + (1-\alpha)^{-1/\alpha} \frac{L^*}{L} \right]^{-1} = R^*(1-\gamma), \tag{12}$$

where we defined $1-\gamma \equiv [1+(1-\alpha)^{-1/\alpha}\frac{L^*}{L}]^{-1}$, so that $\gamma \equiv \frac{(1-\alpha)^{-1/\alpha}\frac{L^*}{L}}{1+(1-\alpha)^{-1/\alpha}\frac{L^*}{L}}$ is the fraction of per period extraction used domestically. Note that the share of oil exports is a positive function of the importer's market size, L.

Equations (9) and (11) can be used to obtain the Keynes-Ramsey rule under uncertainty. First divide (11) throughout by $\frac{\partial V_t}{\partial S_t}$ to get

$$-R_t^* \frac{\partial^2 V_t / \partial S_t^2}{\partial V_t / \partial S_t} + \lambda\left(\frac{\partial V_t^b / \partial S_t}{\partial V_t / \partial S_t} - 1 \right) - \delta = 0. \tag{13}$$

From (9), the numerator of the first term in (13) can be calculated as

$$-R_t^* \frac{\partial^2 V_t}{\partial S_t^2} = \frac{d[u'(c_t^*)\frac{\partial Y}{\partial (R_t^* - R_t)}]}{dt} = u''\big(c_t^*\big)\dot{c}_t^* \frac{\partial Y}{\partial (R_t^* - R_t)} + u'\big(c_t^*\big) \frac{d}{dt}\left[\frac{\partial Y}{\partial (R_t^* - R_t)} \right]$$

$$\begin{aligned}
&= u''\big(c_t^*\big)\dot{c}_t^*(1-\alpha)L^{*\alpha}\big(R^*-R\big)^{-\alpha} \\
&\quad - \alpha u'\big(c_t^*\big)(1-\alpha)L^{*\alpha}\big(R^*-R\big)^{-\alpha-1}\big(\dot{R}^*-\dot{R}\big) \\
&= (1-\alpha)L^{*\alpha}\big(\gamma R^*\big)^{-\alpha}u'\big(c_t^*\big)\left(\frac{u''(c_t^*)\dot{c}_t^*}{u'(c_t^*)} - \alpha\hat{R}^*\right),
\end{aligned}$$

where we made use of (12). This expression can be further simplified by noting that with logarithmic utility we have

$$-\frac{u''(c_t^*)\dot{c}_t^*}{u'(c_t^*)} = \hat{c}_t^*.$$

From the budget constraint (5)

$$\begin{aligned}
c_t^* &= L^{*\alpha}\big(\gamma R^*\big)^{1-\alpha} + (1-\alpha)L^{\alpha}\big[(1-\gamma)R^*\big]^{1-\alpha} \\
&= R^{*1-\alpha}\big[L^{*\alpha}\gamma^{1-\alpha} + (1-\alpha)L^{\alpha}(1-\gamma)^{1-\alpha}\big],
\end{aligned}$$

which implies that $\hat{c}_t^* = (1-\alpha)\hat{R}^*$. Then

$$\begin{aligned}
-R_t^*\frac{\partial^2 V_t}{\partial S_t^2} &= (1-\alpha)L^{*\alpha}\big(\gamma R^*\big)^{-\alpha}u'\big(c_t^*\big)\big[-(1-\alpha)\hat{R}^* - \alpha\hat{R}^*\big] \\
&= -(1-\alpha)L^{*\alpha}\big(\gamma R^*\big)^{-\alpha}u'\big(c_t^*\big)\hat{R}^*.
\end{aligned}$$

Inserting this in (13) yields

$$\begin{aligned}
&-\frac{(1-\alpha)L^{*\alpha}(\gamma R^*)^{-\alpha}u'(c^*)\hat{R}^*}{u'(c^*)(1-\alpha)L^{*\alpha}(\gamma R^*)^{-\alpha}} + \lambda\left[\frac{u'(c^{*b})(1-\alpha)L^{*\alpha}R^{*b-\alpha}}{u'(c^*)(1-\alpha)L^{*\alpha}(\gamma R^*)^{-\alpha}} - 1\right] - \delta = 0 \\
&-\hat{R}^* + \lambda\left[\frac{(c^{*b})^{-1}R^{*b-\alpha}}{(c^*)^{-1}(\gamma R^*)^{-\alpha}} - 1\right] - \delta = 0 \\
&-\hat{R}^* + \lambda\left\{\frac{R^{*1-\alpha}[L^{*\alpha}\gamma^{1-\alpha} + (1-\alpha)L^{\alpha}(1-\gamma)^{1-\alpha}](\gamma R^*)^{\alpha}}{L^{*\alpha}(\delta S_\tau)^{1-\alpha}(\delta S_\tau)^{\alpha}} - 1\right\} - \delta = 0 \\
&-\hat{R}^* + \lambda\left\{R^*\left[\gamma^{1-\alpha} + (1-\alpha)(1-\gamma)^{1-\alpha}\left(\frac{L}{L^*}\right)^{\alpha}\right]\gamma^{\alpha}(\delta S_\tau)^{-1} - 1\right\} - \delta = 0, \\
&\hat{R}^* = \lambda R^*(\delta S_\tau)^{-1} - (\lambda+\delta),
\end{aligned}$$

where $[\gamma^{1-\alpha} + (1-\alpha)(1-\gamma)^{1-\alpha}(\frac{L}{L^*})^{\alpha}]\gamma^{\alpha} = 1$, given the definition of γ. The solution to this non-linear differential equation is given by

$$R_t^* = \frac{(\lambda+\delta)\delta S_\tau e^{-(\lambda+\delta)t}}{\lambda e^{-(\lambda+\delta)t} + C_1(\lambda+\delta)\delta S_\tau},$$

where C_1 is a constant of integration. Evaluating the expression at $t = 0$, we get $C_1 = \frac{(\lambda+\delta)\delta S_\tau - \lambda R_0^*}{(\lambda+\delta)\delta S_\tau R_0^*}$. The complete solution is then

$$R_t^* = \frac{R_0^*(\lambda+\delta)\delta S_\tau}{\lambda R_0^* + [(\lambda+\delta)\delta S_\tau - \lambda R_0^*]e^{(\lambda+\delta)t}}. \tag{14}$$

This solution is then used in Eq. (5) describing the evolution of the resource stock to solve for the initial extraction rate in Phase 1, R_0^*, and the remaining resource stock at the time of the switch, S_τ:

$$\begin{aligned} S_0 &= \int_0^\infty R_t^* dt = \int_0^\infty \frac{R_0^*(\lambda+\delta)\delta S_\tau}{\lambda R_0^* + [(\lambda+\delta)\delta S_\tau - \lambda R_0^*]e^{(\lambda+\delta)t}} dt \\ &= \frac{\delta S_\tau}{\lambda}\{(\lambda+\delta)t - \ln\left([(\lambda+\delta)\delta S_\tau - \lambda R_0^*]e^{(\lambda+\delta)t} + \lambda R_0^*\right)\}\big|_0^\infty \\ &= \frac{\delta S_\tau}{\lambda}\ln\left(\frac{(\lambda+\delta)\delta S_\tau}{(\lambda+\delta)\delta S_\tau - \lambda R_0^*}\right), \end{aligned} \tag{15}$$

$$\begin{aligned} S_\tau &= S_0 - \int_0^\tau R_t^* dt = S_0 - \int_0^\tau \frac{R_0^*(\lambda+\delta)\delta S_\tau}{\lambda R_0^* + [(\lambda+\delta)\delta S_\tau - \lambda R_0^*]e^{(\lambda+\delta)t}} dt \\ &= S_0 - \frac{\delta S_\tau}{\lambda}\{(\lambda+\delta)t - \ln\left([(\lambda+\delta)\delta S_\tau - \lambda R_0^*]e^{(\lambda+\delta)t} + \lambda R_0^*\right)\}\big|_0^\tau \\ &= S_0 - \frac{\delta S_\tau}{\lambda}\left\{(\lambda+\delta)\tau + \ln\left(\frac{(\lambda+\delta)\delta S_\tau}{[(\lambda+\delta)\delta S_\tau - \lambda R_0^*]e^{(\lambda+\delta)\tau} + \lambda R_0^*}\right)\right\}. \end{aligned} \tag{16}$$

Combining (15) and (16), and after some rearrangements, we finally obtain the solution for the initial extraction rate in terms of the exogenous parameters of the model

$$R_0^* = \frac{(\lambda+\delta)S_0[1 - e^{-(\lambda+\delta)\tau}(e^{\lambda/\delta} - 1)^{-1}]}{(\lambda+\delta)\tau + \ln(e^{\lambda/\delta} - 1)}$$

and the remaining resource stock at time τ

$$S_\tau = \frac{\lambda S_0}{\delta[(\lambda+\delta)\tau + \ln(e^{\lambda/\delta} - 1)]}.$$

It can then be clearly established that the extraction rate declines over time with the rate of decline being

$$\begin{aligned} \hat{R}_t^* &= \lambda R_t^*(\delta S_\tau)^{-1} - (\lambda+\delta) = \frac{\lambda R_0^*(\lambda+\delta)}{[(\lambda+\delta)\delta S_\tau - \lambda R_0^*]e^{(\lambda+\delta)t} + \lambda R_0^*} - (\lambda+\delta) \\ &= \frac{\lambda+\delta}{[\frac{(\lambda+\delta)\delta S_\tau}{\lambda R_0^*} - 1]e^{(\lambda+\delta)t} + 1} - (\lambda+\delta) \end{aligned}$$

$$= (\lambda+\delta)\left\{\frac{1}{1+e^{(\lambda+\delta)t}[\frac{e^{\lambda/\delta}-1}{e^{\lambda/\delta}-1-e^{-(\lambda+\delta)\tau}}-1]}-1\right\}$$

$$= -\frac{(\lambda+\delta)e^{(\lambda+\delta)(t-\tau)}}{e^{\lambda/\delta}-1+e^{-(\lambda+\delta)\tau}[e^{(\lambda+\delta)t}-1]} < 0. \tag{17}$$

Consequently the optimal consumption growth rate is also negative:

$$\hat{c}^* = (1-\alpha)\hat{R}^* = -\frac{(1-\alpha)(\lambda+\delta)e^{(\lambda+\delta)(t-\tau)}}{e^{\lambda/\delta}-1+e^{-(\lambda+\delta)\tau}[e^{(\lambda+\delta)t}-1]} < 0.$$

Proposition 2 *At the time of the technological breakthrough, extraction rate may jump either up or down, depending primarily on the relationship between the arrival rate, λ, and the time preference rate, δ.*

This can be seen by comparing the extraction rate the moment just before the invention, R_τ^*, with the extraction rate just after the invention, R_τ^{*b},

$$\begin{aligned} R_\tau^* &\gtrless R_\tau^{*b} \\ \frac{(\lambda+\delta)S_0[e^{\lambda/\delta}-e^{-(\lambda+\delta)\tau}-1]}{[(\lambda+\delta)\tau+\ln(e^{\lambda/\delta}-1)][e^{\lambda/\delta}-e^{-(\lambda+\delta)\tau}]} &\gtrless \frac{\delta\lambda S_0}{\delta[(\lambda+\delta)\tau+\ln(e^{\lambda/\delta}-1)]} \\ \frac{(\lambda+\delta)[e^{\lambda/\delta}-e^{-(\lambda+\delta)\tau}-1]}{e^{\lambda/\delta}-e^{-(\lambda+\delta)\tau}} &\gtrless \lambda \\ (\lambda+\delta)\left[e^{\lambda/\delta}-e^{-(\lambda+\delta)\tau}-1\right] &\gtrless \lambda\left[e^{\lambda/\delta}-e^{-(\lambda+\delta)\tau}\right] \\ \delta\left[e^{\lambda/\delta}-e^{-(\lambda+\delta)\tau}-1\right]-\lambda &\gtrless 0 \\ e^{\lambda/\delta}-e^{-(\lambda+\delta)\tau}-1 &\gtrless \frac{\lambda}{\delta}. \end{aligned} \tag{18}$$

If the chances of the technological breakthrough are slim, say $\lambda \to 0$, then the expression on the left-hand side is negative, equal to $-e^{-\delta\tau}$, while on the right-hand side we have zero. Then $R_\tau^* < R_\tau^{*b}$ and the extraction rate jumps upwards at the time of the invention. The intuition behind the optimal upward jump is the following. When REC believes that the chances of a substitute discovery are low, it optimally spreads the exports over a longer time horizon, and thus "underextracts" at each point in time. When the substitute suddenly arrives, REC finds itself with a resource stock higher than what is optimal and therefore instantaneously adjusts the extraction rate upwards. This is shown in Fig. 2a.

Proposition 3 *The extraction rate in phase 1 may fall faster or slower than in phase 2, depending on the relationship between the arrival rate, λ, and the time preference rate, δ.*

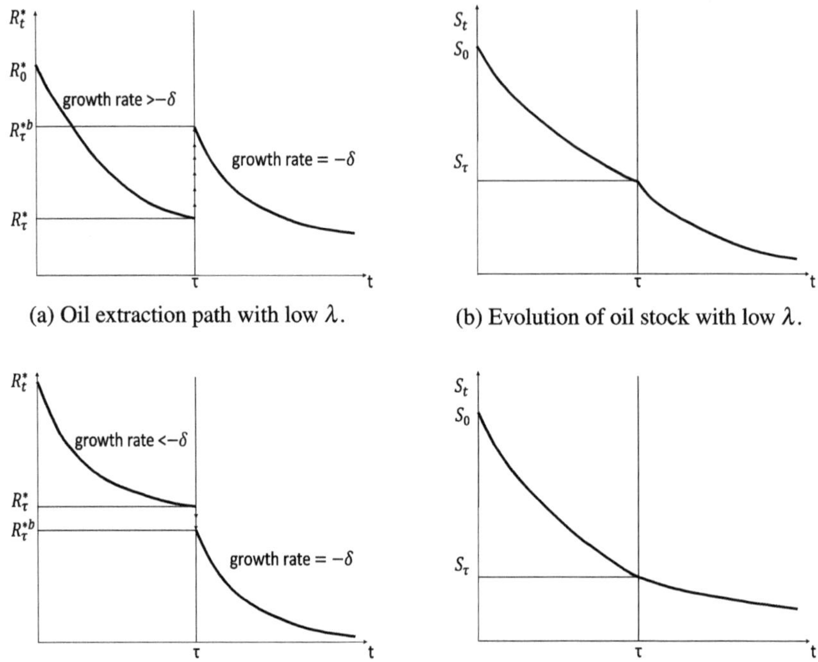

(a) Oil extraction path with low λ. (b) Evolution of oil stock with low λ.

(c) Oil extraction path with high λ. (d) Evolution of oil stock with high λ.

Fig. 2 Oil stock and extraction paths

Recall that $\hat{R}_t^{*b} = -\delta$ while $\hat{R}_t^*$ is given by (17) and compare their absolute values

$$
\begin{aligned}
|\hat{R}_t^*| &\gtrless \delta \\
\frac{(\lambda+\delta)e^{(\lambda+\delta)(t-\tau)}}{e^{\lambda/\delta} - 1 + e^{-(\lambda+\delta)\tau}[e^{(\lambda+\delta)t} - 1]} &\gtrless \delta \\
\left(1 + \frac{\lambda}{\delta}\right)e^{(\lambda+\delta)(t-\tau)} &\gtrless e^{\lambda/\delta} - 1 + e^{-(\lambda+\delta)\tau}\left[e^{(\lambda+\delta)t} - 1\right] \\
\frac{\lambda}{\delta}e^{(\lambda+\delta)(t-\tau)} &\gtrless e^{\lambda/\delta} - 1 - e^{-(\lambda+\delta)\tau}.
\end{aligned}
\tag{19}
$$

The expression on the RHS of (19) is identical to the expression on the LHS of (18). It follows that if $e^{\lambda/\delta} - e^{-(\lambda+\delta)\tau} - 1 \geq \frac{\lambda}{\delta}$ in (18), then $\frac{\lambda}{\delta}e^{(\lambda+\delta)(t-\tau)} < e^{\lambda/\delta} - 1 - e^{-(\lambda+\delta)\tau}$ in (19) because it is relevant only for $t < \tau$, i.e., phase 1. The extraction path then must be as depicted in Fig. 2a. That is, during phase 1 it declines at the rate $|\hat{R}_t^*| < \delta$, at time τ it jumps upwards (a possibility of no jumps at all also exists if (18) is satisfied with strict equality), and afterwards it declines at the rate δ until the end of the planning horizon. It is also easy to show that $R_\tau^{*b} < R_0^*$, i.e., the extraction rate cannot jump above the initial rate in phase 1.

If we now turn to the possibility that the chances of a breakthrough are relatively high, say $\lambda \to \delta$ and $\tau \to 1/\lambda$, then the expression on the LHS of (18) is greater than on the RHS, implying that a downward jump may occur: $R_\tau^* > R_\tau^{*b}$. In this case it is also likely that the absolute value of the rate of decline in the extraction rate is above δ (or, algebraically, the growth rate is smaller than $-\delta$), such as depicted in Fig. 2c. The evolution of the oil stock over time in the case of low and high λ is shown in Figs. 2b and 2d, respectively. If $|\hat{R}_t^*| < \delta$ ($|\hat{R}_t^*| > \delta$), oil stock declines at a slower (faster) rate in phase 1 than it does in phase 2. If there is a jump in the extraction rate at the time of invention, the time path of the oil stock has a kink at $t = \tau$. In Fig. 2b, the slope of the schedule to the left of τ is smaller than the slope to the right of τ. In Fig. 2d the opposite is true.

We examine next the following question: How does the optimal behavior of an oil exporter facing uncertainty compare to his optimal behavior in the case of certainty, i.e., when the date of substitute arrival is known with certainty. Let us assume that the substitute arrives on date τ. Then REC's optimization problem reads

$$\max_{c_t^*, c_t^{*b}, R_t^*} \int_0^\tau u\left(c_t^*\right)e^{-\delta t}dt + \int_\tau^\infty u\left(c_t^{*b}\right)e^{-\delta t}dt$$

subject to

$$\begin{aligned} c_t^* &= Y\left(R_t^* - R_t, L^*\right) + P(R_t)R_t, \quad t \in [0, \tau], \\ c_t^{*b} &= Y\left(R_t^{*b}, L^*\right), \quad t > \tau, \\ \dot{S}_t &= -R_t^*, \quad S_0 \text{ given}, \\ P_t &= (1-\alpha)L^\alpha R_t^{-\alpha}, \quad t \in [0, \tau]. \end{aligned} \tag{20}$$

This problem is solved using the standard dynamic optimization technique. The optimal consumption and extraction paths in phase 2 are described by exactly the same equations as in the case of uncertainty, namely $\hat{R}_t^* = -\delta$, $\hat{c}_t^{*b} = -(1-\alpha)\delta$ for $t > \tau$. The optimal paths in the first phase are, however, different from those under uncertainty. We write the current-value Hamiltonian

$$H = u\left(Y\left(R_t^* - R_t, L^*\right) + P(R_t)R_t\right) - \mu_t R_t^*$$

and the first-order conditions

$$R_t^*: \quad u'\left(c_t^*\right)\frac{\partial Y}{\partial(R_t^* - R_t)} - \mu_t = 0, \tag{21}$$

$$R_t: \quad u'\left(c_t^*\right)\left[-\frac{\partial Y}{\partial(R_t^* - R_t)} + R_t\frac{\partial P(R_t)}{\partial R_t} + P(R_t)\right] = 0, \tag{22}$$

$$S: \quad \delta\mu_t - \dot{\mu}_t = 0. \tag{23}$$

Note that condition (22) is identical to (10), implying that the optimal split between oil exports and domestic use is the same as under uncertainty, $R_t = (1-\gamma)R_t^*$.

Using this in the budget constraint (20) and applying the hat calculus, yields $\hat{c}_t^* = (1-\alpha)R_t^*$. Then, noting that with log utility $\frac{du'(c_t^*)/dt}{u'(c_t^*)} = -\hat{c}_t^*$ and combining (21) with (23), we finally obtain $-\hat{R}_t^* = \delta$. The extraction path is then

$$R_t^* = R_0^* e^{-\delta t}. \tag{24}$$

Contrary to the case of uncertainty, the extraction rate under certainty declines over time at the rate δ *before and after* the invention of the substitute. Recall (14) and set $\lambda = 0$, then we obtain precisely (24). Does the presence of uncertainty imply a faster or slower resource depletion? In other words, how does the path in (14) compare to the path in (24)? We already have the answer to this question in (19). Depending on how likely it is that the substitute is going to be invented, extraction path under uncertainty may happen to be steeper or flatter than in the certainty case. If REC believes that the chances are slim, he will choose a relatively conservationist path, i.e., slower extraction. If it believes that the chances are fairly high, it will choose a relatively fast extraction profile.

If we depart from our assumption of costless extraction and assume instead that oil extraction costs are larger than the backstop price, REC may also want to switch from oil to the backstop. Whether it will do so or not will depend on the structure of the market for the substitute. If the backstop is specific to RIC's geographic location, then there is no scope for sharing it with REC. If, however, the backstop can be easily spread, then it is most likely that RIC will patent the invention. Then, being the monopolist on the substitute market, RIC will try to extract all the rents from the buyer. It will therefore set the price just slightly below the oil extraction cost in order to induce REC to shift from oil to the renewable energy source. Presumably, some of the oil stock will remain unextracted. But then REC's optimization in the first phase must take into account the fact that it may not be optimal to leave oil stock unexploited. On the one hand, leaving some oil in the ground means losing the revenue. On the other hand, extracting everything by the time the substitute arrives means selling more at each point in time and therefore exerting a downward pressure on the price. Whether it is optimal to leave some of the oil stock unexploited or increase current exports depends, of course, on the elasticity of demand for oil. Given the demand function of RIC, this elasticity is simply equal to $-\alpha$, smaller than unity in absolute value. Thus, an increase in supply has a positive effect on total revenue and it is then optimal to increase current exports and aim at extracting everything by the time the backstop arrives.

4 Conclusion

This chapter provides an analysis of the optimal behavior of an oil importer and an oil exporter under uncertainty. We assume that the importer engages in an R&D activity aimed at developing a renewable perfect substitute for the non-renewable fossil resource. The invention is intrinsically uncertain and governed by the Poisson process with the arrival rate being an increasing function of the expenditure devoted to R&D. We cast the importer's and the exporter's problem in the form of the dynamic stochastic optimization in continuous time.

With regard to the oil importer, we show that the invention of the substitute entails a discontinuous upward jump in the consumption rate to a higher level (determined by the quantity of the substitute). The optimal investment expenditure in R&D features multiple solutions, depending on the parameters of the model. It is feasible that the optimal investment is uniquely determined and depends only on the country's rate of time preference and the efficiency of the R&D activity but is completely independent of other parameters, such as, for example, oil price. In this case, there is no scope for the oil exporting country to affect the importer's investment decision. It is also feasible that, under certain conditions, the optimal investment is not uniquely determined, i.e., the solution involves two roots. They do depend on the price of oil, the quantity of the substitute and the structure of the production technology. In this case the oil exporter may reduce the chances of the technological breakthrough by reducing its exports and thus forcing the importer to cut the R&D expenditure.

With regard to the oil extraction, we show that after the arrival of the substitute the total per period extraction falls over time at the rate of time preference. However, before the substitute is available, the extraction declines over time at the rate which may be either smaller or greater than the time preference rate, depending on whether the chances of the technological breakthrough are relatively small or large, respectively. One cannot therefore unambiguously conclude that uncertainty about an eventual discovery of an oil substitute contributes to a faster or slower resource depletion.

In our present model we refrained from introducing an accumulative factor of production, such as physical capital, in order to keep the analysis tractable. However, a possibility of investing into capital accumulation and building a stock of a productive factor which can to some extent substitute for other inputs, opens up a wider range of options for both the oil importer and the exporter. For example, the importer will face a tradeoff between investing into physical capital or investing into R&D, the outcome of which will depend on the interplay between the productivity of capital and efficiency of the research lab. The oil exporter will have to optimally choose how much of revenue from oil sales to consume and how much to invest in capital accumulation. We leave these questions on the agenda for future research.

Appendix

Let us now consider a more general case of CRRA utility, $u(c) = \frac{c^{1-\theta}-1}{1-\theta}$, the limit of which is $\ln(c)$ when θ goes to unity. Parameter θ is the inverse of the elasticity of intertemporal consumption substitution. The new specification for the utility function will affect the derivation of the Keynes-Ramsey rule. Combining (9) and (11), we now obtain:

$$-\frac{(1-\alpha)L^{*\alpha}(\gamma R^*)^{-\alpha}u'(c^*)\hat{R}^*[\alpha+\theta(1-\alpha)]}{u'(c^*)(1-\alpha)L^{*\alpha}(\gamma R^*)^{-\alpha}}$$
$$+\lambda\left[\frac{u'(c^{*b})(1-\alpha)L^{*\alpha}R^{*b-\alpha}}{u'(c^*)(1-\alpha)L^{*\alpha}(\gamma R^*)^{-\alpha}}-1\right]-\delta=0,$$

$$
\begin{aligned}
&-\hat{R}^*[\alpha+\theta(1-\alpha)]+\lambda\left[\gamma^\alpha\left(\frac{c^*}{c^{*b}}\right)^\theta\left(\frac{R^*}{R^{*b}}\right)^\alpha-1\right]-\delta=0,\\
&-\hat{R}^*[\alpha+\theta(1-\alpha)]\\
&\qquad+\lambda\left\{\gamma^\alpha\left\{\frac{R^{*1-\alpha}[L^{*\alpha}\gamma^{1-\alpha}+(1-\alpha)L^\alpha(1-\gamma)^{1-\alpha}]}{L^{*\alpha}R^{*b1-\alpha}}\right\}^\theta\left(\frac{R^*}{R^{*b}}\right)^\alpha-1\right\}-\delta\\
&\quad=0,\\
&-\hat{R}^*[\alpha+\theta(1-\alpha)]\\
&\qquad+\lambda\left\{\gamma^\alpha\left[\gamma^{1-\alpha}+(1-\alpha)\left(\frac{L}{L^*}\right)^\alpha(1-\gamma)^{1-\alpha}\right]^\theta\left(\frac{R^*}{\delta S_\tau}\right)^{\alpha+\theta(1-\alpha)}-1\right\}-\delta\\
&\quad=0,\\
&\hat{R}^*[\alpha+\theta(1-\alpha)]\\
&\quad=\lambda\gamma^\alpha\left[\gamma^{1-\alpha}+(1-\alpha)\left(\frac{L}{L^*}\right)^\alpha(1-\gamma)^{1-\alpha}\right]^\theta\left(\frac{R^*}{\delta S_\tau}\right)^{\alpha+\theta(1-\alpha)}-(\lambda+\delta),\\
&\hat{R}^*\eta=R^{*\eta}\varepsilon-(\lambda+\delta),
\end{aligned}
$$

where $\eta\equiv\alpha+\theta(1-\alpha)$ and $\varepsilon\equiv\lambda\gamma^\alpha[\gamma^{1-\alpha}+(1-\alpha)(\frac{L}{L^*})^\alpha(1-\gamma)^{1-\alpha}]^\theta(\delta S_\tau)^{-\eta}=\lambda\gamma^{\alpha(1-\theta)}(\delta S_\tau)^{-\eta}$. The solution to this differential equation is given by:

$$
R_t^*=\left[\frac{\varepsilon e^{-(\delta+\lambda)t}+C_1(\lambda+\delta)}{\lambda+\delta}\right]^{-1/\eta}e^{-\frac{\lambda+\delta}{\eta}t},
$$

where C_1 is a constant of integration which can be found by evaluating the above expression at $t=0$: $C_1=R_0^{*-\eta}-\frac{\varepsilon}{\lambda+\delta}$. The complete solution for the extraction path is then:

$$
R_t^*=\left\{R_0^{*-\eta}e^{(\lambda+\delta)t}-\frac{\varepsilon[e^{(\lambda+\delta)t}-1]}{\lambda+\delta}\right\}^{-1/\eta}.
$$

We can check that when $\theta=1$, as in our baseline case, then $\eta=1$ and we are back to our solution in (14).

References

Boukas, E. K., Haurie, A., & Michel, P. (1990). An optimal control problem with a random stopping time. *Journal of Optimization Theory and Applications*, *64*, 471–480.

Dasgupta, P., Gilbert, R., & Stiglitz, J. (1983). Strategic considerations in invention and innovation: the case of natural resources. *Econometrica*, *51*, 1439–1448.

Gallini, N., Lewis, T., & Ware, R. (1983). Strategic timing and pricing of a substitute in a cartelized resource market. *Canadian Journal of Economics*, *16*, 429–446.

Gerlagh, R., & Liski, M. (2011). Strategic resource dependence. *Journal of Economic Theory*, *146*, 699–727.
Harris, C., & Vickers, J. (1995). Innovation and natural resources: a dynamic game with uncertainty. *The Rand Journal of Economics*, *26*, 418–430.
Heal, G. (1976). The relationship between price and extraction cost for a resource with a backstop technology. *Bell Journal of Economics*, *7*, 371–378.
Hoel, M. (1978). Resource extraction, substitute production, and monopoly. *Journal of Economic Theory*, *19*, 28–37.
Hung, N. M., & Quyen, N. V. (1993). On R&D timing under uncertainty. *Journal of Economic Dynamics & Control*, *17*, 971–991.
Kamien, M. I., & Schwartz, N. L. (1978). Optimal exhaustible resource depletion with endogenous technical change. *Review of Economic Studies*, *45*, 179–196.
Olsen, T. E. (1993). Perfect equilibrium timing of a backstop technology. *Journal of Economic Dynamics & Control*, *17*, 123–151.
Tsur, Y., & Zemel, A. (2003). Optimal transition to backstop substitutes for non-renewable resources. *Journal of Economic Dynamics & Control*, *27*, 551–572.
Tsur, Y., & Zemel, A. (2005). Scarcity, growth and R&D. *Journal of Environmental Economics and Management*, *49*, 484–499.
Valente, S. (2011). Endogenous growth, backstop technology, and optimal jumps. *Macroeconomic Dynamics*, *15*, 293–325.
Van der Ploeg, F., & Withagen, C. A. (2012). Is there really a green paradox? *Journal of Environmental Economics and Management*, *64*, 342–363.
Vinogradova, A. (2012). Investment in an uncertain backstop: optimal strategy for an open economy. Mimeo, ETH Zurich.

Robust Control of a Spatially Distributed Commercial Fishery

William A. Brock, Anastasios Xepapadeas, and Athanasios N. Yannacopoulos

Abstract We consider a robust control model for a spatially distributed commercial fishery under uncertainty, and in particular a tracking problem, i.e. the problem of robust stabilization of a chosen deterministic benchmark state in the presence of model uncertainty. The problem is expressed in the form of a stochastic linear quadratic robust optimal control problem, which is solved analytically. We focus on the emergence of breakdown from the robust stabilization policy, called hot spots, and comment upon their significance concerning the spatiotemporal behaviour of the system.

1 Introduction

An important issue in understanding ecosystems and designing efficient management rules with the purpose of preventing collapse and secure long-term sustainable productivity, is their spatial and temporal structure. The study of the emergence and the properties of regular spatial or spatiotemporal patterns which can be found in abundance in nature, such as for example stripes or spots on animal coats, ripples in sandy desserts, vegetation patterns in arid grazing systems or spatial patterns of fish species, has drawn much attention in natural sciences (e.g. Murray 2003).

W.A. Brock
Department of Economics, University of Wisconsin, Madisson, Wisconsin, USA
e-mail: wbrock@ssc.wisc.edu

W.A. Brock
Department of Economics, University of Missouri, Columbia, Columbia

A. Xepapadeas (✉)
Department of International and European Economic Studies, Athens University of Economics and Business, Athens, Greece
e-mail: xepapad@aueb.gr

A.N. Yannacopoulos
Department of Statistics, Athens University of Economics and Business, Athens, Greece
e-mail: ayannaco@aueb.gr

E. Moser et al. (eds.), *Dynamic Optimization in Environmental Economics*, Dynamic Modeling and Econometrics in Economics and Finance 15,
DOI 10.1007/978-3-642-54086-8_10,

Thus, in the management of natural resources and the regulation of pollution it seems natural to analyze mechanisms causing spatiotemporal patterns to arise, and to design regulatory policies with spatial characteristics. In renewable resource economics, modelling with spatial-dynamic processes Smith et al. (2009) has been used to study issues such as harvesting in metapopulation models governed by discrete spatial-dynamic processes, design of optimal policies in a spatiotemporal domain, marine or terrestial reserve policies, or bioinvasions (see, e.g., Sanchirico and Wilen 1999; Wilen 2007). Pattern formation and spatially dependent policies in renewable resource management have been also studied in the context of optimal control of reaction diffusion spatiotemporal systems (Brock and Xepapadeas 2008, 2010). In spatial pollution regulation the main objective is the internalization of the pollution externality through spatially dependent taxes (see, e.g., Goetz and Zilberman 2000), while spatial analysis has also been used to study water pricing in which the concept of a spatial distribution is combined with a two-stage optimal control problem (Xabadia et al. 2004).

Another issue which is of considerable interest in resource management is decision making when the decision maker is trying to make good choices when she regards her model not as the correct one but as an approximation of the correct one, or to put it differently, when the decision maker has concerns about possible misspecifications of the correct model and wants to incorporate these concerns into the decision-making rules (e.g., Salmon 2002; Hansen and Sargent 2001, 2008; Hansen et al. 2006; JET 2006).

The purpose of the present paper is to study the regulation of a commercial fishery following the classic model of commercial fishing (Smith 1969) with explicit spatial dependence where spatial interconnections in economic and biological variables are captured by local and non-local spatial effects. In this model the regulator has concerns about possible misspecifications of the spatiotemporal evolution of the phenomenon. That is, the regulator regards her model as an approximation of the correct spatiotemporal dynamics and seeks spatially dependent regulation that performs well under the approximating model. In this context we try to study how a regulator could design optimal spatiotemporal robust control for this fishery, how hot spots, which are location where the qualitative properties of the system change along with the structure of the regulation, may emerge, and what implications they might have for regulation.

The contribution of this approach is that it allows us to study in a unified model the optimal regulation of spatially interconnected distributed parameter fishery when concerns about model misspecification vary across the spatial domain. We follow Hansen et al. (2006) or Hansen and Sargent (2008), and regard concerns about model misspecification to imply that the regulator distrusts her model and wants robust decisions over a set of possible models that surround the regulator's approximating or benchmark model, and which are difficult to distinguish with finite data sets. The robust decisions are obtained by introducing Nature, a fictitious "adversarial agent". Nature promotes robust decision rules by forcing the regulator, who seeks to maximize profits from the commercial fishery over an entire spatial domain, to explore the fragility of decision rules with respect to departures from the benchmark model. A robust decision rule to model misspecification means that lower bounds to the rule's performance are determined by Nature—the adversarial agent—who acts as a minimizing agent when constructing these lower bounds. Hansen et al. (2006) show that robust control theory can be interpreted as a recursive version of max-min expected utility theory (Gilboa and Schmeidler 1989).

In our model, considering the spatial domain of the fishery as a ring of cells, the regulator is trying to determine an optimal level of harvesting per vessel in each spacial cell. This harvesting level can be used, for example, to set up a quota system in each site of the fishing area. The regulator's objective could be either the maximization of discounted profits over the whole ring, or the minimization deviations (or the cost of deviations) from target harvesting and biomass levels in each ring, by taking into account biomass diffusion as well as stock, congestion, and productivity externalities

The regulator is however uncertain regarding the true statistical distribution of the state of the system. This means that the regulator has concerns regarding the specification of biomass dynamics in each cell, and depending on her scientific knowledge, she might trust a benchmark model of the fishery more or less depending on the specific cell. For a large enough ring, this assumption—which implies spatially differentiated degrees of model uncertainty—seems plausible, and it is related to a localized in space entropy constraint of the spatially varying interconnected systems. In this context we derive optimal robust harvesting rules for each site and identify conditions under which concerns about model misspecification at specific site(s) could cause regulation to break down or to be very costly. We call sites associated with these phenomena hot spots. We are also able to identify spatial hot spots where the need to apply robust control induces spatial agglomerations and breaks down spatial symmetry. From the theory point of view this is a new source for generating spatial patterns as compared to the classic Turing diffusion induced instability (Turing 1952) which belongs to the recently identified family of optimal diffusion or spatial-spillover-induced instabilities (Brock and Xepapadeas 2008, 2010; Brock et al. 2012).

Distributed parameter models result in optimal control problems in infinite dimensional spaces. By using Fourier methods and exploiting the property of spatial invariance of a class of linear quadratic problems, we are able to obtain closed form solutions to these infinite dimensional problems which reveal important information on the qualitative features of the optimal policy, possible deviations from it or breakdowns as well as its dependence on the choice of model. Furthermore, by obtaining a linear quadratic approximation around a deterministic optimal trajectory of a nonlinear distributed parameter robust control problem of a commercial fishery, the tracking problem of keeping the controlled trajectory under uncertainty and concerns about model misspecification close to the optimal deterministic trajectory, representing the ideal benchmark model of the fishery manager, and comment upon hot spot formation and their importance.

Fig. 1 The circular fishery and the relevant state variables

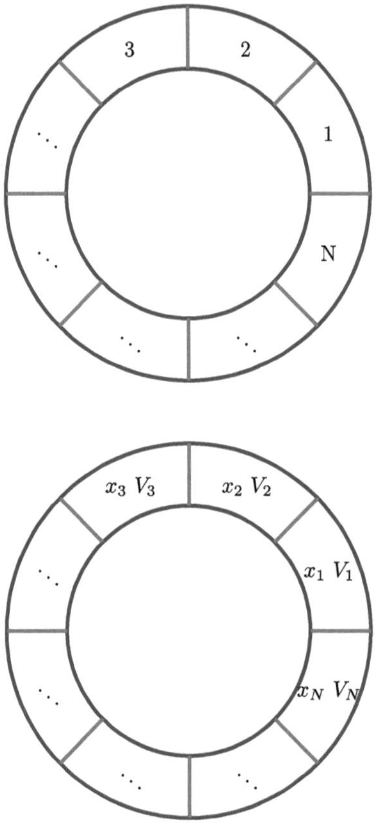

2 Modelling a Fishery with Spatial Interactions

2.1 A Spatial Profit Maximization Fishery Model

We consider a commercial fishery occupying an area that consists of a circular ring of N cells or sites on a finite lattice, so that space can be considered as the finite group of integers modulo N, $\mathbb{Z}_N$. The state of the system is quantified in terms of the fish biomass in each cell, x_n, and the number of vessels or firms fishing in each cell V_n, $n \in \mathbb{Z}_N$ (see Fig. 1).

Let $x_n(t)$ denote biomass at time $t \geq 0$ and cell $n \in \mathbb{Z}_N$. Fish biomass moves from cell to cell. The movements if there are strictly local can be described by classic diffusion with diffusion coefficient $D > 0$, which means that fish move from cells of high biomass concentration to adjacent cells of low biomass concentration. In this case the spatial movement can be modelled using the discrete Laplacian by a term $D[x_{n+1}(t) - 2x_n(t) + x_{n-1}(t)]$. More general spatial interactions across locations can be modelled by an influence "kernel" (or rather a discretized version of an influence kernel) which can be represented in terms of a matrix $\mathsf{A} = (\alpha_{nm}) \in \mathbb{R}^{N \times N}$. The entry α_{nm} provides a measure of the influence of the biomass of the system at point m to the biomass concentration of the system at point n. If there is no movement of biomass across cells then $\mathsf{A} = \alpha_{nm} = \delta_{n,m}$ where $\delta_{n,m}$ is the Kronecker delta. If only next neighborhood movements are possible then α_{nm} is non-zero only if m is a neighbor of n. Such an example is the discrete Laplacian, and matrix A in this case has a general form

$$\mathsf{A} = D \begin{pmatrix} 1 & -2 & 1 & 0 & \cdots & 0 & 0 & 0 \\ 0 & 1 & -2 & 1 & \cdots & 0 & 0 & 0 \\ \vdots & \vdots & \vdots & \vdots & \ddots & \vdots & \vdots & \vdots \\ 0 & 0 & 0 & 0 & \cdots & 1 & -2 & 1 \end{pmatrix}.$$

This can be considered as the discretization of the Laplace operator $\mathsf{A} = D\frac{\partial^2}{\partial z^2}$, when the space is considered as continuous e.g. the interval $[-\pi, \pi]$.

Let $V_n(t)$ denote the number of identical vessels or firms operating at cell n of the ring, and $h_n(t)$ the harvest rate at cell n per unit time. Thus total harvesting at cell n is $h_n(t)V_n(t)$. The temporal evolution of biomass of the fishery is subject to statistical fluctuations (noise), which is introduced into the model via stochastic factors (sources),[1] modelled in terms of a stochastic process $w = \{w_n\}$, $n \in \mathbb{Z}_N$, which is considered as a vector valued Wiener process on a suitable filtered probability space $(\Omega, \{\mathscr{F}_t\}_{t\in\mathbb{R}_+}, \mathscr{F}, \mathbb{P})$ (see e.g., Karatzas and Shreve 1991). The introduction of noise turns the biomass for a fixed time t into an $\mathbb{R}^N$-valued random variable, thus $x(\cdot)$ is an $\mathbb{R}^N$-valued stochastic process. We assume that this stochastic process is the solution of a stochastic differential equation:

$$dx_n(t) = \Big[f\big(x_n(t)\big) + \sum_m \alpha_{nm} x_m(t) - h_n(t)V_n(t) \Big] dt + \sum_m s_{nm} dw_m, \tag{1}$$
$$x_n(0) = x_{0,n}, \quad n, m \in \mathbb{Z}_N.$$

In the above equation $f(x)$, $x \geq 0$, is the recruitment rate or growth function for the fishery. This function has the properties that there exist three values $\underline{x}$, $\bar{x}$ and x^0 with $0 \leq \underline{x} < x^0 < \bar{x}$, such that $f(\underline{x}) = f(\bar{x}) = 0$, $f'(x^0) = 0$, $f''(x^0) < 0$. An example of such a function is a quadratic function which models logistic growth. It is assumed that the parameters of model (1) are chosen so that positivity of solutions is guaranteed (i.e. noise levels are assumed to be small and have a weak effect). Furthermore, for the rest of the paper $\sum_m$ will be used as a shorthand for $\sum_{m\in\mathbb{Z}_N}$.

[1]There is uncertainty concerning the state of the system (i.e. the true figures for the biomass) which is represented in terms of the vector valued stochastic process w. These common factors affect the state of the biomass x at the different sites. Each factor has a different effect on the state of the biomass on each particular site; this will be modelled by a suitable correlation matrix. It is not of course necessary that the number of factors is the same as the number of sites in the system however, without loss of generality we will make this assumption and assume that there is one factor or source of uncertainty related to each site. This assumption can be easily relaxed.

The last term of (1), describing the fluctuations of the biomass due to the stochasticity, is understood in the sense of the Itô theory of stochastic integration. In compact form it can be represented by a finite matrix $\mathsf{S} = (s_{nm})$ with elements s_{nm} indicating how the uncertainty at site m is affecting the uncertainty concerning the biomass of the fishery at site n. The matrix $\mathsf{S} = (s_{nm})$ can be thought of as the spatial autocorrelation operator for the system. Thus the evolution of the system can be written in a compact form as:

$$dx = \left[\mathsf{F}(x) + \mathsf{A}x - y\right]dt + \mathsf{S}dw, \tag{2}$$

where we have used the vector notation

$$\begin{aligned} x &= (x_1, \ldots, x_N)^{tr}, \\ w &= (w_1, \ldots, w_n)^{tr}, \\ y &= (h_1 V_1, \ldots, h_N V_N)^{tr}, \\ \mathsf{F}(x) &= \left(f(x_1), \ldots, f(x_N)\right)^{tr}, \end{aligned}$$

and $\mathsf{A}, \mathsf{S} : \mathbb{R}^N \to \mathbb{R}^N$ are linear operators, representable by finite matrices with elements $\{\alpha_{nm}\}$, $\{s_{nm}\}$, respectively. We will also use the notation $y = h \otimes V$ for the vector which is defined by componentwise multiplication of the vectors h, V.

The cost per vessel operating at a cell n for harvesting rate h is determined by a cost function $c(h_n(t), x_n(t), C_n(t), P_n(t))$. This is a function of the harvesting rate; the biomass level at the specific cell, $x_n(t)$ which reflects recourse stock externalities; and the number of other vessels operating in the neighborhood of the cell n, which reflect two types of externalities: crowding or congestion externalities and productivity or knowledge externalities. Crowding externalities, which are negative (cost increasing), and productivity externalities, which are positive (cost reducing), are non-local effects, which are modeled by spatial kernels as:

$$\begin{aligned} C_n(t) &= \sum_m c_{nm} V_m(t) =: (\mathsf{C}V)_n(t), \\ P_n(t) &= \sum_m \gamma_{nm} h_m(t) =: (\Gamma h)_n(t), \end{aligned} \tag{3}$$

where $\mathsf{C}, \Gamma : \mathbb{R}^N \to \mathbb{R}^N$ are linear operators, representable by finite matrices with elements c_{nm}, γ_{nm}, respectively. We assume that: (i) $\frac{\partial c}{\partial h} > 0$, $\frac{\partial^2 c}{\partial h^2} \geq 0$; (ii) $\frac{\partial c}{\partial x} < 0$, which implies resource stock externalities; (iii) $\frac{\partial c}{\partial C} > 0$, which implies crowding externalities due to congestion effects. We assume that an increase in vessels in a given cell will always increase costs, that is $\frac{\partial c}{\partial C} > 0$. This kernel formulation in the cost function means that vessels not only in cell n but also near cell n could create congestion effects and increase operating costs of the vessels operating in cell n; and (iv) $\frac{\partial c}{\partial P} < 0$, which implies knowledge or productivity externalities because harvesting that takes place near cell n helps the development of harvesting knowledge in n and reduces operating costs.

Assuming that harvested fish is sold at an exogenous price $\mathscr{P}$, which is homogeneous over the whole ring, profit per vessel at n is defined as:

$$\pi_n(t) = \mathscr{P} h_n(t) - c\big(h_n(t), x_n(t), (\mathsf{C}V)_n(t), (\Gamma h)_n(t)\big). \tag{4}$$

Vessels are attracted to cell n if profits per vessel at this site are higher than the average profit over the whole spatial domain. Vessels can be attracted to the ring from locations outside the ring if profits are positive in cells of the ring, so the number of vessels in the ring does not need to be conserved.[2] Assuming that the rate of growth of vessels in each cite is proportional to the difference between the profit per vessel at n with the average profit per vessel over the whole lattice, the evolution of vessels in each site is described by:

$$\begin{aligned} &\frac{d}{dt} V_n(t) = \phi\Big(\pi_n(t) - \frac{1}{N}\sum_m \pi_m(t)\Big) V_n(t), \\ &V_n(0) = V_{0,n}, \end{aligned} \tag{5}$$

where $\phi > 0$ measures the speed of adjustment and is set equal to one without loss of generality. Note that Eq. (5), though not an Itô stochastic differential equation, is now a random differential equation since x is a stochastic process.

A regulator is trying to determine in each cell an optimal level of harvesting per vessel, h_n. This harvesting level can be used, for example, to set up a quota system in each cell of the ring. The regulator's objective is the maximization of discounted profits over the whole ring by taking into account biomass diffusion as well as stock, congestion and knowledge externalities.[3] The regulator's objective is therefore

$$\begin{aligned} &\max_{\{h_n(t)\}} \mathbb{E}\Big[\int_0^\infty e^{-rt}\Big(\sum_n V_n(t)\pi_n(t)\Big)dt\Big], \\ &\text{subject to} \quad (2) \text{ and } (5), \end{aligned} \tag{6}$$

where the per vessel profit π_n is given by (4). It is clear that the state of the system is characterized by the biomass x and the vessel distribution V, and we will use the notation $X = (x, V)^{tr}$ where $X \in \mathbb{R}^{2N\times 1}$.

2.2 Misspecification Concerns

We now assume that the regulator has concerns regarding the specification of biomass dynamics in each cell, which can be modelled as follows: Assume that

---

[2]To simplify we ignore transportation costs.

[3]To simplify the interpretation of results and the analysis, we do not include existence values for the biomass.

there is some uncertainty concerning the "true" statistical distribution of the state of the system. This corresponds to a family of probability measures $\mathcal{Q}$ such that each $Q \in \mathcal{Q}$ corresponds to an alternative stochastic model (scenario) concerning the state of the system. From Girsanov's theorem $\bar{w}_n(t) = w_n(t) - \int_0^t v_n(s)ds$ is a Q-Brownian motion for all $n \in \mathbb{Z}$, where the drift term v_n may be considered as a measure of the model misspecification at lattice site n, where $v = (v_1, \ldots, v_N)^{tr}$ is an $\mathbb{R}^N$-valued stochastic process which is measurable with respect to the filtration $\{\mathcal{F}_t\}$ satisfying the Novikov condition $\mathbb{E}[\exp(\int_0^T \sum_n v_n^2(t)dt)] < \infty$. Thus, Girsanov's theorem (see e.g. Karatzas and Shreve 1991) shows that the adoption of the family $\mathcal{Q}$ of alternative measures concerning the state of the system, leads to a family of differential equations for the biomass

$$\begin{aligned} dx_n(t) &= \left[f\left(x_n(t)\right) + \sum_m \alpha_{nm} x_m(t) - h_n(t)V_n(t) + \sum_m s_{nm} v_m \right] dt \\ &\quad + \sum_m s_{nm} d\bar{w}_m, \quad n, m \in \mathbb{Z}_N, \\ x_n(0) &= x_{0,n}, \end{aligned} \tag{7}$$

parameterized by the information drift v. In (7) x indicates the state of the system when the measure[4] Q corresponding to the information drift v and the control procedure $h = (h_1, \ldots, h_N)^{tr}$, which will be denoted by Q_v, is adopted. This is an Ornstein Uhlenbeck equation which in compact form can be expressed as

$$dx = \left[\mathsf{F}(x) + \mathsf{A}x - h \otimes V + \mathsf{S}v\right]dt + \mathsf{S}d\bar{w}. \tag{8}$$

The regulator's problem when there are concerns about model misspecification is solved under the adoption of the measure Q, related to the drift v, i.e. it is solved under the dynamic constraints (7) and (5). This will provide a solution leading to a value function $\mathcal{V}(X; v)$; corresponding to the maximum discounted profits over the whole spatial domain obtained for the model Q_v under the optimal harvesting effort, given that the system had initial state $X = (x(0), V(0)) = (x, V)$. Being uncertain about the true model, the decision maker will opt to choose the strategy that will work in the worst case scenario; this being the one that minimizes $V(X; v)$—the maximum over all h having chosen v—over all possible choices for v. Therefore, the robust control problem to be solved is of the general form

$$\begin{aligned} \mathcal{V}(X) &= \max_h \min_v J(h, v), \\ &\text{subject to} \quad (7) \text{ and } (5), \end{aligned} \tag{9}$$

where

$$J(h, v) = \mathbb{E}_{Q_v} \left\{ \int_0^\infty e^{-rt} \left[\sum_n V_n(t)\pi_n(t) + \sum_n \theta_n \left(v_n(t)\right)^2 \right] dt \right\}.$$

[4]We will identify a model by a probability measure.

The vector $\theta = (\theta_1, \ldots, \theta_N)^{tr}$ corresponds to the weight assigned to concerns related to model misspecification in a local sense (differentially in space). To clarify this point, we refer to Brock et al. (2012), as by a simple modification of the arguments in this work it can be shown that robust optimization problems of the form

$$\begin{aligned} &\sup_h \inf_{Q \in \mathscr{Q}} \mathbb{E}_Q \left[\int_0^\infty \mathrm{e}^{-rt} \sum_n V_n(t) \pi_n(t) dt \right], \\ &\text{subject to} \quad \mathscr{H}(\mathbb{P}_n \mid Q_n) < H_n, \; n \in \mathbb{Z}_N, \end{aligned} \tag{10}$$

and the dynamic constraints (7) and (5), can be written as equivalent to (9) where now the vector $\theta \in \mathbb{R}_+^N$ plays the role of a Lagrange multiplier associated with the constraints in (10). In (10) by $\mathscr{H}(\mathbb{P}_n \mid Q_n)$ we denote the Kullback-Leibler entropy of the marginal probability measures $\mathbb{P}_n$ and Q_n (i.e. the probability measures $\mathbb{P}$ and Q respectively, averaged over all possible states of the noise over the remaining sites). The localized entropic constraints mean that the regulator is only considering models in each cell (i.e., measures Q_n) whose deviation in terms of the relative entropy from the "true" model in the cell (i.e., the measure $\mathbb{P}_n$) is less than H_n.

The introduction of the local entropic constraints means that the concern of the policy maker about uncertainty on site n is quantified by H_n, the smaller H_n is the less model uncertainty she is willing to accept for site n, given her information about this site. This assumption is not unreasonable as certain cells may be considered as more crucial than others therefore specific care should be taken for them.

In the robust control problem the minimizing adversarial agent—Nature—chooses a $\{v_n(t)\}$ while $\theta_n \in (\underline{\theta}_n, +\infty]$, $\underline{\theta}_n > 0$, is a penalty parameter restraining the maximizing choice of the decision maker. As noted above θ_n is associated with the Lagrange multiplier of the entropy constraint at each site. In the entropy constraint H_n is the maximum misspecification error that the decision maker is willing to consider given the existing information about the system at site n.[5] The lower bound $\underline{\theta}_n$ is a so-called breakdown point beyond which it is fruitless to seek more robustness because the adversarial (i.e. the minimizing) agent is sufficiently unconstrained so that she/he can push the criterion function to $-\infty$ despite the best response of the maximizing agent. Thus when $\theta_n < \underline{\theta}_n$ for a specific site robust control rules cannot be attained. In our terminology this site is a candidate for a "nucleus" of a hot spot since misspecification concerns for this site will break down robust control for the whole spatial domain. On the other hand when $\theta_m \to \infty$ or equivalently $H_m = 0$ there are no misspecification concerns for this site and the benchmark model can be used. The effects of spatial connectivity can be seen in this extreme example. The spatial relation of site m with site n could break down regulation for both sites. If site m was spatially isolated from n there would have been no problem with regulation at m.

[5] If the decision maker can use physical principles and statistical analysis to formulate bounds on the relative entropy of plausible probabilistic deviations from her/his benchmark model, these bounds can be used to calibrate the parameters H_n (Athanassoglou and Xepapadeas 2012).

2.3 Robust Stabilization of a Desired Optimal State

Problem (9) is a non linear robust control problem. The full nonlinear problem, eventhough accessible to either abstract analysis or numerical treatment, will not allow an analysis in terms of closed form expressions and as such will obscure our main interest in this paper, which is to show the existence of hot spots and spatial pattern formation. To illustrate these points we will instead choose to work in terms of a linear quadratic approximation of the full nonlinear problem, which allows a rather detailed analytical treatment. However, rather than taking a linear quadratic local approximation of the full problem (9) we choose an alternative approach. This alternative approach is related to a tracking problem, which allows the decision maker to "correct" her benchmark policy in such a way as to optimally make up for possible misspecifications of the model. Tracking problems have been addressed by the control theory community and find important applications in a variety of problem in mathematical, environmental and financial economics (see e.g., Leizarowitz 1985, 1986; Artstein and Leizarowitz 1985).

In this section we formulate a related linear quadratic robust control problem which is associated with a stabilization policy, under the effect of noise and uncertainty with respect to the nature of this noise, which allows the decision maker to keep the system as close as possible to a desired optimal state of (9). We assume that the desired optimal state is the one that corresponds to the deterministic version of the model, i.e. the case where there is no noise present. Let us call this state $(x^{(0)}, V^{(0)})$ and assume that it is supported by the optimal control procedure $h^{(0)}$. The triple $(x^{(0)}, V^{(0)}, h^{(0)})$ is thus the solution of the deterministic optimal control problem

$$\max_h \int_0^\infty e^{-rt} \sum_n V_n(t)\pi_n(t)dt,$$

subject to

$$\dot{x}_n = \sum_m a_{nm}x_m - f(x_n) - V_n h_n, \quad n \in \mathbb{Z}_N \tag{11}$$

$$\dot{V}_n = \phi\left(\pi_n - \frac{1}{N}\sum_m \pi_m\right)V_n, \quad n \in \mathbb{Z}_N.$$

This is the idealized problem that the fishery manager wants to solve. The solution of that, furnishes the "best" she can do to optimize her profit, given the capabilities of the fishery, in the absence of unforeseen event (i.e. noise).

The solution $x^{(0)}$ is determined by the deterministic Pontryagin principle, associated with the Hamiltonian function

$$H(x, V, \mathfrak{p}, \bar{\mathfrak{p}}; h) = \sum_n V_n\pi_n + \sum_n \mathfrak{p}_n\left(\sum_m a_{nm}x_m - f(x_n) - V_n h_n\right)$$
$$+ \sum_n \bar{\mathfrak{p}}_n\left[\phi\left(\pi_n - \frac{1}{N}\sum_m \pi_m\right)V_n\right],$$

where $\mathfrak{p} = (\mathfrak{p}_1, \ldots, \mathfrak{p}_N)^{tr}$, $\bar{\mathfrak{p}} = (\bar{\mathfrak{p}}_1, \ldots, \bar{\mathfrak{p}}_N)^{tr}$ are the adjoint variables associated with the state variables $x = (x_1, \ldots, x_N)^{tr}$ and $V = (V_1, \ldots, V_N)^{tr}$ respectively. The solution of the benchmark optimal control problem is reduced to the solution of the system of differential equations

$$\begin{aligned}
&\frac{\partial H}{\partial x_n}(x, V, \mathfrak{p}, \bar{\mathfrak{p}}; h) - \dot{\mathfrak{p}}_n - r\mathfrak{p}_n = 0, \quad n \in \mathbb{Z}_N, \\
&\frac{\partial H}{\partial V_n}(x, V, \mathfrak{p}, \bar{\mathfrak{p}}; h) - \dot{\bar{\mathfrak{p}}}_n - r\bar{\mathfrak{p}}_n = 0, \quad n \in \mathbb{Z}_N, \\
&\frac{\partial H}{\partial \mathfrak{p}_n}(x, V, \mathfrak{p}, \mathfrak{p}; h) - \dot{x}_n = 0, \quad n \in \mathbb{Z}_N, \\
&\frac{\partial H}{\partial \bar{\mathfrak{p}}_n}(x, V, \mathfrak{p}, \bar{\mathfrak{p}}; h) - \dot{V}_n = 0, \quad n \in \mathbb{Z}_N, \\
&\frac{\partial H}{\partial h_n}(x, V, \mathfrak{p}, \bar{\mathfrak{p}}; h) = 0, \quad n \in \mathbb{Z}_N,
\end{aligned} \tag{12}$$

where the last set of equations is an optimality condition. The solution $(x^{(0)}, V^{(0)}, h^{(0)})$ of this system gives the optimal benchmark path. In general this system has a solution which is spatially dependent, i.e., $x_n(t) \neq x_m(t)$ for $n \neq m$. However, it may also have solutions which are uniform in space. For example in the case of diffusive coupling, $\sum_m a_{nm} = 0$, $\sum_m \beta_{nm} = 0$ and $\sum_m \gamma_{nm} = 0$ the system (12) may admit a solution which is uniform in space, i.e. a solution $\{x_n^0(t), V_n^0(t)\}$ such that $x_n^0(t) = x^0(t)$, $V_n^0(t) = V^0(t)$ for all $n \in \mathbb{Z}$. Similarly for the optimal control $h^{(0)}$. Furthermore, we may assume that these equations have a stationary uniform in space solution, i.e., a solution that is time independent. While this assumption is not necessary for the development of the proposed model, it simplifies the exposition and will be adopted. It should be stressed though that a general theory for time dependent as spatially nonhomogeneous $x^{(0)}$, $V^{(0)}$, $h^{(0)}$ can be formulated and the necessary modifications are technical.

However, true life is often far from the idealized model, that the manager has in mind. This means that the manager should be adept to sidetrack from the idealized optimal control procedure $h^{(0)}$ as an effect of unforeseen circumstances, modeled here by noise. An important question is the following: Can we design optimally a corrective policy $h^{(1)}$ which will take into account the effects of noise so that the true system keeps as close as possible to the idealized optimal state $(x^{(0)}, V^{(0)})$ as provided by the solution of the optimal control problem (11)?

Assume that we have the nonlinear problem (1), subject to weak additive noise. The problem is subject to model uncertainty (with respect to the nature of the noise term) which may be modelled in terms of a drift v so that applying Girsanov's theorem we obtain the family of models

$$
\begin{aligned}
dx_n &= \Big[f(x_n) + \sum_m \alpha_{nm} x_n - h_n V_n + \varepsilon \sum_m s_{nm} v_n \Big] dt + \varepsilon \sum_m s_{nm} dw_m, \\
dV_n &= \phi\Big(\pi_n - \frac{1}{N} \sum_m \pi_m \Big) V_n dt, \quad n \in \mathbb{Z}_N.
\end{aligned} \tag{13}
$$

This family of models will give the "observed" state of the system (x, V). The system is still subject to a control procedure h, and it is our aim to choose h so that the actual state of the system (x, V) is kept as close as possible to the ideal profit maximizing state $(x^{(0)}, V^{(0)})$ with h as close as possible to $h^{(0)}$.

Since the noise is assumed to be weak we may consider as a zeroth order approximation to (13) (i.e., the solution setting $\varepsilon = 0$) the deterministic optimal path $(x^{(0)}, V^{(0)})$. Let us consider perturbations of $\{x, V, h, v\}$ around this reference solution, i.e. let us consider solutions of the above problem of the form

$$
\{x, V, h, v\} = \{x^0, V^0, h^0, 0\} + \varepsilon\{x^1, V^1, h^1, v^1\},
$$

where now $\{x, V, h, v\}$ are subject to uncertainty and are solutions of the stochastic biomass equation (13) with ε a small parameter. The terms $(x^{(1)}, V^{(1)})$ quantify the divergence of the actual state of the system from the ideal profit maximizing optimal state $(x^{(0)}, V^{(0)})$, the fishery manager would like to follow. This deviation is in general going to be spatially dependent; this spatial dependence will depend on the interaction between the dynamics of the system and noise. We still allow the manager a control procedure $(h^{(1)}, v^{(1)})$, this is considered as the correction procedure on top of the pre-planned ideal optimal control procedure $h^{(0)}$ where $v^{(1)}$ takes care of model uncertainty which will be chosen so as to minimize deviation from the ideal plan of action $(x^{(0)}, V^{(0)}, h^{(0)})$. As we shall see this correction procedure can be chosen in terms of a feedback control procedure, whereby the corrections are determined upon observation of the deviation from the ideal desired state.

We linearize the state equations around the state $s^{(0)} := \{x^{(0)}, V^{(0)}, h^{(0)}, v^{(0)}\}$ to obtain to first order in ε that

$$
\begin{aligned}
dx^{(1)} &= \big[A^{(1)} x^{(1)} + A^{(2)} V^{(1)} + B^{(1)} h^{(1)} + S v^{(0)}\big] dt + S d\bar{w}, \\
dV^{(1)} &= \big[A^{(3)} x^{(1)} + A^{(4)} V^{(1)} + B^{(2)} h^{(1)}\big] dt,
\end{aligned}
$$

where $x^{(1)}, V^{(1)}, h^{(1)}, v \in \mathbb{R}^N$ and $A^{(i)}, B^{(j)}, i = 1, \dots, 4, j = 1, 2$ are $\mathbb{R}^{N \times N}$ matrices with elements

$$
\begin{aligned}
A^{(1)}_{nm} &= f'\big(x^{(0)}_n\big) \delta_{nm} + a_{nm}, \\
A^{(2)}_{nm} &= -h^{(0)}_n \delta_{nm}, \\
A^{(3)}_{nm} &= -\phi V^{(0)}_n \left(\frac{\partial c_0}{\partial x} \right)_n \delta_{nm} + \frac{1}{N} \phi V^{(0)}_n \left(\frac{\partial c_0}{\partial x} \right)_m,
\end{aligned}
$$

$$A^{(4)}_{nm} = \phi\left(\pi^{(0)}_n - \frac{1}{N}\sum_k \pi^{(0)}_k\right)\delta_{nm} - \phi V^{(0)}_n \left(\frac{\partial c_0}{\partial C}\right)_n \beta_{nm}$$
$$+ \frac{1}{N}\phi V^{(0)}_n \left(\sum_k \beta_{km}\left(\frac{\partial c_0}{\partial C}\right)_k\right),$$
$$B^{(1)}_{nm} = -V^{(0)}_n \delta_{nm},$$
$$B^{(2)}_{nm} = \phi V^{(0)}_n \left(\mathscr{P} - \left(\frac{\partial c_0}{\partial h}\right)_n\right)\delta_{nm} - \phi V^{(0)}_n \gamma_{nm}\left(\frac{\partial c_0}{\partial P}\right)_n - \frac{1}{N}\phi \mathscr{P} V^{(0)}_n$$
$$+ \frac{1}{N}\phi V^{(0)}_n \left(\frac{\partial c_0}{\partial h}\right)_m + \frac{1}{N}\phi V^{(0)}_n \left(\sum_k \gamma_{km}\left(\frac{\partial c_0}{\partial P}\right)_k\right),$$

where by $(\frac{\partial c_0}{\partial z})_n$, $z = h, x, C, P$, in the above we mean that the respective partial derivatives are calculated at the state $s^{(0)}$ and at site n. Assuming that the zeroth order state is spatially uniform, and assuming also that the interaction kernels have the property that $\sum_m \beta_{nm} = 0$, $\sum_m \gamma_{nm} = 0$ (diffusive coupling) the above expressions can simplify considerably to

$$A^{(1)}_{nm} = f'(x^{(0)}_n)\delta_{nm} + a_{nm},$$
$$A^{(2)}_{nm} = -h^{(0)}_n \delta_{nm},$$
$$A^{(3)}_{nm} = -\phi V^{(0)}_n \left(\frac{\partial c_0}{\partial x}\right)\delta_{nm} + \frac{1}{N}\phi V^{(0)}_n \left(\frac{\partial c_0}{\partial x}\right),$$
$$A^{(4)}_{nm} = \phi\left(\pi^{(0)}_n - \frac{1}{N}\sum_k \pi^{(0)}_k\right)\delta_{nm} - \phi V^{(0)}_n \left(\frac{\partial c_0}{\partial C}\right)\beta_{nm},$$
$$B^{(1)}_{nm} = -V^{(0)}_n \delta_{nm},$$
$$B^{(2)}_{nm} = \phi V^{(0)}_n \left(\mathscr{P} - \left(\frac{\partial c_0}{\partial h}\right)_n\right)\delta_{nm} - \phi V^{(0)}_n \gamma_{nm}\left(\frac{\partial c_0}{\partial P}\right)$$
$$- \frac{1}{N}\phi \mathscr{P} V^{(0)}_n + \frac{1}{N}\phi V^{(0)}_n \left(\frac{\partial c_0}{\partial h}\right).$$

We may now express the linearized system in compact form as the stochastic control system

$$dX = [\mathbb{A}X + \mathbb{B}u + \mathbb{S}v]dt + \mathbb{S}d\bar{w}, \tag{14}$$

where

$$\mathbb{A} := \begin{pmatrix} A^{(1)} & A^{(2)} \\ A^{(3)} & A^{(4)} \end{pmatrix}, \qquad \mathbb{B} := \begin{pmatrix} B^{(1)} \\ B^{(2)} \end{pmatrix}, \qquad \mathbb{S} := \begin{pmatrix} S \\ 0 \end{pmatrix},$$

where 0 is the $N \times N$ zero matrix and $X = (x^{(1)}, V^{(1)})^{tr}$, $u = h^{(1)}$, $v = v^{(1)}$. It is clear that $X \in \mathbb{R}^{2N\times 1}$, $u, v, w \in \mathbb{R}^{N\times 1}$, $\mathbb{A} \in \mathbb{R}^{2N\times 2N}$, $\mathbb{B}, \mathbb{S} \in \mathbb{R}^{2N\times N}$. It should be noted that matrices $\mathbb{A}$ and $\mathbb{B}$ incorporate stock, congestion and productivity externalities in the linearized dynamics

We now consider the problem of controlling the linearized system by proper choice of the control procedure u so that the system is kept as close as possible and at the minimum possible cost at the zeroth order desired steady state $s^{(0)}$. Ideally, we would like to choose $u = 0$ and keep $X = 0$ at all times, as this would correspond to keeping the system to the profit maximizing state $(x^{(0)}, V^{(0)}, h^{(0)})$. However, this is not possible and we choose the less ambitious task of minimizing the deviation of X from 0 at the minimum possible cost. This is equivalent to the robust optimal control problem[6]

$$\begin{aligned} &\min_u \max_v \bar{J}(u, v), \\ &\text{subject to} \quad (14), \end{aligned} \tag{15}$$

where

$$\bar{J}(u, v) := \mathbb{E}\left[\frac{1}{2}\int_0^\infty e^{-rt}\big(X^{tr}(t)\mathsf{P}X(t) + u^{tr}(t)\mathsf{Q}u(t) - v^{tr}(t)\mathsf{R}v(t)\big)dt\right].$$

The solution to problem (15) guarantees that we get as close as possible to the desired state, at the worst possible deviation from our ideal model (11). The matrices $\mathsf{P} \in \mathbb{R}^{2N\times 2N}$, $\mathsf{Q} \in \mathbb{R}^{N\times N}$ and $\mathsf{R} \in \mathbb{R}^{N\times N}$ are positive definite and invertible and without loss of generality can be considered to be copies of the identity matrix, i.e.

$$\mathsf{P} = \begin{pmatrix} pI & 0 \\ 0 & \bar{p}I \end{pmatrix}, \qquad \mathsf{Q} = qI, \qquad \mathsf{R} = \theta I,$$

where I is the $N \times N$ identity matrix. For this particular case the objective functional becomes

$$\bar{J}(u, v) = E\left[\frac{1}{2}\int_0^\infty e^{-rt}\sum_n\big(p(x_n^{(1)}(t))^2 + \bar{p}(V_n^{(1)}(t))^2 + q(h_n^{(1)}(t))^2 + \theta(v_n^{(1)}(t))^2\big)dt\right]. \tag{16}$$

In objective (16) the coefficients $(p, \bar{p}, q, \theta)$ reflect the relative importance attache by the regulator to deviations from the optimal deterministic path, with r expressing the cost or being robust. Without loss of generality and to simplify the expressions we may choose $p = \bar{p}$. The parameter θ in (16) should be interpreted as the parameter associated with the global entropic constraint. If we are dealing with local

[6]It is obvious that upon setting $\hat{J} = -\bar{J}$ the $\min_u \max_v \bar{J}$ problem becomes equivalent to the $\max_u \min_v \hat{J}$ problem, which is in a form similar to the robust control problem (9), where now the profit functional is a replaced by the negative of a loss functional quantifying costs of deviation from a target.

entropic constraints matrix R should be defined as:

$$\mathsf{R} = \begin{pmatrix} \theta_1 & \cdots & 0 \\ 0 & \ddots & 0 \\ 0 & \cdots & \theta_n \end{pmatrix}.$$

This a more complicated case which can be dealt with methods appropriate for the solution of the general linear quadratic robust control problem presented in Brock et al. (2012). Because of its relative simplicity functional (16) allows us to use the Fourier space solution of the problem as we will see in the next section.

Note that this problem is different from the problem treated in Magill (1977a, 1977b) where a linear quadratic approximation of a nonlinear stochastic optimal control problem is proposed. Here instead, we propose an exact linear quadratic procedure, which minimizes the tracking error from the optimal solution of a nonlinear idealized deterministic profit maximization problem. Our approach differs in spirit, however, correspond to a realistic situation. Most policy is designed upon ideal and simplified models (as for instance model (11)). It is important for the policy maker to have guidelines concerning the necessary corrections needed when the true state of the system deviates from the ideal state (as for instance under model (13)), so as to correct her policy in order to minimize deviations from the target. However, the generalization of Magill's procedure to a robust control problem is of interest in its own right, and will be treated elsewhere.

3 Robust Stabilization of the Benchmark Solution

Problem (15) can now be treated using the Hamilton-Jacobi-Belman-Isaacs equation. This is expressed in terms of the generator $\mathscr{L}$ of the Ornstein-Uhlenbeck process (14), defined through its action on a twice continuously differentiable function $\mathscr{V} : \mathbb{R}^{2N} \to \mathbb{R}$

$$\mathscr{L}\mathscr{V} = (\mathbb{A}X + \mathbb{B}u + \mathbb{S}v) D_X \mathscr{V} + \frac{1}{2} \mathbb{S}\mathbb{S}^{tr} D_X^2 \mathscr{V},$$

where $D_X \mathscr{V}$ is the gradient of $\mathscr{V}$ with respect to the coordinates of the vector X and $D_X^2 \mathscr{V}$ is the Hessian matrix of the function $\mathscr{V}$ with respect to the coordinates of the vector X. The above are shorthands for the relevant expressions in coordinate form, e.g.,

$$D_X \mathscr{V} = \left(\frac{\partial \mathscr{V}}{\partial x_1^{(1)}}, \ldots, \frac{\partial \mathscr{V}}{\partial x_N^{(1)}}, \frac{\partial \mathscr{V}}{\partial V_1^{(1)}}, \ldots, \frac{\partial \mathscr{V}}{\partial V_N^{(1)}} \right)^{tr},$$

and similarly for the Hessian. Using the generator we may define the Hamiltonian function

$$H(\mathscr{V}; X, u, v) = \mathscr{L}\mathscr{V} + \langle \mathsf{P}X, X \rangle + \langle \mathsf{Q}u, u \rangle - \theta \langle \mathsf{R}v, v \rangle,$$

where by $\langle\cdot,\cdot\rangle$ we denote the inner product in $\mathbb{R}^{2N}$. The value function $\mathscr{V}$ is the solution of the Hamilton-Jacobi-Belman-Isaacs (HJBI) equation

$$r\mathscr{V} + \min_u \max_v H(\mathscr{V}; X, u, v) = 0$$

(since by the saddle point theorem we may interchange the order of the $\min_u$ and $\max_v$ operations). The optimal policy is then related to the solution of the optimization problem for the Hamiltonian function. The HJBI equation is a fully nonlinear PDE, but on account of the linear quadratic nature of the system it can be solved in terms of the matrix Ricatti equation. Adapting the general results of Brock et al. (2012) to the model under consideration we find that the optimal correction policy is given by

$$u = -\mathsf{Q}^{-1}\mathbb{B}^{tr}\mathsf{H}X, \tag{17}$$

where $\mathsf{H} \in \mathbb{R}^{2N\times 2N}$ is the symmetric solution of the matrix Ricatti equation

$$\mathsf{H}\mathbb{A} + \mathbb{A}^{tr}\mathsf{H} - \mathsf{H}\mathsf{E}^{s}\mathsf{H} - r\mathsf{H} + \mathsf{P} = 0$$

and

$$\mathsf{E}^{s} = \frac{1}{2}(\mathsf{E} + \mathsf{E}^{tr}),$$

$$\mathsf{E} = \mathbb{B}\mathsf{Q}^{-1}\mathbb{B}^{tr} - \frac{1}{\theta}\mathbb{S}\mathsf{R}^{-1}\mathbb{S}^{tr}.$$

Once the matrix H computed, in principle numerically, the correction $u = h^{(1)}$ needed to modify the benchmark control procedure $h^{(0)}$ so as to keep the true state of the system as close as possible to the benchmark optimal state $(x^{(0)}, V^{(0)})$ is given by the feedback rule (17). This rule is very easy to apply as it only requires the manager to monitor the current value of the state $X(t)$, i.e. the current deviations $(x^{(1)}(t), V^{(1)}(t))$ of the true state of the system from the benchmark optimal state $(x^{(0)}, V^{(0)})$. We remark that our approach through the Ricatti equation does not necessarily require the benchmark state to be time independent not spatially homogeneous. However, even in the benchmark state enjoys both these properties, the deviations from this state, $X(t)$ will not necessarily satisfy them; it is in general expected to be both time varying and is expected to display spatial patterns. Furthermore, the optimal state X (i.e. the optimal deviations from the benchmark state once the optimal correction policy $u = h^{(1)}$ is adopted) is given by the solution of the Ornstein-Uhlenbeck equation

$$dX = \left(\mathbb{A} - \mathbb{B}\mathsf{Q}^{-1}\mathbb{B}^{tr}\mathsf{H} + \frac{1}{\theta}\mathbb{S}\mathsf{R}^{-1}\mathbb{S}^{tr}\mathsf{H}\right)X dt + \mathbb{S}dw.$$

The matrix Ricatti equation can be treated through a multitude of analytic or numerical methods leading to either interesting qualitative features of its solution, or to accurate computations. Therefore the above analysis provides a general and computationally feasible approach to the problem of correcting the benchmark optimal strategy in order to lead the realistic system to the desired state. Here, in order to provide some qualitative results with the less possible technicalities involved, we treat the simple, yet realistic case where the operators related to the matrices $\mathbb{A}$, $\mathbb{B}$, $\mathbb{S}$ are translation invariant, the fishery is situated on a ring (i.e. periodic boundary conditions $x_1(t) = x_N(t)$ and $V_1(t) = V_N(t)$ for all t are imposed) and the loss functional related to the deviations of the system from the benchmark model is given in the form (16). We note that operators such as the discrete Laplacian often employed in models concerning the transport of biomass enjoy the translation invariant property. Furthermore, for this particular approach we have to assume that the benchmark state is spatially invariant, while the analysis is simplified considerably if it is also a steady state.

When all the above assumptions are satisfied, we may treat the robust control problem (15) with the choice of objective functional as in (16) by using the discrete Fourier transform. Importantly, the problem decouples[7] in Fourier space, a fact that allows us to obtain closed form solutions in terms of the Fourier transform of X.

For a vector $x = \{x_n\} = (x_1, \ldots, x_N)$ defined on the spatial domain $\mathbb{Z}_N$, we may define a vector $\hat{x} = \{\hat{x}_k\} = (\hat{x}_1, \ldots, \hat{x}_N)$, by

$$\hat{x}_k := \sum_{n=1}^{N} x_n \exp\left(-i2\pi k \frac{n-1}{N} \right), \quad k \in \mathbb{Z}_N.$$

The k coordinates of the vector $\hat{x}$ are considered as taking values in a dual space, often called the Pontryagin dual space or simply Fourier space, which in this simple case coincides with $\mathbb{Z}_N$. The discrete Fourier transform of $X = (x, V)$ where x and V are vectors defined on the spatial domain $\mathbb{Z}_N$ is defined by $\hat{X} = (\hat{x}, \hat{V})$. The discrete Fourier transform has very interesting properties, one of which is very important in the simplification of problem (15) with the choice of objective functional as in (16). The Fourier transform turns a convolution to a product, in the sense that the Fourier transform of $\mathsf{A}x$ is equal to $\hat{\mathsf{A}}\hat{x}$ as long as A is translation invariant, i.e. commutes for all m with the translation operators T_m defined by $(T_m x)_n = x_{n-m}$ where of course periodicity is taken into account. Matrices such as those corresponding to the discrete Laplacian have this form. This property leads to a decoupled set on equations for the state variables, when treated in Fourier space. Furthermore, by the special form of the objective functional (16), the use of the Parceval identity allows us to rewrite the objective functional in essentially identical form but now interpreted in Fourier space. This leads to a decoupling of the full problem into N scalar problems which are amenable to full analytic consideration.

Denoting by $\hat{X}$ the Fourier transform of X it can be shown (see Brock et al. 2012) that the optimal state is the solution of the Ornstein-Uhlenbeck equation

$$d\hat{X}_k = R_k \hat{X}_k + \hat{\sigma}_k dw_k, \quad k \in \mathbb{Z}_N,$$

[7]Essentially turning the matrix Ricatti equation to a set of scalar, uncoupled Ricatti equations, amenable to analytic solution.

where $\hat{\sigma}_k$ is a constant whose exact expression is not needed for what follows,

$$R_k := \hat{a}_k - \frac{\hat{b}_k^2 M_{2,k}}{2q} + \frac{\hat{c}_k^2 M_{2,k}}{2\theta},$$

and $M_{2,k}$ is the solution of

$$\left(\frac{\hat{c}_k^2}{2\theta} - \frac{\hat{b}_k^2}{2q}\right) M_{2,k}^2 + (2\hat{a}_k - r) M_{2,k} + 2p = 0. \quad (18)$$

The terms $\hat{a}_k$, $\hat{b}_k$ and $\hat{c}_k$ are related to the Fourier transform of the matrices $\mathbb{A}$, $\mathbb{B}$ and $\mathbb{S}$. Furthermore, the optimal controls are given by the feedback laws

$$\hat{u}_k = -\frac{\hat{b}_k M_{2,k}}{2q} \hat{x}_k, \qquad \hat{v}_k = \frac{\hat{c}_k M_{2,k}}{2\theta} \hat{x}_k.$$

4 Hot Spot Formation

In this section we study the validity and the qualitative behavior of the controlled system (15). We will call the qualitative changes of the behavior of the system **hot spots**. In the present context, hot spots will correspond to important deviations of the stabilization procedure presented in the previous section, that will have as consequence important quantitative and qualitative deviations of the true controlled system from the desired ideal benchmark model, no matter what the decision maker does in order to correct her policy by proper adjustment procedures. We may thus consider hot spots as possible failures of the adjustment procedure, which may have important consequences on the true state of the controlled fishery.

We will define three types of hot spots:

Hot spot of type I: This is a breakdown of the solution procedure, i.e., a set of parameters where a solution to the above problem does not exist.

Hot spot of type II: This corresponds to the case where the solution exists but may lead to spatial pattern formation, i.e., to spatial instability similar to the Turing instability.

Hot spot of type III: This corresponds to the case where the cost of robustness becomes more that what is offering us, i.e., where the relative cost of robustness may become very large.

In what follows, for simplicity, we discuss the formation of hot spots under the assumption that the tracking problem (15) with the choice of objective functional as in (16) is translation invariant (which requires certain symmetry conditions). The results are stated in terms of a number of propositions, providing relevant parameter values for the formation of the different types of hotspots, the proofs of which may be found in Brock et al. (2012). However, similar results hold for the general case of non-translation invariant systems, by the full treatment of the matrix Ricatti equation (see Brock et al. 2012).

4.1 Hot Spots of Type I

The breakdown of the solution procedure can be seen quite easily by the following simple argument. The value function assumes a simple quadratic form, as long as the algebraic quadratic equation

$$\left(\frac{\hat{c}_k^2}{2\theta}-\frac{\hat{b}_k^2}{2q}\right)M_{2,k}^2+(2\hat{a}_k-r)M_{2,k}+2p=0, \tag{19}$$

admits real valued solutions, at least one of which is positive. The positivity of the real root is needed since, by general considerations in optimal control, the value function must be convex. If the above algebraic quadratic equation does not admit at least one positive real valued solution this is an indication of breakdown of the existence of a solution to the robust control problem which will be called a hot spot of Type I.

Proposition 1 (Type I hot spot creation) *Hot spots of Type I may be created in one of the following two cases*:

(I_A) *Either*,

$$(2\hat{a}_k-r)^2<8p\left(\frac{\hat{c}_k^2}{2\theta}-\frac{\hat{b}_k^2}{2q}\right). \tag{20}$$

(I_B) *Or*,

$$(2\hat{a}_k-r)^2>8p\left(\frac{\hat{c}_k^2}{2\theta}-\frac{\hat{b}_k^2}{2q}\right),\qquad \left(\frac{\hat{c}_k^2}{2\theta}-\frac{\hat{b}_k^2}{2q}\right)>0,\quad 2\hat{a}_k-r>0. \tag{21}$$

Hot spots of this type may arise either due to low values of θ, or due to high values of q or low values of r. For example, they may arise either if

$$\theta<\frac{p\hat{c}_k^2}{(\hat{a}_k-\frac{r}{2})^2+\frac{p}{q}\hat{b}_k^2},\qquad k\in\mathbb{Z}_N,$$

or if

$$\theta>\frac{p\hat{c}_k^2}{(\hat{a}_k-\frac{r}{2})^2+\frac{p}{q}\hat{b}_k^2},\qquad \frac{q}{\theta}>\frac{\hat{b}_k^2}{\hat{c}_k^2},\qquad r<2\hat{a}_k,\ k\in\mathbb{Z}_N.$$

In particular hot spots are expected to occur in the limit as $\theta\to 0$ while they are not expected to occur in the limit as $\theta\to\infty$.

As mentioned above, a hot spot of Type I represents breakdown of the solvability of the optimal control problem. We argue that this represents some sort of loss of convexity of the problem thus leading to non existence of solution. To illustrate this point more clearly let us take the limit as $\theta\to 0$ which corresponds to hot spot formation. For such values of θ, the particular ansatz employed for the solution breaks down and in fact as $\theta \to 0$ we expect $M_{2,k} \to 0$ so that the quadratic term in the value function will disappear. This leads to loss of strict concavity of the functional, which may be seen as follows: The functional contains a contribution from $\hat{v}_k$ through the dependence of $\hat{x}_k$ on $\hat{v}_k$ which contributes a quadratic term of positive sign in $\hat{v}_k$. The robustness term, which is proportional to $-\theta$ contributes a quadratic term of negative sign in $\hat{v}_k$. For large enough values of θ the latter term dominates in the functional and guarantees the strict concavity, therefore, leading to a well defined maximization problem. In the limit of small θ the former term dominates and thus turn the functional into a convex functional leading to problems with respect to the maximization problem over $\{\hat{v}_k\}$. We call this breakdown of concavity in v, which lead to loss of convexity of the value function in x, for small values of θ a hot spot of type I. When this happens, there is a duality gap, since the assumptions of the min-max theorem do not hold. In terms or regulatory objectives this means that concerns about model misspecification make regulation impossible.

The effect of the parameters of the fishery model employed on the formation of hot spots, can quantified by the results of Proposition 1 through the dependence of the Fourier transformed operators $\mathbb{A}$, $\mathbb{B}$, $\mathbb{S}$ on the model parameters. For instance, if prices $\mathscr{P}$ increase, whereas the rest of the parameters remain fixed, then $\hat{b}_k$ will increase with respect to the other parameters $\hat{a}_k$ and $\hat{c}_k$. This will result to a decrease of the right hand side of e.g., Eq. (20) thus leading to a suppression of such a hot spot. Due to the large number of parameters of the model, extreme care should be taken when interpreting qualitatively the above conditions. However, having chosen a particular model and having estimated some of the parameters, the decision maker may investigate numerically the above analytic conditions and provide parameter regimes for creation or suppression of the various type of hot spots. Since our major interest here is the formulation of a general methodology, rather than a detailed treatment of a particular model, we provide two simplified examples that allow us to provide a qualitative understanding of hot spot formation as a result of the various interacting "forces" that influence the system and comment upon their relative importance. These examples are provided here, for lack of space, to hot spots of type I only, but they can be extended to the study of the other hot spots as well.

The following examples show some interesting limiting situations, in terms of simplifications of the operators $\mathbb{A}$, $\mathbb{B}$ and $\mathbb{S}$:

Example 1 Assume that $\mathbb{A}$ is the discrete Laplacian whereas $\mathbb{B}$ and $\mathbb{S}$ are copies of the identity operator. This corresponds to the case that there is diffusive coupling in the state equation but controls as well as the uncertainty have purely localized effects. A quick calculation shows that in this case $a_k = \alpha(1 + 2\cos(\frac{2\pi k}{N}))$ where α is the diffusion coefficient whereas $b_k = \beta$ and $c_k = \gamma$ for every $k \in \mathbb{Z}_N$ where β and γ is a measure for the control and the uncertainty respectively. In this particular case, the quadratic equation becomes

$$\left(\frac{\gamma^2}{2\theta} - \frac{\beta^2}{2q}\right) M_{2,k}^2 + \left(2\alpha\left(1 + 2\cos\left(\frac{2\pi k}{N}\right)\right) - r\right) M_{2,k} + 2p = 0,$$

which must have a real valued solution for every k. There will not exist real valued solutions if

$$\Delta := \left(2\alpha\left(1+2\cos\left(\frac{2\pi k}{N}\right)\right)-r\right)^2 - 8p\left(\frac{\gamma^2}{2\theta}-\frac{\beta^2}{2q}\right) < 0$$

or equivalently after some algebra

$$\left(\left(1+2\cos\left(\frac{2\pi k}{N}\right)\right)^2 - \frac{r}{2\alpha}\right) < \frac{p}{\alpha^2}\left(\frac{\gamma^2}{\theta}-\frac{\beta^2}{q}\right).$$

This is the condition for generation of a hot spot of Type I in this particular example. If this condition holds for some $k \in \mathbb{Z}_N$, this particular k is a candidate for such a hot spot. We may spot directly that this cannot hold for any $k \in \mathbb{Z}_N$ if the right hand side of this inequality is negative, i.e., when $\theta > \theta_{cr} := q\frac{\gamma^2}{\beta^2}$, therefore hot spots of this type will never occur for large enough values of θ. The critical value of θ for the formation of such hot spots will depend on the relative magnitude of uncertainty over control. For $\theta < \theta_{cr}$ then a hot spot of Type I may occur for the modes k such that

$$\left(1+2\cos\left(\frac{2\pi k}{N}\right)\right)^2 \leq \frac{r}{2\alpha}+\rho$$

or equivalently for k such that

$$\left(1+2\cos\left(\frac{2\pi k}{N}\right)\right)^2 \leq \left(\frac{r}{2\alpha}+\rho\right)^{\frac{1}{2}},$$

where $\rho^2 = \frac{p}{\alpha^2}(\frac{\gamma^2}{\theta}-\frac{\beta^2}{q})$.

Example 2 The opposite case is when $\mathbb{A}$ is again the discrete Laplacian while $\mathbb{B}$ and $\mathbb{S}$ are multiples of matrices containing 1 in the diagonal and the same entry ν in every other position. This means that the controls as well as the uncertainty has a globalized effect to all lattice points, in the sense that the controls even at remote lattice sites have an effect at each lattice point. Then $\hat{b}_k = \beta\delta_{k,0}$, $\hat{c}_k = \gamma\delta_{k,0}$, i.e., the Fourier transform is fully localized and is a delta function. Then, for $k = 0$ the quadratic equation becomes

$$\left(\frac{\gamma^2}{2\theta}-\frac{\beta^2}{2q}\right)M_{2,0}^2 - (6\alpha - r)M_{2,0} + 2p = 0,$$

while for $k \neq 0$ the quadratic term vanishes yielding

$$-\left(2\alpha\left(1+2\cos\left(\frac{2\pi k}{N}\right)\right)-r\right)M_{2,0} + 2p = 0.$$

4.2 Hot Spots of Type II

We now consider the spatial behavior of the optimal path, as given by the Itô stochastic differential equation

$$d\hat{x}_k^* = R_k \hat{x}_k^* dt + \hat{c}_k d\hat{w}_k.$$

The optimal path is a random field, thus leading to random patterns in space, some of which may be short lived and generated simply by the fluctuations of the Wiener process. We thus look for the spatial behavior of the mean field as described by the expectation $\hat{X}_k := \mathbb{E}_Q[\hat{x}_k^*]$. By standard linear theory $\hat{X}_k(t) = \hat{X}_k(0)\exp(R_k t)$ and this means that for the modes $k \in \mathbb{Z}_N$ such that $R_k \geq 0$ we have temporal growth and these modes will dominate the long term temporal behavior. On the contrary modes k such that $R_k < 0$ decay as $t \to \infty$ therefore such modes correspond to (short term) transient temporal behavior, not likely to be observable in the long term temporal behavior. The above discussion implies that the long time asymptotic of the solution in Fourier space will be given by

$$\hat{X}_k(t) \simeq \begin{cases} \hat{x}_k(0)\exp(R_k t), & k \in \mathscr{P} := \{k \in \mathbb{Z}_N : R_k \geq 0\}, \\ 0, & \text{otherwise.} \end{cases}$$

To see what this pattern will look like in real space, we simply need to invert the Fourier transform, thus obtaining a spatial pattern of the form

$$X_n(t) := \mathbb{E}_Q\big[x_n(t)\big] = \sum_{k \in \mathscr{P}} \hat{x}_k(0)\exp(R_k t)\cos\left(2\pi \frac{k}{N} n\right). \tag{22}$$

The above discussion therefore leads us to a very important conclusion, which is of importance to economic theory of spatially interconnected systems:

> If as an effect of the robust optimal control procedure exerted on the system there exist modes $k \in \mathbb{Z}_N$ such that $R_k > 0$, then this will lead to spatial pattern formation which will create spatial patterns of the form (22). As we will see there are cases what such patterns will not exist in the uncontrolled system and will appear as an effect of the control procedure. We will call such patterns an **optimal robustness induced spatial instability** or hot spot of Type II.

The economic significance of this result should be stressed. We show the emergence of a spatial pattern formation instability, which can be triggered by the optimal control procedures exerted on the system; in other words emergence of spatial clustering and agglomerations in the fishery (as observed in the spatial distribution of the biomass and the number of vessels) caused by uncertainty aversion and robust control. This observation can further be extended in the case of nonlinear dynamics, in the weakly nonlinear case. When the dynamics are nonlinear in the state the emergence of hot spots of Type II and optimal robustness induced spatial instability should be linked to the spatial instability of a spatially uniform steady state corresponding to the linear quadratic approximation of a nonlinear system. This instability which can be thought as pattern formation precursor will induce the emergence of spatial clustering. As time progresses and the linearized solution (22) grows beyond a certain critical value (in terms of a relevant norm) then the deviation from the homogeneous steady state is so large that the linearized dynamics are no longer a valid approximation. Then the nonlinear dynamics will take over and as an effect of that some of the exponentially growing modes could be balanced thus leading to more complicated stable patterns. At any rate even in the nonlinear case the mechanism described here will be a Turing type pattern formation mechanism explaining the onset of spatial patterns in the fishery.

The next proposition identifies which modes can lead to hot spot of Type II formation (optimal robustness induced spatial instability) and in this way through Eq. (22) identifies possible spatial patterns that can emerge in the fishery.

Proposition 2 (Pattern formation for the primal problem) *There exist pattern formation behavior for the primal problem if there exist modes k such that $R_k > 0$, i.e., if there exist modes k such that*

$$\frac{1}{2}\left(r-\sqrt{r^2+8p\left(\frac{\hat{c}_k^2}{2\theta}-\frac{\hat{b}_k^2}{2q}\right)}\right) \leq \hat{a}_k \leq \frac{1}{2}\left(r+\sqrt{r^2+8p\left(\frac{\hat{c}_k^2}{2\theta}-\frac{\hat{b}_k^2}{2q}\right)}\right),$$

$$r^2+8p\left(\frac{\hat{c}_k^2}{2\theta}-\frac{\hat{b}_k^2}{2q}\right) \geq 0. \tag{23}$$

It is interesting to see what is the behavior of the system as a function of parameters with respect to pattern formation and the qualitative behavior of the optimal path.

Note that this pattern formation behavior is in full accordance with the fact that our state equation is the optimal path for the linear quadratic control problem. Since it solves this problem it is guaranteed that $I := \mathbb{E}_Q[\int_0^\infty e^{-rt}\hat{x}_k^2(t)dt]$ is finite[8] therefore $\hat{x}_k(t)$ can at most grow as $e^{\frac{r}{2}t}$, otherwise the quantity I would be infinite. This is verified explicitly by the observation that $R_k \leq \frac{r}{2}$ for every $k \in \mathbb{Z}_N$. Therefore, all possible patterns may at most exhibit growth rates less or equal to $r/2$. In the limit as $r \to 0$ i.e. in the limit of small discount rates pattern formation is becoming increasingly difficult in the linear quadratic model since growing patterns will be suppressed by the control procedures.

Proposition 3 (Stabilizing or destabilizing effects of control) *The robust control procedure may either have a stabilizing or destabilizing effect with respect to pattern formation. in the sense that it may either stabilize an unstable mode of the uncontrolled system or on the contrary facilitate the onset of instabilities.*

In particular,

(i) *If* $\frac{q}{\theta} < \frac{\hat{b}_k^2}{\hat{c}_k^2}$ *then the robust control procedure has a stabilizing effect*

[8]This is in fact equivalent to the assertion that the optimal path satisfies temporal transversality conditions at infinity.

(ii) *If* $\frac{q}{\theta} > \frac{\hat{b}_k^2}{\hat{c}_k^2}$ *then the robust control procedure has a destabilizing effect.*

Case (ii) *suggests robust control caused pattern formation, in the sense that we obtain a growing mode leading to a pattern which would not have appeared in the uncontrolled system.*

As seen by Proposition 3 in the $\theta \to \infty$ limit, the control has a stabilizing effect on unstable modes of the uncontrolled system. Similarly, by Proposition 3 in the $\theta \to 0$ limit, the robust control has a destabilizing effect on modes of the uncontrolled system which are "marginal" to be stable i.e. with $\hat{\alpha}_k$ negative but close to zero.

In closing this discussion we wish to ponder upon some similarities and differences of Type II hot spots with the occurrence of the celebrated Turing instability; Formation of hot spots of type II is similar to Turing instability leading to pattern formation but with a very important difference! In contrast to Turing instability which is observed in an uncontrolled forward Cauchy problem, this instability is created in an optimally controlled problem in the infinite horizon. This has important consequences and repercussions both from the conceptual as well as from the practical point of view. On the conceptual level, a controlled system is related to a system that somehow its final state (at $t \to \infty$ in our case) is predescribed. Therefore, our result is an "extension" of Turing instability in a forward-backward system and not just to a forward Cauchy problem, as is the case for the Turing instability. On the practical point of view, the optimal control nature of the problem we study here induces serious constraints on the growth rate of the allowed patterns which has a strict upper bound is related only to the discount factor of the model and not on the operator A. This is not the case for the standard Turing pattern formation mechanism, in which the growth rate upper bound is simply related to the spectrum of the operator A.

4.3 Hot Spots of Type III: The Cost of Robustness

The value function is of the form $V_k = \frac{M_{2,k}}{2}\hat{x}_k^2 + \frac{\hat{c}_k^2 M_{2,k}}{2r}$. This gives us the total cost of the minimum possible deviation from the desired goal and it is made up from contributions by three terms:

- the term proportional to p in the cost functional which corresponds to the cost related to the deviation from the desired target,
- the term proportional to q in the cost functional which corresponds to the cost related to the cost of the control u needed to drive the system to the desired target and
- the term proportional to θ in the cost functional which corresponds to the cost of robustness (which is the cost incurred by the regulator because she wants to be robust when she has concerns about the misspecification of the model).

The value function depends on all these three contributions and this may be clearly seen since $M_{2,k}$ is in fact a function of the parameters p, q, θ.

An interesting question is which is the relevant importance of each of these contributions in the overall value function. Does one term dominates over the others or not?

A simple answer to this question will be given by the elasticity of the value function with respect to these parameters, i.e., by the calculation of the quantities $\frac{1}{V}\frac{\partial V}{\partial p}$, $\frac{1}{V}\frac{\partial V}{\partial q}$ and $\frac{1}{V}\frac{\partial V}{\partial \theta}$. It is easily seen that these elasticities are independent of $\hat{x}_k$ and reduce to $\frac{1}{M_{2,k}}\frac{\partial M_{2,k}}{\partial p}$, $\frac{1}{M_{2,k}}\frac{\partial M_{2,k}}{\partial q}$ and $\frac{1}{M_{2,k}}\frac{\partial M_{2,k}}{\partial \theta}$, respectively. Whenever one of these quantities tends to infinity, that means that the contribution of the relevant procedure dominates the control problem.[9]

In particular whenever $\frac{1}{M_{2,k}}\frac{\partial M_{2,k}}{\partial \theta} \to \infty$, then we say that the cost of robustness becomes more expensive than what it offers, and we will call that a hot spot of type III. This quantity can be calculated directly from the solution of the quadratic equation (18) through straightforward but tedious algebraic manipulations, which we choose not to reproduce here.

However, an illustrative partial case, which allows some insight on the nature of hot spots of type III is the following:

Differentiating (18) with respect to θ yields

$$-\frac{\hat{c}_k^2}{2\theta^2}M_{2,k}^2 + 2\left(\frac{\hat{c}_k^2}{2\theta} - \frac{\hat{b}_k^2}{2q}\right)M_{2,k}\frac{\partial M_{2,k}}{\partial \theta} + (2\hat{a}_k - r)\frac{\partial M_{2,k}}{\partial \theta} = 0.$$

Dividing by $M_{2,k}^2$ we obtain

$$-\frac{\hat{c}_k^2}{2\theta^2} + 2\left(\frac{\hat{c}_k^2}{2\theta} - \frac{\hat{b}_k^2}{2q}\right)\frac{1}{M_{2,k}}\frac{\partial M_{2,k}}{\partial \theta} + (2\hat{a}_k - r)\frac{1}{M_{2,k}^2}\frac{\partial M_{2,k}}{\partial \theta} = 0.$$

Let us now take the particular case where $2\hat{a}_k = r$, so that

$$\frac{1}{M_{2,k}}\frac{\partial M_{2,k}}{\partial \theta} = \frac{\hat{c}_k^2}{4\theta\left(\frac{\hat{c}_k^2}{2\theta} - \frac{\hat{b}_k^2}{2q}\right)},$$

which becomes infinite for values of θ such that $\theta \to \frac{q\hat{c}_k^2}{\hat{b}_k^2}$. The general case $2\hat{a}_k \neq r$ may present similar phenomena.

---

[9]This interpretation arises from observation that close to a point (p_0, q_0, θ_0) the value function behaves as

$$V_k \simeq \left.\frac{\partial V_k}{\partial p}\right|_{p=p_0}(p - p_0) + \left.\frac{\partial V_k}{\partial q}\right|_{q=q_0}(q - q_0) + \left.\frac{\partial V_k}{\partial \theta}\right|_{\theta=\theta_0}(\theta - \theta_0).$$

5 Concluding Remarks

In this paper we studied the optimal management of a commercial fishery which is distributed over a finite spatial domain, is characterized by stock, congestion and productivity externalities, and the fishery manager has concerns about model misspecification.

We solve this problem as a robust control linear quadratic distributed parameter model. The linear quadratic approximation is formulated as a tracking problem where stochastic dynamics indicating model uncertainty are linearized around a deterministic optimal path, and the control process aims at keeping the system close to the optimal path. Harvesting rules are obtained as robust tracking rules which can be used by the manager to set policy such as quotas on each site of the spatial domain.

An important result of our paper is the identification of spatial hot spots, which are sites of special interest emerging from the interactions between concerns about model uncertainty, spatial interactions and the structure of the fishery. In such hot spots optimal robust regulation may be impossible and the inability to regulate is extended to the whole domain (type I hot spot); regulation may lead to spatial nonhomogeneity in the harvesting rules, implying spatially differentiated quotas (type II hot spots); or misspecification concerns may lead to very costly regulation, indicating excessive cost of robust regulation (type III hot spot).

These results although qualitative in nature provide insights to regulation of a commercial fishery under model uncertainty and under explicit spatial interactions. Future research may include solution of a linear quadratic approximation in the sense of Magill approximations, instead of solution of the optimal tracking problem, or attempts to characterize the solution of the full nonlinear problem.

Acknowledgements This research has been co-financed by the European Union (European Social Fund—ESF) and Greek national funds through the Operational Program "Education and Lifelong Learning" of the National Strategic Reference Framework (NSRF)—Research Funding Program: Thalis—Athens University of Economics and Business—Optimal Management of Dynamical Systems of the Economy and the Environment.

References

Artstein, Z., & Leizarowitz, A. (1985). Tracking periodic signals with the overtaking criterion. *IEEE Transactions on Automatic Control*, *30*(11), 1123–1126.

Athanassoglou, S., & Xepapadeas, A. (2012). Pollution control with uncertain stock dynamics: when, and how, to be precautious. *Journal of Environmental Economics and Management*, *63*, 304–320.

Brock, W., & Xepapadeas, A. (2008). Diffusion-induced instability and pattern formation in infinite horizon recursive optimal control. *Journal of Economic Dynamics & Control*, *32*(9), 2745–2787.

Brock, W., & Xepapadeas, A. (2010). Pattern formation, spatial externalities and regulation in coupled economic-ecological systems. *Journal of Environmental Economics and Management*, *58*, 149–164.

Brock, W., Xepapadeas, A., & Yannacopoulos, A. N. (2012). Robust control and hot spots in spatiotemporal economic systems. Athens University of Economics and Business. Working Paper (DIES), 1223 pp.
Gilboa, I., & Schmeidler, D. (1989). Maxmin expected utility with non-unique prior. *Journal of Mathematical Economics*, *18*, 141–153.
Goetz, R. U., & Zilberman, D. (2000). The dynamics of spatial pollution: the case of phosphorus runoff from agricultural land. *Journal of Economic Dynamics & Control*, *24*(1), 143–163.
Hansen, L. P., & Sargent, T. J. (2001). Robust control and model uncertainty. *The American Economic Review*, *91*(2), 60–66.
Hansen, L. P., & Sargent, T. (2008). *Robustness in economic dynamics*. Princeton: Princeton university press.
Hansen, L. P., Sargent, T. J., Turmuhambetova, G., & Williams, N. (2006). Robust control and model misspecification. *Journal of Economic Theory*, *128*, 45–90.
JET (2006). Symposium on model uncertainty and robustness. *Journal of Economic Theory*, *128*, 1–163.
Karatzas, I., & Shreve, S. E. (1991). *Brownian motion and stochastic calculus*. Berlin: Springer.
Leizarowitz, A. (1985). Infinite horizon autonomous systems with unbounded cost. *Applied Mathematics & Optimization*, *13*(1), 19–43.
Leizarowitz, A. (1986). Tracking nonperiodic trajectories with the overtaking criterion. *Applied Mathematics & Optimization*, *14*(1), 155–171.
Magill, M. (1977a). A local analysis of N-sector capital accumulation under uncertainty. *Journal of Economic Theory*, *15*(1), 211–219.
Magill, M. J. P. (1977b). Some new results on the local stability of the process of capital accumulation. *Journal of Economic Theory*, *15*(1), 174–210.
Murray, J. (2003). *Mathematical biology* (3rd ed., Vols. I and II). Berlin: Springer.
Salmon, M. (2002). *Special issue on robust and risk sensitive decision theory. Macroeconomic dynamics*. Cambridge: Cambridge University Press.
Sanchirico, J. N., & Wilen, J. E. (1999). Bioeconomics of spatial exploitation in a patchy environment. *Journal of Environmental Economics and Management*, *37*(2), 129–150.
Smith, V. L. (1969). On models of commercial fishing. *Journal of Political Economy*, *77*(2), 181–198.
Smith, M. D., Sanchirico, J. N., & Wilen, J. E. (2009). The economics of spatial-dynamic processes: applications to renewable resources. *Journal of Environmental Economics and Management*, *57*(1), 104–121.
Turing, A. (1952). The chemical basis of morphogenesis. *Philosophical Transactions of the Royal Society of London*, *237*, 37–72.
Wilen, J. E. (2007). Economics of spatial-dynamic processes. *American Journal of Agricultural Economics*, *89*(5), 1134–1144.
Xabadia, A., Goetz, R., & Zilberman, D. (2004). Optimal dynamic pricing of water in the presence of waterlogging and spatial heterogeneity of land. *Water Resources Research*. doi:10.1029/2003WR002215.

On the Effect of Resource Exploitation on Growth: Domestic Innovation vs. Technological Diffusion Through Trade

Francisco Cabo, Guiomar Martín-Herrán, and María Pilar Martínez-García

Abstract The economic growth in a developing country endowed with a natural resource and with a resource-dependent economy can be based on its own investments in new technology. Conversely, it can rely on trade as a channel for technology diffusion from a technologically advanced country. The existence, uniqueness and stability of a sustainable growth path are proved under both assumptions. Our second concern is on the resource curse hypothesis. When the developing country does not export the natural resource but uses it as an essential input in the production of a final good, resource bounty is not a curse. Resource abundance increases long-run growth in the closed-economy scenario, and it is growth-neutral but consumption-enhancing when technology is transmitted from abroad through international trade.

1 Introduction

A significant number of developing economies are linked to the extraction of natural resources, which can be either exported or transformed in the productive process. One of the main challenges for these economies is to achieve unlimited economic growth with a finite natural-resource sector. The finite character of the resource sector refers to the fact that, although a renewable resource is considered, its availability as a productive input at each time is bounded from above. The debate on the effect of resource richness on the growth rate of the economy is closely connected to this aim of sustainability. While resource abundance seems to slow economic growth

F. Cabo · G. Martín-Herrán (✉)
Departamento de Economía Aplicada (Matemáticas), IMUVA, Universidad de Valladolid, Avda. Valle de Esgueva, 6, 47011 Valladolid, Spain
e-mail: guiomar@eco.uva.es

F. Cabo
e-mail: pcabo@eco.uva.es

M.P. Martínez-García
Departamento de Métodos Cuantitativos para la Economía, Universidad de Murcia, Campus de Espinardo, 30100 Murcia, Spain
e-mail: pilarmg@um.es

E. Moser et al. (eds.), *Dynamic Optimization in Environmental Economics*, Dynamic Modeling and Econometrics in Economics and Finance 15,
DOI 10.1007/978-3-642-54086-8_11,

down in many countries, others take advantage of this resource bounty to improve their own growth rate (see, for example, the excellent surveys by van der Ploeg 2011 and Frankel 2012). The conditions that cause a natural resource to be a curse or a blessing are at the core of our analysis. Moreover, our paper deals with the challenge of sustainability of resource-dependent economies within a context of international trade.

Since the 90s, authors like Grossman and Helpman (1991), Smulders (1995), Bovenberg and Smulders (1995) and Elbasa and Roe (1996) among others, suggest that the chances of achieving sustainable growth are critically dependent on maintaining a steady flow of technological innovation, a conclusion that is roughly consistent with the historical experience of industrialized countries. However, many developing economies rely on underdeveloped, or even non-existent R&D sectors.

To understand the mechanism through which a developing economy endowed with a renewable natural resource can attain sustainability, two models are studied. A closed economy that invests in innovation and an open economy which imports new technology from abroad. The first model follows the literature on endogenous growth and the environment developed during the 90s. Investment in R&D is the base for sustainable growth, although it may not completely explain the realities of developing countries with a weak or null investment sector. The major source of innovation in developing countries is not domestic but foreign R&D investment, which can be transferred by international trade.[1]

However, most endogenous growth models that tackle environmental problems study an isolated country and do not take into account trade relationships as a transmission channel for economic growth.[2] The technology transfer through international trade is analyzed in the second model in this paper, where technological improvements occur in a technologically leading country as an increase in the number of varieties of intermediate goods. These new varieties are adapted to the production process in the developing economy, which buys them from their producers in the leading country. Technology diffusion through international trade is presented as the key to sustainability in industrializing economies. Given the conditions for sustainable growth, the second question at stake is the effect of resource wealth on the growth rate of the economy.

A large body of empirical work shows a negative relationship between resource abundance and economic performance (see, for example, Gylfason 2001 and Sachs and Warner 2001). However, history asserts that some countries have managed to take advantage of their natural wealth and received a blessing. Many authors think

[1]Coe et al. (1997) reports that, in 1990, industrial countries accounted for 96 % of the world's R&D expenditure. Ninety percent of the world's patents originate from countries like the United States, Japan, Germany, France and the UK. The rest of the countries in the world are considered technological followers. For a recent study see Coe et al. (2009).

[2]An exception is Cabo et al. (2005), where a natural resource extracted in one country is traded in exchange for consumption goods. Although no technology is transferred through this trade channel, the consumption growth in the resource-dependent economy is a direct consequence of the economic growth in the industrialized country.

that the positive economic development of countries like Australia, Canada, Iceland, United States, New Zealand and the Scandinavian countries was stimulated by their resource wealth (see, for example, Stevens 2003; Mehlum et al. 2006 and references therein). There is no single explanation of which conditions create a curse rather than a blessing. Auty (2001), for example, cites several causes: structuralist policies, Dutch disease, policy failure, inefficient investment and rent seeking. Recently, there has been a growing evidence that institutions are a key element in avoiding the curse (see, for example, Holder 2006 and Mehlum et al. 2006).

Within this debate, our model, which is to some extent based on Elíasson and Turnovsky (2004), is in contrast to their conclusion of a resource curse. Within an international trade framework, Elíasson and Turnovsky (2004) model a small open economy in which the renewable resource is used to purchase imports of a consumption good. They prove the existence of a sustainable growth path with production being independent of the harvested resource. One of their main conclusions is that resource abundance reduces the long-run growth rate. More labor is allocated in the resource sector at the expense of less employment in the final output sector, and consequently, the long-run growth rate is shortened. This result is obtained without invoking other explanatory variables given in the literature, such as rent seeking, sub-optimal allocation of resources, terms of trade or political incentives (see Stevens 2003; Papyrakis and Gerlagh 2004; Robinson et al. 2006).

In Cabo et al. (2008) the harvested resource is not exported, but used as an essential input to produce an elaborated final consumption good, which constitutes the exports of the developing economy. Two scenarios are compared depending on whether innovation only occurs in the technologically leading country or also in the resource-abundant developing economy. In either case resource abundance affects the real exchange rate and hence consumption, however it has no effect on the long run growth rate. The idea that the natural resource is not a curse if it is transformed before exporting, is empirically supported by Brunnschweiler and Bulte (2008). The authors find a natural resource curse when considering primary export, but a resource blessing if resource wealth data do not take exports into account.

In the present paper the modelization of the technology diffusion through trade is inherited from Cabo et al. (2008), giving entrance to a more general harvesting function. Therefore, resource abundance does not affect the long-run growth rate of a resource-abundant open economy which imports technology from abroad. Moreover, we also analyze the case of a closed economy which carries out innovative activities, and which growth rate will be dependent on resource bounty.

This paper throws light on two of the main challenges faced by developing economies. First, sustainable economic growth can be attained even if the country does not invest in technological progress but innovation is imported from a technologically leading country. Secondly, the growth rate of the economy does not necessarily shrink with the resource wealth if the harvested resource is not directly exported but it is transformed into a final consumption good. Resource curse is turned into a blessing when considering a closed developing country that carries out R&D activities. Alternatively, resource abundance is growth-neutral when technology improvements are internationally traded from the technologically leader country.

The rest of the paper is organized as follows. In Sect. 2, we present a closed economy endowed with a natural resource that also invests in technological innovation. In Sect. 3, the economy does not carry out R&D activities, but relies on technology diffusion through international trade. In both sections we study the existence, uniqueness and stability of a steady-state equilibrium. We provide a sensitivity analysis of the steady-state equilibrium. In Sect. 4, we compare the long-run growth rates and the consumer welfare obtained with domestic innovation and technology diffusion through international trade. Section 5 gives our conclusions.

2 Sustainable Growth with Domestic Innovation

In this section we deal with a closed economy endowed with a stock of a renewable natural resource, which is harvested and used as an essential input in the production of final output, combined with labor and intermediate nondurable goods. The total labor force is allocated between the harvesting of the natural resource and the production of final output. Intermediate goods are invented and produced by monopolistic entrepreneurs. Population is assumed to be constant, $\bar{L}$.

In the resource sector, the property rights associated with the natural resource are equally distributed among identical consumer-owner agents. Each agent initially owns a portion s_0 of the natural resource.[3] The net growth rate of each agent's resource share is given by its natural reproduction minus the harvesting, that is,[4]

$$\dot{s} = g(s) - h = \tilde{g}s(1 - s/\kappa) - h, \quad s(0) = s_0, \tag{1}$$

where s is the stock of the consumer-owned natural resource, $g(s)$ describes its gross reproduction rate of the logistic or Verlhust type (see, for example, Clark 1990) and h is the rate of harvest. Parameters $\tilde{g}$ and κ denote the intrinsic growth rate and the carrying capacity or saturation level of each agent's resource share.

A representative consumer is also endowed with one unit of labor per unit of time. At each time, the consumer supplies a fraction v of its labor to produce final output and a fraction $1 - v$ to harvest its natural resource share, with $v \in [0, 1]$.

[3]Historically, the distribution of communal resources among the users is one of the solutions that the economic literature has proposed to avoid the overexploitation of open-access resources. This approach relies on an external authority, who distributes the property rights. However, researchers have recently proved that private property rights may emerge internally as a result of individual agents' desire to avoid cost externalities. See Birdyshaw and Ellis (2007) and the real examples therein. Cabo et al. (2012) analyze how the ownership and distribution of the exploitation rights upon the natural resource may affect the sustainable growth rate for the two trading economies and the resource conservation.

Although the distribution of property rights among users can be easily established is same cases, like forestry, partitioning is unfeasible for other resources such as fisheries. Nevertheless, results hereafter remain valid if the resource is managed by some institution like a fishers' association or cooperative that regulates the optimal exploitation of the resource.

[4]The time argument is eliminated when no confusion can arise.

The harvesting of the resource h depends on labor and on the size of the renewable resource (its stock). The harvesting function presents decreasing marginal returns to the effort (in our case, labor) and the stock level. Thus, the per-capita harvest rate is:

$$h(v, s) = b(1 - v)^{1-\delta} s^{\theta}, \quad b > 0,\ 0 < \delta < 1,\ 0 \leq s \leq \kappa,\ 0 \leq \theta \leq 1. \tag{2}$$

In particular, $\theta = 0$ implies that harvesting is independent of the stock size; while $\theta = 1$, corresponds to the well-known Schaefer pattern. In this case, the harvest is proportional to the stock of the renewable resource.[5] In what follows we shall name the harvest flow h, omitting the arguments v and s.

A representative consumer sells the extracted natural resource to final output producers, who use it as a productive input. A consumer receives the income derived from the exploitation of the resource, which is sold at a price p_h, and a wage income derived from its labor services in the final output sector, where w is the wage rate. In addition, a consumer accumulates assets and receives financial interest income from them. Thus, the per-capita budget constraint for a representative consumer is

$$\dot{a} = ra + vw + p_h h - c, \quad a(0) = a_0, \tag{3}$$

where a are the per-capita assets, r is the rate of return on assets, c is the per-capita consumption of final good, and a_0 is the initial amount of per-capita assets.

A representative consumer has to decide consumption, c, and the fraction of labor, v and $1 - v$, employed either in the final-output production or in harvesting, to maximize discounted utility:[6]

$$\max_{c,v} \int_0^{\infty} \ln(c) e^{-\rho t} dt, \quad \rho > 0, \tag{4}$$

subject to (1) and (3).[7]

The final output sector comprises a large number of identical firms. Output production demands labor, natural resources and intermediate goods. Thus, the output-production function of a representative firm is given by

$$Y = A(v\bar{L})^{1-\alpha-\beta} \sum_{j=1}^{N} X_j^{\alpha} (h\bar{L})^{\beta}, \quad A > 0,\ 0 < \alpha,\ \beta,\ \alpha + \beta < 1, \tag{5}$$

where N is the number of intermediate good varieties and X_j is the amount of the jth type of intermediate goods, $j \in \{1, \ldots, N\}$. This is the standard production function with an increasing variety of inputs, (see, for example, Barro and Sala-i Martin

[5]The hypothesis $\theta = 0$ is appropriate for forests or fish living close to the surface; whereas, $\theta = 1$ is suitable for bottom-dwelling fish (see Elíasson and Turnovsky 2004 and references therein).

[6]Our results are upheld for any iso-elastic utility function. For the sake of simplicity we have chosen the logarithmic expression.

[7]Under the assumption of perfect property rights, each consumer bears the full cost of their actions and the resource is used more efficiently than under open-access.

1999, Chap. 6). Additionally, the natural resource is a necessary factor for production in (5). Output production has diminishing marginal productivity for each input $(v\bar{L}, X_j, h\bar{L})$, and constant returns to scale for all inputs taken together. The final good sector is competitive and firms take prices as given. Therefore, the problem of a representative firm in the final-good sector is given by

$$\max_{v,h,X_j,j=1,\ldots,N} A(v\bar{L})^{1-\alpha-\beta} \sum_{j=1}^{N} X_j^{\alpha}(h\bar{L})^{\beta} - w(v\bar{L}) - \sum_{j=1}^{N} p_j X_j - p_h h\bar{L}, \qquad (6)$$

where p_j is the price of intermediate good j and the price of the final output is normalized to one.

Technological progress takes place in an innovative sector. In this sector there is a changing number of firms, each of which monopolizes the production of a specific intermediate good. At a given point in time, the existing technology allows the production of N varieties of intermediate goods. Technological progress takes the form of an expansion in this number of varieties.[8] Once invented, an intermediate good of type j costs σ units of Y to produce. The monopolist sets the price p_j, at each date, to maximize instantaneous profits, $\pi_j = (p_j - \sigma)X_j$, taking the demand function for the intermediate good derived from (6) as given.

The cost of inventing a new type of product is fixed at η times the production cost, that is, $\eta\sigma$ units of output Y. We assume free entry into the business of being an inventor so, in equilibrium, the present value of the profits for each intermediate good must equal $\eta\sigma$, that is,

$$\eta\sigma = \int_t^{\infty} \pi_j e^{-\bar{r}(z,t)(z-t)} dz, \qquad (7)$$

with $\bar{r}(z,t) = [1/(z-t)] \int_t^z r(u)du$ the average interest rate between times t and z.

2.1 *Steady-State Equilibrium*[9]

The economy faces a problem of environmental shortage when consumers harvest at rates below the harvesting rate under an open-access regime. The following proposition shows the fraction of labor that consumers devote to output production and to harvesting in the case of an open-access natural resource.

Proposition 1 *If the natural resource is an open-access resource, the representative consumer would allocate a fraction of labor* $v^{oa} = 1/\phi$, *to output production*

[8]Technology increases productivity in the final output sector. It does not, however affect the resource sector.

[9]The proofs of the propositions are available from the authors upon request.

(*and correspondingly*, $1 - 1/\phi$, *to harvesting*), *where*

$$\phi = \frac{1-\alpha-\delta\beta}{1-\alpha-\beta} > 1. \tag{8}$$

Under open access consumers do not take into account the dynamics of the natural resource (1). They solve the maximization problem (4) subject to their budget constraint given by (3). From the necessary conditions for optimality, the result immediately follows. Consequently, a natural resource is scarce if $v > v^{oa} = 1/\phi$, that is, the harvesting rates are below $h^{oa} = b(1-1/\phi)^{1-\delta}$, the open-access rate.

The maximization of consumers' utility needs the standard Euler equation describing the evolution of per-capita consumption

$$\frac{\dot{c}}{c} = r - \rho. \tag{9}$$

Firms maximize benefits by equalizing net marginal products to factor prices:

$$w = (1-\alpha-\beta)\frac{Y}{v\bar{L}}, \qquad p_h = \beta\frac{Y}{h\bar{L}}, \qquad X_j = \left(\frac{\alpha A}{p_j}\right)^{\frac{1}{1-\alpha}} \bar{L} v^{\frac{1-\alpha-\beta}{1-\alpha}} h^{\frac{\beta}{1-\alpha}}. \tag{10}$$

Taking the demand function for an intermediate good j as given in the third expression in (10), the monopoly that produces it maximizes profits at price $p_j = \sigma/\alpha > \sigma$. Using this price in the demand we obtain

$$\begin{aligned} X_j = X &= \left(\frac{A}{\sigma}\right)^{\frac{1}{1-\alpha}} \alpha^{\frac{2}{1-\alpha}} \bar{L} v^{\frac{1-\alpha-\beta}{1-\alpha}} h^{\frac{\beta}{1-\alpha}}, \\ Y &= A(v\bar{L})^{1-\alpha-\beta} N X^{\alpha} (h\bar{L})^{\beta} = \frac{\sigma}{\alpha^2} N X. \end{aligned} \tag{11}$$

Note that the amount of the intermediate good X_j is the same for all $j \in \{1, \dots, N\}$ and depends on variables v and s. Using the first expression in (11) in (7) and differentiating the latter with respect to t it follows that

$$r = \frac{1}{\eta}\frac{1-\alpha}{\alpha} X = \frac{1}{\eta}\frac{1-\alpha}{\alpha}\left(\frac{A}{\sigma}\right)^{\frac{1}{1-\alpha}} \alpha^{\frac{2}{1-\alpha}} \bar{L} v^{\frac{1-\alpha-\beta}{1-\alpha}} h^{\frac{\beta}{1-\alpha}}. \tag{12}$$

Since the economy is closed to international asset exchange, total households' assets, $a\bar{L}$, equal the market value of the firms that produce the intermediate goods, $\eta\sigma N$. Taking into account (3), (10), (11) and (12), the dynamics of the number of intermediate goods, N, is

$$\dot{N} = \frac{1}{\eta}\left[\frac{Y - c\bar{L}}{\sigma} - NX\right], \qquad N(0) = N_0. \tag{13}$$

Definition 1 Given N_0 and s_0, an equilibrium consists of time paths for N, s, c and v that maximize the utility of a representative consumer who is subject to (1) and (3), where w, p_h, and X, are given by (10) and (11). A steady-state equilibrium would be an equilibrium where all variables grow at constant rates (which could be zero for some variables).

Note that v and s cannot grow indefinitely at a non-zero constant rate (they are lower and upper bounded). The constancy of v and s implies that h and r are also constant at a steady-state equilibrium. From (9), (10), (11) and (13), the following proposition, which characterizes a steady-state equilibrium, immediately follows.

Proposition 2 *If a steady-state equilibrium exists, along this path*:

- s, v, h *and* r *remain constant*;
- Y, c, p_h *and* w, *all grow at the same rate as* N.

A steady-state equilibrium can be seen as a sustainable growth path. In such a solution, the economy will grow continuously, while maintaining a constant stock of the renewable resource. Note that a steady-state equilibrium, as described in Proposition 2, will exist if and only if variables v, s and $\tilde{c} = c/N$ remain constant. The dynamics of these three variables are given in the following lemma.

Lemma 1 *Any steady-state equilibrium for the model previously described corresponds to a steady state of the following three differential equations*:

$$\dot{\tilde{c}} = \tilde{c}\left\{\frac{\bar{L}}{\eta}\left[\frac{\tilde{c}}{\sigma} - \left(\frac{1-\alpha}{\alpha}\right)\alpha^{\frac{1+\alpha}{1-\alpha}}\left(\frac{A}{\sigma}\right)^{\frac{1}{1-\alpha}} v^{\frac{1-\alpha-\beta}{1-\alpha}} h^{\frac{\beta}{1-\alpha}}\right] - \rho\right\}, \tag{14}$$

$$\dot{v} = \Omega(v)\left\{\Theta(v,s) + (\phi v - 1)\frac{\dot{\tilde{c}}}{\tilde{c}}\right\}, \tag{15}$$

$$\dot{s} = g(s) - h(v,s), \tag{16}$$

where $\tilde{c} = c/N$, ϕ *is given in* (8) *and*

$$\Omega(v) = \frac{(1-\alpha)v(1-v)}{(1-\alpha)(1-v) + (1-\alpha-\beta)(1-\delta v)(\phi v - 1)},$$

$$\Theta(v,s) = \left[\rho - \tilde{g}\left(1 - \frac{2s}{\kappa}\right) + \theta\frac{1-\alpha-\beta}{1-\alpha}\frac{\dot{s}}{s}\right](\phi v - 1) - \theta(1-v)\frac{h}{s}.$$

Expressions (14)–(16) immediately arise from the necessary conditions for an interior solution of the consumer's maximization problem (4) subject to (1) and (3), taking into account (2), (10) and (13). The following proposition collects all the hypotheses needed to guarantee the existence of a unique steady state.

Proposition 3 *The existence and uniqueness of a steady-state equilibrium with* $\tilde{c}^* > 0$, $1/\phi < v^* < 1$ *and* $0 < s^* < \kappa$ *have been proven:*[10]

- *for* $\theta = 1$, *under sufficient condition* $\tilde{g} \geq \rho$;
- *for* $\theta = 0$, *under necessary and sufficient condition* $\tilde{g} \in (\rho, \tilde{g}^+)$, *where*

$$\tilde{g}^+ = \frac{h^{oa}}{\kappa/2} + \sqrt{\left(\frac{h^{oa}}{\kappa/2}\right)^2 + \rho^2},$$

and h^{oa} *is the harvesting rate under an open-access regime.*

When the stock of the resource does not affect harvesting, $\theta = 0$, steady-state values s^* and v^* can be explicitly found:

$$s^* = \frac{\tilde{g} - \rho}{2\tilde{g}}\kappa < \frac{\kappa}{2}, \qquad v^* = 1 - \left[\frac{(\tilde{g}^2 - \rho^2)\kappa}{4\tilde{g}b}\right]^{\frac{1}{1-\delta}}. \tag{17}$$

Thus, a necessary condition for the positivity of s^* is $\tilde{g} > \rho$, which also guarantees $v^* < 1$. That is, the intrinsic growth rate of the resource must be greater than the rate of temporal discount, in order for a feasible interior steady state to exist.

Recall that we assume a scarce natural resource, i.e., we focus on equilibria with a harvesting rate, h^* below the open-access' $h^{oa} = b(1 - 1/\phi)^{1-\delta}$. This condition is guaranteed if and only if $\tilde{g} < \tilde{g}^+$, for $\theta = 0$, which also ensures $v^* > 1/\phi$.

When the stock of the resource linearly affects harvesting (i.e. $\theta = 1$), a closed form for the stock of the resource and the labor share in the final output sector at the steady state cannot be found. However, the assumption $\tilde{g} \geq \rho$, ensures the existence of a unique equilibrium with $h^* \leq h^{oa}$.

The lack of complete stability is a typical property of balanced paths in endogenous growth models when transversality conditions are satisfied (see Martínez-García 2003). This is also the case for our model. The following proposition proves conditional stability when $\theta = 0$ or $\theta = 1$ because the conditions in Martínez-García (2003) are satisfied.

Proposition 4 *The steady-state equilibrium is a saddle point with a one-dimensional stable manifold.*

Below we shall concentrate on the particular cases $\theta = 0$ and $\theta = 1$, where the existence, uniqueness, and saddle-point stability are proved.

The following proposition presents the responses of the steady-state equilibrium values of the natural resource stock and of the labor allocated to each sector, to changes in the environmental parameters. They can be easily obtained taking partial derivatives in expressions (17), when $\theta = 0$, and by implicit differentiation when $\theta = 1$.

---

[10]Transversality conditions together with the concavity of functions guarantee the optimality of the unique steady-state equilibrium.

Proposition 5 *When a unique steady-state equilibrium exists, the stock of the resource, s^*, increases while the labor share in the final output sector, v^*, decreases, with the intrinsic growth rate of the resource, $\tilde{g}$. Likewise, s^* increases with the carrying capacity, κ, while its effect on v^* is negative for $\theta = 0$ but is null for $\theta = 1$.*

A higher $\tilde{g}$ leads consumers to devote a larger labor share to the resource sector, $1 - v^*$, which pushes up harvesting. Nevertheless, since the resource grows faster, this situation is compatible with a larger stock of the resource at the steady state.

The carrying capacity of the natural resource, κ, has a positive effect on the equilibrium resource stock, s^*. Nevertheless, the effect of κ on the labor share in each sector depends on the value of θ. When the stock of the natural resource does not affect harvesting, $\theta = 0$, a greater carrying capacity requires an increment in harvesting to maintain a constant resource stock, which requires a higher extraction effort, $1 - v^*$. However, when the stock of the natural resource does affect harvesting, $\theta = 1$, the increment in κ also raises the stationary resource stock, s^*. The increment in κ increases harvesting in the same proportion as the increment in s^*, which renders unnecessary a higher extraction effort, $1 - v^*$. Thus, for $\theta = 1$, the labor share in each sector is unaffected by the carrying capacity.

From the consumption growth rate in (9) and the rate of return in (12), the growth rate of the economy, γ, along a steady-state equilibrium is as follows:

$$\gamma = \frac{1}{\eta}\frac{1-\alpha}{\alpha}X - \rho = \frac{1}{\eta}\frac{1-\alpha}{\alpha}\alpha^{\frac{2}{1-\alpha}}\bar{L}\left(\frac{A}{\sigma}\right)^{\frac{1}{1-\alpha}} v^{*\frac{1-\alpha-\beta}{1-\alpha}} h\left(v^*, s^*\right)^{\frac{\beta}{1-\alpha}} - \rho. \quad (18)$$

The following proposition presents the responses of the long-run growth rate of the economy to changes in the cost of innovation and environmental parameters. The results are immediate by taking into account (18) and the results in Proposition 5.

Proposition 6 *The long-term growth rate of the economy, γ, decreases with η, and increases with κ and $\tilde{g}$.*

An increment in the cost of innovation, η, reduces the rate of return on assets for investors, which lessens the growth rate of the economy.

As Eq. (18) states, the stock of the resource at the steady state, s^*, only affects the growth rate through its effect on harvesting. Therefore, resource abundance, as measured by the carrying capacity, κ, or by the intrinsic growth rate, $\tilde{g}$, influences the economic growth rate in the long term through its effect on v^*, when $\theta = 0$, or on both v^* and s^*, when $\theta = 1$.

When $\theta = 0$, a higher κ (or a higher $\tilde{g}$) reduces the labor share in the final output sector, v^*, increasing the labor share in the resource sector, $1 - v^*$. The decrease in v^* has two opposite effects on the growth rate in the long term. On the one hand, a direct negative effect on the growth rate in the long run. This is the effect in Elíasson and Turnovsky (2004), where the extracted resource is traded to obtain foreign consumption goods and harvesting has no effect on the production of final output. Then, if the economy has access to a more bountiful natural resource, it will choose more consumption today, at the cost of slower growth in the long run. However, in our model, harvesting the resource has a positive influence on the final-output production. Thus, the decrease in v^* associated with a higher κ (or a higher $\tilde{g}$) presents an indirect effect on the growth rate: a larger labor share in the resource sector increases harvesting, which is then used to augment the final-output production, which in turn raises the economy's growth rate. These two effects have opposite signs. Under open access, consumers would choose the labor share, $v^{oa} = 1/\phi$, where these two effects exactly cancel out. However, the assumption of perfect property rights over the resource leads consumers to devote a lower share of labor to harvesting the resource, $h(v^*, s^*) < h^{oa}$, and consequently $v^* > v^{oa}$. In this situation, the environmental restriction is forcing an overlarge labor force in final output and an underutilization of the resource. Resource abundance relaxes the environmental restriction, and the labor share in the final output sector moves down towards v^{oa}. The indirect effect of a higher harvesting is stronger than the direct effect of less labor in the final output sector. Thus, we can conclude that an economy that has access to a more bountiful natural resource will grow faster.

When $\theta = 1$, a larger intrinsic growth rate, $\tilde{g}$, lowers the labor share in the final-output sector, and raises the labor share devoted to harvesting, which pushes extraction up. In addition, the increment in the stationary stock of the resource, s^*, which is associated with a higher $\tilde{g}$, pushes extraction up further. Thus, the net effect of a higher intrinsic growth rate in the economy's growth rate is positive, as for $\theta = 0$. As for the carrying capacity, it has no effect on the labor share devoted to each sector. However, the stock of the natural resource, which now has a positive effect on the growth rate, will be greater, leading to a higher growth rate.

It is easy to show that the effect on $\tilde{c}^*$ of changes in the environmental parameters κ and $\tilde{g}$ is the same as that on the economy's growth rate, γ. This therefore leads to the following corollary:

Corollary 1 *The steady-state equilibrium of consumption per variety of intermediate good, $\tilde{c}^*$, depends positively on η, κ, and $\tilde{g}$.*

Since an increment in the cost of innovation, η, reduces the rate of return on assets, consumers tend to increase their consumption with respect to investment, augmenting the ratio of consumption per variety of intermediate good, $\tilde{c}^*$. Moreover, a more bountiful resource increases harvesting, and so, consumers attain a larger income from their extraction activities in the resource sector, which increases consumption and the ratio $\tilde{c}^*$.

3 Sustainable Growth with Technological Diffusion Through Trade

Developing economies tend to rely on foreign innovation rather than on domestic innovation as the source of technological development. Following this idea, we present a model of bilateral trade between a technological leading country (country L) and a country endowed with a renewable natural resource, which is a technological follower (country F). We assume that final output producers in country F import new intermediate inputs from country L, whereas consumers in the latter buy final output produced in country F. This model is an extension of the two-country endogenous growth model of Barro and Sala-i Martin (1999) (Chap. 8). We shall see how, despite the fact that no technological investments are carried out in the country endowed with the natural resource, the trade relationship with a technological leader enables sustained economic growth that allows the conservation of the natural resource. This possibility is also analyzed in Cabo et al. (2008).

We assume that the countries' decisions determine pricing. This situation can arise when one country is the only supplier and its counterpart, the only demander of the exchanged goods. The terms of trade are determined by their actions when L (resp. F) is a representative economy of many clone technologically leading (resp. follower) economies. We shall describe the problem each sector faces.

As in the model of the previous section, a representative consumer in country F manages a natural resource whose net growth rate is given by (1)–(2). Since no innovative activity exists in this country, and there is no international trade on financial assets, consumers from country F do not accumulate assets. Thus, the per-capita budget constraint for a representative consumer is

$$vw_F + p_h h = c_F, \tag{19}$$

where w_F is the wage rate and c_F is per capita consumption in country F.

The only asset that this country's consumers can hold is the ownership of the natural resource. Thus, a representative consumer has to decide the consumption c_F and the fraction of labor, v and $1-v$, employed either in the final-output production or in harvesting, to maximize utility:

$$\int_0^{\infty} \ln(c_F)e^{-\rho t}dt, \tag{20}$$

subject to (1) and (19).

Firms in country F import existing intermediate goods from country L. Final output production presents the same functional form as (5):[11]

$$Y_F = A_F(vL_F)^{1-\alpha-\beta}\sum_{j=1}^{N} X_{Fj}^{\alpha}(hL_F)^{\beta}.$$

Consumers of the innovating country consume, together with domestic final output, a final good imported from country F. In contrast to consumers in country F, they accumulate assets in the form of ownership claims on innovative firms and

[11] Subscript F denotes variables corresponding to the follower country.

receive financial interest from them. Thus, the per-capita budget constraint for a representative consumer is

$$\dot{a}_L = ra_L + w_L - c_L - p_F c_{LF}, \quad a_L(0) = a_{L0}, \tag{21}$$

where a_L is the per-capita assets, r is the assets-return rate, w_L is the wage rate, c_L is the per-capita consumption of the domestic final good, and c_{LF} is the per-capita consumption of the good imported from country F at price p_F. A representative consumer has to decide the consumptions c_L and c_{LF} to maximize utility:

$$\int_0^\infty \left[\ln(c_L) + \ln(c_{LF})\right] e^{-\rho t} dt, \tag{22}$$

subject to (21).[12]

We consider the price of the domestic final good as a numeraire, $p_L = 1$. Consequently, p_F not only represents the price of the good imported from F, but also the follower's real exchange rate, i.e., the value of one unit of consumption imported from country F in units of country L's output. Production of final output by a representative firm in country L is described by[13]

$$Y_L = A_L L_L^{1-\alpha} \sum_{j=1}^{N} X_{Lj}^{\alpha}.$$

Production of intermediate goods is carried out in the leading country. This situation applies as long as intellectual-property rights are protected both domestically and internationally. Once invented, an intermediate good of type j costs σ_L units of Y_L to produce, while the innovator who produces this intermediate good obtains p_j unit of Y_L. For simplicity, we normalize $\sigma_L = 1$. The monopolist decides the price p_j to maximize instantaneous profits from sales to final-output producers in L and F, given by $\pi_j = (p_j - 1)(X_{Lj} + X_{Fj})$, where X_{Lj} and X_{Fj} are the demand functions of intermediate good j in countries L and F, respectively.

As in the closed economy, the cost of creating a new intermediate is supposed to be η times the cost of producing it, that is, η units of Y_L. However, an innovator must pay a cost beyond the initial R&D outlay to transfer and adapt his product for use in country F. This cost is represented by ν and is lower than η assuming that the innovator is better suited than other entrepreneurs to the process of adapting a discovery for use in other country. We assume also that the cost ν is low enough to ensure this adaptation is immediately worthwhile. Once more, the free-entry assumption

---

[12]Our model assumes that families in country L consume final output imported from country F. However, families in country F do not import consumption goods from abroad. None of the properties and conclusions of the paper would be affected if consumers in country F are allowed to import final output.

[13]Subscript L denotes variables corresponding to the leading country.

equates the present value of the profits for each intermediate to $\eta + \nu$,

$$\eta + \nu = \int_t^{\infty} \pi_j e^{-\bar{r}(z,t)(z-t)} dz. \tag{23}$$

3.1 Steady-State Equilibrium

By solving the profit-maximization problem of firms in the final output sector in country F we obtain that net marginal products are equated to factor prices:

$$w_F = (1-\alpha-\beta)\frac{Y_F}{vL_F}, \qquad p_h = \beta\frac{Y_F}{hL_F}, \qquad X_{Fj} = vL_F\left(\frac{\alpha A_F}{p_j^F}\right)^{\frac{1}{1-\alpha}}\left(\frac{h}{v}\right)^{\frac{\beta}{1-\alpha}}, \tag{24}$$

with p_j^F the price paid for the intermediate goods to the leading country's entrepreneurs.

A firm producing final-output in country L solves its profit-maximization problem to yield

$$w_L = (1-\alpha)\frac{Y_L}{L_L}, \qquad X_{Lj} = L_L\left(\frac{\alpha A_L}{p_j}\right)^{\frac{1}{1-\alpha}}, \tag{25}$$

where p_j is the price of intermediate input j in this country.

The maximization problem (21) and (22) leads to the necessary condition for an interior solution

$$c_L = p_F c_{LF} \tag{26}$$

and the following consumption growth rates

$$\frac{\dot{c}_L}{c_L} = r - \rho, \qquad \frac{\dot{c}_{LF}}{c_{LF}} = r - \rho - \frac{\dot{p}_F}{p_F}. \tag{27}$$

The growth rate of the domestic-good consumption is again as in (9), with a possibly different value for the interest rate, r. The difference between this rate and the growth rate of the follower's real exchange rate gives the growth rate of the imported-good consumption.

Given the demand functions for intermediate good j in countries F and L in (24) and (25), the firm that produces it solves its profit-maximization problem to yield $p_j = 1/\alpha > 1$. Then, $p_j^F = 1/(\alpha p_F)$, and consequently, the amount of every intermediate in each country is

$$X_{Lj} = X_L = L_L A_L^{\frac{1}{1-\alpha}} \alpha^{\frac{2}{1-\alpha}}, \qquad X_{Fj} = X_F = L_F (p_F A_F)^{\frac{1}{1-\alpha}} \alpha^{\frac{2}{1-\alpha}} v^{\frac{1-\alpha-\beta}{1-\alpha}} h^{\frac{\beta}{1-\alpha}}. \tag{28}$$

Note that while X_L is constant, the quantity X_F depends on p_F, v and h (which is, in turn, a function of v and s).

Using (28) in (23) and following the same reasoning carried out for the closed economy, we obtain the rate of return on innovation and foreign investment- that is, adaptation of products for use in country F—that would be given by

$$r = \frac{1}{\eta+\nu}\frac{1-\alpha}{\alpha}(X_L + X_F). \tag{29}$$

Investment returns in the technologically leading country are linked to the monopolistic benefits in the intermediate-goods sector. If we consider an economy that is closed to international asset exchange, total households' assets, $a_L L_L$, are equal to the market value of the firms that produce these intermediate goods, $(\eta+\nu)N$. From $a_L L_L = (\eta+\nu)N$ and taking into account (25), as well as the relationship $\alpha^2 Y_L = N X_L$, and the dynamics of the assets in (21), the dynamics of N is obtained:

$$\dot{N} = \frac{1}{\eta+\nu}\left[Y_L - (c_L + p_F c_{LF})L_L - N\left(X_L - \frac{1-\alpha}{\alpha}X_F\right)\right], \quad N(0) = N_0.$$

As we will show, the permanent increment in this number fuels production growth in the final-output sector, not only in the technologically leading country, but also in the follower, without reducing the stock of the natural resource.

Before defining an equilibrium for the two trading economies described above, let us consider the problem to be solved in each country. The problem for the leader country, PL: a representative consumer of country L chooses c_L and c_{LF} to maximize (22) subject to (21). The salary w_L and the rate of return r will be given by (25) and (29). In a symmetric fashion, the problem for the follower country, PF: a representative consumer of country F chooses v to maximize (20) subject to (1) and (19). The wage rate, w_F, and the price of the resource, p_h, will be given by (24).

The price, p_F, is determined by equating the value of the final good traded from F to L, to the value of the intermediate goods sold from innovators in L to producers in F:

$$L_L p_F c_{LF} = p_j N X_F. \tag{30}$$

Definition 2 The equilibria of the model are such that given N_0 and s_0, and considering time paths for N, s, c_L, c_{LF} and v, problems PL and PF are solved and p_F is endogenously determined from Eq. (30).

Below we concentrate exclusively on the steady-state equilibria. Following the same reasoning as in Proposition 2 it can be shown that the behavior of the different variables along a steady-state equilibrium is characterized as follows.

Proposition 7 *If a steady-state equilibrium exists, along this path,*

- v, s, h, p_F *and* r *remain constant;*
- $Y_L, Y_F, c_L, c_{LF}, c_F, p_h, w_L,$ *and* w_F *grow at the same rate as* N.

The steady-state equilibrium corresponds to a constant growth path in the leading country. Furthermore, although the follower country does not invest in technological improvements, the trade relationship with the leader allows a sustainable growth path in this country. Both trading economies grow at the same constant rate.

As in Sect. 2, the steady-state equilibrium corresponds with a steady state of variables $\tilde{c}_L \equiv c_L/N$, v and s. Following the same reasoning as in Lemma 1, expressions (31)–(33) can be obtained from the necessary conditions for an interior solution of the consumer's maximization problems (20) and (22).

Lemma 2 *Any steady-state equilibrium for the trade model described by the dynamic problems for countries L and F, corresponds to a steady state of the following three differential equations:*[14]

$$\dot{\tilde{c}}_L = \tilde{c}_L \left\{ \frac{1}{\eta + v} \left[2L_L \tilde{c}_L - (1-\alpha) L_L A_L^{\frac{1}{1-\alpha}} \alpha^{\frac{2\alpha}{1-\alpha}} \right] - \rho \right\}, \tag{31}$$

$$\dot{v} = \Omega^{oe}(v) \Theta^{oe}(v, s), \tag{32}$$

$$\dot{s} = g(s) - h(v, s), \tag{33}$$

where

$$\Omega^{oe}(v) = \frac{v(1-v)}{1 - v + (1-\delta) v (\phi v - 1)},$$

$$\Theta^{oe}(v, s) = \left(\rho - g'(s) + \theta \frac{\dot{s}}{s} \right) (\phi v - 1) - \theta (1 - v) \frac{h(v, s)}{s}.$$

Notice that we can study the existence of the steady state by isolating the two last dynamic equations, which is equivalent to the system solved in Proposition 3. Moreover, the negative sign of the determinant of the Jacobian matrix associated with the system ensures the saddle-point property.

Proposition 8 *Under the conditions in Proposition* 3, *there exists a unique steady-state equilibrium with* $\tilde{c}_L^* > 0$, $1/\phi < v^* < 1$ *and* $0 < s^* < \kappa$. *Furthermore, values* v^* *and* s^* *coincide with those obtained for the closed economy. The steady-state equilibrium is a saddle point with a one-dimensional stable manifold.*[15]

We should note that the steady-state values of the stock of the resource, s^*, and the labor share in the final output sector, v^*, are solutions of the same equation system as those obtained for the closed economy. Thus, the effect of changes in the carrying capacity and in the intrinsic growth rate, as collected in Proposition 5 remains valid. However, the effect of resource abundance on the growth rate is not

[14] From now on superscript *oe* denotes open economy scenario.

[15] Transversality conditions together with the concavity of functions guarantee the optimality of the unique steady state equilibrium.

the same. The reason is the follower's real exchange rate, which has a significant influence on the economic-growth rate, and is determined by the balanced-trade condition (30) in the case of open economies.

Proposition 9 *When a unique steady-state equilibrium exists, the follower's real exchange rate along this equilibrium, p_F^*, increases with η and ν; and, it decreases with κ and $\tilde{g}$.*

The previous result is directly obtained from the expression of p_F^* as a function of υ^*, s^* and the parameters. This expression can be explicitly obtained from (30), (26), (28) and the expression of the steady-state value $\tilde{c}_L^*$. An increment in either the cost of innovation, η, or the cost of adaptation, ν, implies a reduction in the rate of return on assets for investors in the leading country, r. Lower returns lead consumers to increase their consumption (domestic and imported) with respect to investment, augmenting the ratio of foreign consumption per variety of intermediate good, $\tilde{c}_{LF} = c_{LF}/N$, in the leading country at the steady state. As long as η and ν do not affect the demand for intermediate inputs in F, the bilateral trade equilibrium leads to a gain for the follower's real exchange rate, p_F^*.

An increment in either the carrying capacity, κ, or the intrinsic growth rate, $\tilde{g}$, leads consumers in the follower country, who own the resource, to reduce the labor share in the final output sector in favor of a higher harvesting, which pushes up the final-output production. The second effect is stronger both when s^* does not affect harvesting ($\theta = 0$), and when the increment in s^* further fuels harvesting and final-output production ($\theta = 1$). A higher final-output production in the follower country requires higher imports of intermediate inputs. As long as κ and $\tilde{g}$ do not affect the demand for foreign consumption in the leading country, the equilibrium in bilateral trade leads to a drop in the follower's real exchange rate.

These changes in the follower's real exchange rate may also affect consumption. The next proposition can be easily shown and studies the effects upon consumption per variety of intermediate good, along the steady-state equilibrium.

Proposition 10 *When a unique steady-state equilibrium exists, along this equilibrium, $\tilde{c}_L^*$, $\tilde{c}_{LF}^*$ and $\tilde{c}_F^*$, (with $\tilde{c}_i = c_i/N, i \in \{L, LF, F\}$) increase with η and ν.*

An increment in κ or $\tilde{g}$ would increase $\tilde{c}_{LF}^*$ and $\tilde{c}_F^*$, while $\tilde{c}_L^*$ remains constant.

As previously explained, an increment in either η or ν leads consumers in the leading country to increase their consumption (domestic and imported) with respect to investment. Therefore, η and ν have positive effects on $\tilde{c}_L^* = c_L^*/N$ and p_F^*, which reduces imported consumption in country L. This reduction cuts down, but does not fully offset the rise in $\tilde{c}_{LF}^* = c_{LF}^*/N$.

As long as η and ν do not affect the demand for intermediate inputs in F, their effect on consumer income in this country is null. However, η and ν lead to a gain in the follower country's real exchange rate, p_F^*, increasing its income and consumption, $\tilde{c}_F^*$.

The relative price for the follower country drops with κ and $\tilde{g}$, increasing the consumption of imported goods in the leading country, as stated in Proposition 10.

The effect of resource bounty on the consumption per variety of intermediate good in the follower country, $\tilde{c}_F^*$, is twofold. On the one hand, a higher κ or $\tilde{g}$ increases harvesting, $h(v^*, s^*)$, and so, consumers attain larger incomes from their extraction activities. Conversely, resource abundance also means a lower relative price for F, pushing down net revenues from bilateral trade. The first effect, which boosts consumption in F, is greater than the negative effect of a lower real exchange rate.

Finally, the next proposition states the long-term growth rates for open economies and shows the results of the sensitivity analysis.

Proposition 11 *Along a steady-state equilibrium, the economies in both the technological-leader and -follower countries grow at the rate given by*

$$\gamma^{oe} = (1+\alpha)\left[\frac{(1-\alpha)\alpha^{\frac{2\alpha}{1-\alpha}}}{2(\eta+\nu)} L_L A_L^{\frac{1}{1-\alpha}} - \rho\right].$$

γ^{oe} *decreases with* η *and* ν. *Furthermore,* γ^{oe} *is independent of* κ *and* $\tilde{g}$.

An increment in the cost of innovation, η, or the cost of adaptation, ν, reduces the net benefits of innovators and then the rate of return on assets for investors in the leading country, r. Thus, by the usual definition of the growth rate presented in (27), the negative effect on γ^{oe} follows.

The resource bounty, described either by the carrying capacity or the intrinsic growth rate, has no effect on the growth rate of open economies.

Resource bounty fuels final-output production in country F. A higher final-output production requires more imports of each type of intermediate input. However, resource bounty lessens the real exchange rate, p_F, pushing down the imports of intermediate goods in country F. This negative effect exactly compensates for the previous pressure to increase the import of intermediates to maintain the value of consumption imports in L invariant, $p_F^* \tilde{c}_{LF}^* = \tilde{c}_L^*$. Since resource bounty has no influence on the amount of intermediate goods traded, neither does it affect the rate of return in L or the growth rate of both countries. Empirical evidence supports this result (see, for example, Evans 1996).

4 Domestic Innovation vs. Technological Diffusion Through Trade

We have proved that both domestic innovation and technological diffusion through trade with a leading country (TDT hereafter) allow sustainable economic growth in a country that is endowed with a renewable natural resource that has a limited regeneration rate and a bounded carrying capacity. The main question asked in this section is whether the country is better off when innovation is carried out within its borders or when technology is imported from abroad. This section compares the long-run growth rates and the representative consumer's consumption and utility under both scenarios.

In the case of two open economies with TDT, innovators in the leading country pay one unit of Y_L to produce an already-invented intermediate good. If there were inventors in the follower country, we assume that they would face the same production cost, that is, one unit of Y_L, or equivalently, $1/p_F^*$ units of the good produced in this country, Y_F. Thus, to compare the domestic-innovation and TDT scenarios, parameter σ equates $1/p_F^*$ in the former. Moreover, for comparison purposes, $\overline{L} = L_F$, $A = A_F$, and thus, $X(v^*, s^*) = X_F(v^*, s^*)$.

Proposition 12 *The long-run growth rate under domestic innovation, γ, is higher than that under TDT, γ^{oe}, if and only if $\nu/\eta > X_L/X_F(v^*, s^*)$.*

From Eqs. (12) and (29), the rates of return on assets under domestic innovation and under TDT coincide if and only if $X_F(v^*, s^*)/\eta = [X_L + X_F(v^*, s^*)]/(\eta + \nu)$. That is, for each intermediate good, the employed amount over the cost of innovation in the domestic scenario matches the employed amount over the costs of innovation and adaptation under TDT. Under this condition, the return to asset holders is the same under both scenarios. Since this rate of return determines the growth rate of consumption both under domestic innovation in (9) and under TDT in (27) the economies grow at the same rate.

Proposition 12 shows that the higher the cost of adaptation in terms of the cost of innovation, the stronger is the incentive to switch from TDT to domestic innovation. Equivalently, the greater the amount of each intermediate good in the country that has to decide whether to innovate or to import intermediate goods from the leading country, the stronger is again its incentive to innovate.

Corollary 2 *For an open economy, the shift from TDT to domestic innovation enhances the long-run growth rate if the ratio ν/η is not lower than $2\alpha/(1-\alpha)$. Conversely, the long-run growth rate decreases if the output elasticity of the intermediate good is sufficiently large, specifically, $\alpha \geq 2/3$.*

Corollary 2 establishes two sufficient conditions to ensure that it is more (or less) profitable for the economy to innovate rather than to import new intermediate goods. The economy grows faster with domestic innovation if the cost of adaptation with respect to the cost of innovation surpasses a lower bound, which depends positively on the output elasticity of the intermediate goods, α. Since $\nu < \eta$ this first sufficient condition can only occur if $\alpha < 1/3$, and it is more likely the smaller α gets. The smaller the output elasticity of the intermediate goods, the less beneficial it is to import them from abroad. Conversely, domestic innovation slows down growth when the output elasticity of the intermediate good is large enough.

Resource bounty affects the economic growth rate and the consumption per variety of intermediate goods differently under domestic innovation and under TDT. These effects are collected in the two propositions below.

Proposition 13 *A switch from TDT to domestic innovation leads to a greater consumption per variety of intermediate goods at the steady state. This increment is larger the more abundant the natural resource (the lower the follower's real exchange rate).*

To know whether resource wealth is or it is not an incentive for an economy to switch from TDT to domestic innovation, we need to analyze its effect on the growth rate of the number of these varieties.

Proposition 14 *If the long-run growth rate is greater when technology is traded from abroad, $\gamma < \gamma^{oe}$, this gap is shortened by resource abundance. Conversely, resource abundance will widen the gap if the long-run growth rate is greater under domestic innovation.*

In consequence, resource abundance either makes more profitable domestic innovation, or makes it less attractive to import the technology developed abroad.

5 Concluding Remarks

For a country that is endowed with a natural resource and that has a resource-dependent economy, two models have been analyzed depending on whether the economy invests in new technology or imports technology developed abroad. The first concern of the paper is to analyse the sustainability of the economic growth, in both models. Furthermore, we have focused on the effect of resource abundance on the growth rate of the economy, the follower's real exchange rate, the resource stock, and the consumers' welfare along the balanced path.

Our findings are compared for the two scenarios of domestic innovation and technological diffusion through trade. Under both scenarios, we prove the existence, uniqueness and saddle-point stability of a steady-state equilibrium that allows sustained economic growth, while maintaining the natural-resource stock constant.

The results contest the literature of the resource curse. It differs from the negative relationship between an economy's resource wealth and its long-term growth rate, established by Elíasson and Turnovsky (2004). In their model, the stock of the resource does not influence extraction, and the extracted resource is used to purchase imports of a foreign consumption good. In our model, the natural resource is an essential input in the production of a final good, and technological innovation enhances its productivity. In this study we assert that if technological improvements enhance the resource returns on income, the economy will avoid the resource curse. The resource curse turns into a blessing under domestic innovation; and the resource wealth is growth-neutral but consumption-enhancing under TDT.

The first model assumes a resource-dependent economy that develops its own R&D sector. Resource wealth, measured either by the carrying capacity or the intrinsic growth rate, enhances the long-run growth rate of the economy. A more bountiful natural resource also increases the consumption per variety of intermediate good. Both effects lead to a higher consumer welfare.

In the second model, the economy endowed with the natural resource can obtain new technology from abroad. A technologically leading country invests in technological innovation, which is exported to the technological follower. To the best of our knowledge, this is one of the first attempts to simultaneously tackle trade, technology transfer and natural-resource management, in the context of endogenous growth economies. Our approach allows us to study the existence of sustainable growth in some economies of developing countries, which are linked to the extraction of a natural resource, and with an underdeveloped or non-existent R&D sector.

Our results prove that technological innovation in the leading country guarantees sustainable economic growth in both countries. The diffusion of technology through trade permits the reconciliation of unlimited economic growth with bounded natural resources in developing countries.

For open economies, a more bountiful natural resource reduces the real exchange rate of the country owning the natural resource, canceling out the positive effect on the growth rate and making growth independent of the resource wealth. Empirical evidence in Evans (1996) supports this result.

Consumption per variety of intermediate good increases with resource wealth in the follower country. This abundance also affects consumption in the leading country, where an increment in imports is associated with lower real exchange rate in the follower.

The adaptation and innovation costs, and the amounts of intermediate goods employed in each country establish a condition that determines if the long-run growth rate is larger under domestic innovation or TDT. The better the economy is supplied with natural resources, the greater is the increment (resp. the lower the decrement) in welfare associated with a change from TDT to domestic or a non-dependent policy of technology innovation. Thus, the incentive to carry out R&D investment activities is strengthened by resource bounty.

Acknowledgements We thank an anonymous reviewer for his/her helpful comments. The authors have been partially supported by MEC projects ECO2008-01551/ECON and ECO2011-24352. The projects are co-financed by FEDER funds. The third author acknowledges the support by COST Action IS1104 "The EU in the new economic complex geography: models, tools and policy evaluation".

References

Auty, R. M. (Ed.) (2001). *Resource abundance and economic development*. London: Oxford University Press.

Barro, R., & Sala-i-Martin, X. (1999). *Economic growth*. New York: McGraw-Hill.

Birdyshaw, E., & Ellis, C. (2007). Privatizing an open-access resource and environmental degradation. *Ecological Economics*, *61*, 469–477.

Bovenberg, A. L., & Smulders, S. (1995). Environmental quality and pollution-augmenting technological change in a two-sector endogenous growth model. *Journal of Public Economics*, *57*, 369–391.

Brunnschweiler, C. N., & Bulte, E. H. (2008). The resource curse revisited and revised: a tale of paradoxes and red herrings. *Journal of Environmental Economics and Management*, *55*(3), 248–264.
Cabo, F., Martín-Herrán, G., & Martínez-García, M. P. (2005). North-South trade and the sustainability of economic growth: a model with environmental constraints. In R. Loulou, J. F. Waaub, & G. Zaccour (Eds.), *Energy and environment* (pp. 1–25). New York: Springer.
Cabo, F., Martín-Herrán, G., & Martínez-García, M. P. (2008). Technological leadership and sustainable growth in a bilateral trade model. *International Game Theory Review*, *10*(1), 73–100.
Cabo, F., Martín-Herrán, G., & Martínez-García, M. P. (2012). Property rights for natural resources and sustainable growth in a two-country trade model. *Decisions in Economics and Finance*. doi:10.1007/s10203-012-0135-5.
Clark, C. W. (1990). *Mathematical bioeconomics: the optimal management of environmental resources*. New York: Wiley.
Coe, D. T., Helpman, E., & Hoffmaister, A. W. (1997). North-South R&D spillovers. *The Economic Journal, Royal Economic Society*, *107*(440), 134–149.
Coe, D. T., Helpman, E., & Hoffmaister, A. W. (2009). International R&D spillovers and institutions. *European Economic Review*, *53*(7), 723–741.
Elbasa, E. H., & Roe, T. L. (1996). On endogenous growth: the implications of environmental externalities. *Journal of Environmental Economics and Management*, *31*, 240–268.
Elíasson, L., & Turnovsky, S. J. (2004). Renewable resources in an endogenous growing economy: balanced growth and transitional dynamics. *Journal of Environmental Economics and Management*, *48*, 1018–1049.
Evans, P. (1996). Using cross-country variances to evaluate growth theories. *Journal of Economic Dynamics & Control*, *20*, 1027–1049.
Frankel, J. A. (2012). The natural resource curse: a survey of diagnoses and some prescriptions. In R. Arezki, C. A. Pattillo, M. Quintyn, & M. Zhu (Eds.), *Commodity price volatility and inclusive growth in low-income countries*, Washington: International Monetary Fund.
Grossman, G. M., & Helpman, E. (1991). *Innovation and growth in the global economy*. Cambridge: MIT Press.
Gylfason, T. (2001). Natural resources, education and economic development. *European Economic Review*, *45*, 847–859.
Holder, R. (2006). The curse of natural resources in fractionalized countries. *European Economic Review*, *50*, 1367–1386.
Martínez-García, M. P. (2003). The general instability of balanced paths in endogenous growth models: the role of transversality conditions. *Journal of Economic Dynamics & Control*, *27*, 599–618.
Mehlum, H., Moene, K., & Torvik, R. (2006). Institutions and the resource curse. *The Economic Journal*, *116*, 1–20.
Papyrakis, E., & Gerlagh, R. (2004). The resource curse hypothesis and its transmission channels. *Journal of Comparative Economics*, *32*, 181–193.
Robinson, J. A., Torvik, R., & Verdier, T. (2006). Political foundations of the resource curse. *Journal of Development Economics*, *79*, 447–468.
Sachs, J. D., & Warner, A. M. (2001). The curse of natural resources. *European Economic Review*, *45*, 827–838.
Smulders, S. (1995). Environmental policy and sustainable economic growth. *De Economist*, *143*, 163–195.
Stevens, P. (2003). Resource impact-curse or blessing? A literature survey. *Journal of Energy Literature*, *9*(1), 3–42.
van der Ploeg, F. (2011). Natural resources: curse or blessing? *Journal of Economic Literature*, *49*(2), 366–420.

Forest Management and Biodiversity in Size-Structured Forests Under Climate Change

Renan Goetz, Carme Cañizares, Joan Pujol, and Angels Xabadia

Abstract Climate change is threatening biodiversity conservation at a global scale, urging the need for action in order to prevent current and future losses. In forestry, the consideration of some stand features such as requiring a certain volume of deadwood and/or large trees as a part of the management regime may help to preserve and enhance biodiversity. However, it is likely to lead to a decrease in the benefits obtained from timber sales. This chapter presents a bioeconomic model that allows the optimal selective logging regime of a size-distributed forest to be determined, while taking climate change and biodiversity into account. It analyzes to what extent structural targets related to biodiversity affect the optimal forest management regime and the profitability of forests. For this purpose, an empirical analysis under various climate change scenarios is conducted for two diameter-distributed stands of *Pinus sylvestris* in Catalonia. The results show that the costs of biodiversity conservation in terms of reduced profitability can be significant, and augment with climate change.

1 Introduction

The effects of climate change on forests have been analyzed over the last 20 years. Initially academic studies focused on changes in forest growth and its economic consequences (Solberg et al. 2003; Perez-Garcia et al. 2002; Sohngen et al. 2001; Shugart et al. 2003). Soon this effort was widened by also looking at the importance of carbon sequestration in forest and agriculture to mitigate climate

R. Goetz (✉) · C. Cañizares · J. Pujol · A. Xabadia
University of Girona, Campus de Montilivi, 17071 Girona, Spain
e-mail: renan.goetz@udg.edu

C. Cañizares
e-mail: carme.canyizares@udg.edu

J. Pujol
e-mail: joan.pujol@udg.edu

A. Xabadia
e-mail: angels.xabadia@udg.edu

E. Moser et al. (eds.), *Dynamic Optimization in Environmental Economics*,
Dynamic Modeling and Econometrics in Economics and Finance 15,
DOI 10.1007/978-3-642-54086-8_12,

change (UNFCCC, http://unfccc.int/essential_background/items/6031.php accessed 28.02.2013). The interest is based on the fact that biological carbon sequestration can be considered as a policy option to reduce the impact of climate change. To evaluate its competitiveness with policy options outside the land-use sector, most studies aimed to determine the cost of forest carbon sequestration. The specification of these costs allows the cost of very different climate change mitigation policies to be compared (Alig et al. 2002; Murray et al. 2005; White et al. 2010).

Over the last ten years, however, the predominant focus on timber and forest carbon sequestration has been abandoned. At the international level for example the United Nations Forum of Forests, UNFF, adopted a multi-year program of work (2007–2015) which centers its interest on the forest ecosystem with respect to economic and biological parameters. More specifically it calls for sustainable forest management in the presence of climate change and it additionally aims for the conservation of biodiversity. The European Union forest policy also reenacted this shift of paradigm. In 2006 the European Union Commission approved the Forest Action Plan (EU 2006) which provides a coherent framework for forest-related initiatives at European Union level. The different formulated key actions bring together aspects such as competitiveness of the forest sector, compliance with international obligations on climate change mitigation (UNFCCC 1997, Kyoto protocol) and preservation of biodiversity. To respond to these different demands all at the same time it is necessary to model this complex decision making problem by taking account of sustainability and the multifunctional role of the forest.

The mere maximization of timber benefits has resulted in clear-cutting becoming the prevailing logging technique in many regions of the world. Traditionally, forest economists determined the optimal rotation period of the forest stand with the Faustmann formula (Conrad and Clark 1987). However, apart from timber, forests have a multifunctional role, for instance they are a source of important by-products such as mushrooms or cork (Croitoru 2007) and present scenic and recreational values (Scarpa et al. 2000). Likewise, a diversified size structure of trees and canopy is the basis for biological diversity (Whittam et al. 2002) as it provides habitat for a wide range of species (Doyon et al. 2005; Sawadogo et al. 2005). Finally, forests also provide important environmental services, such as protection of floods, avalanches and landslides, the enhancement of the water buffering capacity and the sequestration of carbon (van Kooten and Sohngen 2007). If the multifunctional role of forests were taken into account, clear-felling would most likely not be the optimal logging regime.

Responding to social demand, the literature has begun to study the adoption of management regimes which take the multifunctional role of forests into account. Two prominent examples of these different management regimes are modified clear-felling known as green-tree retention (GTR) and selective-logging (Koskela et al. 2007; St-Laurent et al. 2008; Tahvonen et al. 2010). Summerville and Crist (2002) for instance, analyzed the effects of timber harvest on the presence of Lepidoptera (order of insects including moths and butterflies) and found that the diversity and presence of these species were significantly lower in clear-cut stands, but were identical between selectively logged and unlogged stands. Rosenvald and Lõhmus

(2008) reviewed a large number of studies to determine the effect of different management regimes on biodiversity. They found that GTR improves the habitat for a number of species, but the selective logging regime seems to preserve biodiversity to a larger extent. Thus, a selective-logging regime is especially indicated for stands with species that have a high biodiversity value and poor survival rate under GTR.

In this chapter we present a theoretical model to determine the optimal selective-logging regime of a size-distributed forest under changing climatic conditions when biodiversity conservation is taken into account. The law of motion of the economic model is governed by a partial integrodifferential equation that describes the evolution of the forest stock over time. Given the complexity of the problem, it is not possible to obtain an analytical solution. To solve the problem numerically we employ a technique known as the "Escalator Boxcar Train", previously employed by Goetz et al. (2008, 2011).

The empirical part of the chapter determines the selective-logging regime that maximizes the discounted net benefits from timber production of a stand of *Pinus sylvestris* (Scots Pine), and compares it with the optimal selective-logging regime when biodiversity is accounted for. The results show that the costs of biodiversity conservation in terms of reduced profitability can be significant when stringent targets are implemented. Moreover, we found that these costs are accentuated by climate change. Our results show that targeting biodiversity in terms of deadwood volume and large trees is substantially cheaper in mature stands than in young stands. Hence, it is important to determine the link between the chosen indicators for biodiversity and the type of forest in order to determine the costs of biodiversity.

The chapter is organized as follows. The next section describes the features of the bioeconomic model and explains the different components of the biological processes. Section 3 presents the empirical study which determines the optimal selective-cutting regime of two particular stands, and conducts a sensitivity analysis of the previous results with respect to various biological parameters. The chapter closes with a summary and discussion of the results.

2 Bioeconomic Model

In order to specify the economic model, one needs to characterize the underlying biological model that describes the dynamics of the forest. The description of the evolution of the stand is quite brief; a more detailed explanation can be found in Goetz et al. (2010).

2.1 Stand Dynamics

In forest sciences the size of a tree is usually measured by the diameter at breast height, that is, the diameter of the trunk at height of 1.30 meters above the ground.

Given this definition, time, denoted by t, and diameter, denoted by l, are incorporated as the domain of the control and state variables. The lower boundary of the diameter domain, l_0, indicates the diameter of the seedlings and the upper boundary of the domain, l_m, can be interpreted as the maximum diameter a tree can reach under perfect environmental conditions. It is assumed that a diameter-distributed forest can be fully characterized by the number of trees and by the distribution of the diameter of the trees. In other words, the spatial distribution or particular location of the trees is not accounted for. It is assumed that all individuals have the same environmental conditions and the same amount of space. Moreover, given that the diameter of a tree lies in the interval $[l_0, l_m)$, and that the number of trees is sufficiently large, the forest can be represented by a density function. This function is denoted by $x(t, l)$ and indicates the population distribution with respect to the structuring variable, l, at time t. Thus, the number of trees in the forest at time t is given by

$$X(t) = \int_{l_0}^{l_m} x(t, l)dl.$$

In order to model the dynamics of the forest, the processes of growth, reproduction and mortality are determined in the following paragraphs; where the influence of the individual tree on the vital functions of other individuals (intra-specific competition) is taken into consideration.

In order to express the biotic or abiotic factors which influence the life cycle of the individual, biologists use the term environment. The maximum height of a tree is determined only by its genetic information. However, its diameter is a function of time and environmental conditions.

Let $g(E(t, l), l)$ define the change in diameter of a tree over time as a function of its current diameter, and of a collection of environmental characteristics, $E(t, l)$, which affect the individual life cycle. In a context where the atmospheric and soil conditions are given, and in the absence of diseases and pests, these characteristics are given by the local environmental conditions of the tree, and by the pressure of competition between trees from the same stand for space, light and nutrients. Since our model does not consider the exact location of each tree, $E(t, l)$, measures exclusively the intra-specific competition. Thus, the change in the diameter of a single tree over time is given by

$$\frac{dl}{dt} = g\big(E(t, l), l\big).$$

The instantaneous death rate is denoted by $\mu(E(t, l), l)$. It describes the rate at which the probability of survival of a tree with diameter l, given the environmental characteristics $E(t, l)$, decreases with time.

Finally, as far as reproduction is concerned, we assume assisted natural regeneration.[1] The seedlings with diameter l_0 are the result of natural reproduction and

[1] Assisted natural regeneration (ANR) is a cost-effective regeneration method that facilitates forest growth. It is based on the natural regeneration of forest trees, and aims to accelerate natural succes-

posterior selection for upgrowth by the forest manager. Thus, the control variables of the model, $u(t,l)$ and $p(t,l_0)$, denote the density of cutting in time t with diameter l, and the flow of trees selected for upgrowth in time t with diameter l_0, respectively.

These processes allow us to model the forest dynamics based on the equations described by de Roos (1997), and Metz and Diekmann (1986). It can be modeled as:

$$\frac{\partial x(t,l)}{\partial t} + \frac{\partial (g(E(t,l),l)x(t,l))}{\partial l} = -\mu\big(E(t,l),l\big)x(t,l) - u(t,l). \tag{1}$$

2.2 Biodiversity Considerations

The Convention on Biological Diversity defines biodiversity as the variability among living organisms from all sources, including diversity within genes, between species and of ecosystems. In this study we focus on the second type. According to the literature, biodiversity of forest ecosystems is measured either directly, by the number of species for a given territory, or indirectly, by assessing the presence of different factors that favor biodiversity (Norddahl-Kirsch and Bradshaw 2004; Torras and Saura 2008; Mäkelä et al. 2012).

Within the indirect factors, the volume of deadwood accumulated in the forest ecosystem, and the number of large-diameter trees are considered fundamental for biodiversity (Harmon et al. 1986; McComb and Lindenmayer 1999; Nilsson et al. 2001). Deadwood, including fallen trees and standing dead trees, is of vital importance for the survival of many different species. It has also been proposed under the initiative "Streamlining European Biodiversity Indicators" (http://biodiversity.europa.eu/topics/sebi-indicators accessed 03.01.2013) which is supported by, among others, the United Nations Environmental Program, the European Environmental Agency and the European Commission. Steele (1972) stresses that 20 % of animal species associated with wood are associated with deadwood. For instance, wood-consuming insects which in turn are the prey for insectivorous birds. Bütler et al. (2004) point out that deadwood is essential for the development of the three-toed woodpecker. Brin et al. (2009) found significant correlations between the volume of deadwood and the variety and number of saproxylic beetle species. Moreover, deadwood also allows, among other things, the development of flora such as lichens and mosses (Ódor et al. 2006; Moning et al. 2009).

Besides deadwood, high and old trees also provide important ecological functions. They offer birds a diverse and abundant supply of natural cavities in the trunk

---

sion by removing or reducing barriers, such as intra-specific competition, and forest disturbances (Shono et al. 2007). We assume that the ingrowth of trees is sufficiently large, and thus the effect of ANR is limited to obtaining the optimal number of trees. Although it represents a simplification, the effect is not decisive since insufficient ingrowth could be resolved through enrichment planting, that is, by planting additional trees to reach the desired number of trees.

and branches for nesting and refuge (Camprodon et al. 2008; Vaillancourt et al. 2008). Old trees also accommodate a great diversity of mycorrhizae, which are absent or very scarce in younger trees. In the case of Scots pine, for instance, the cavities are normally found in trees older than 150 years. Since trees are typically cut within the range of 80–120 years, most managed forests do not provide these habitats for birds.

The volume of deadwood and the number of old trees are also employed as indicators in the ongoing Fourth National Forest Inventory to determine the status of the biodiversity of Spanish forest ecosystems. For these reasons, we will focus on these two structural elements as indicators for biodiversity in forest ecosystems. In particular we consider trees as old once their diameter at breast height exceeds 50 cm, since they start presenting cavities (Alberdi et al. 2005). For the sake of brevity we will denote these trees as large-diameter trees throughout the chapter.

The most immediate approach to introduce biodiversity in the economic model would be to express the social benefits of biodiversity in economic terms and include them in the objective function of a mathematical optimization problem. The resulting solution of the optimization problem would determine the optimal structure of standing and dead trees, and of young and old trees. Given the complexity and vagueness of the concept of biodiversity, it becomes very difficult to define a function which expresses biodiversity in economic terms.

Since the former approach is not feasible, an alternative option is to include biodiversity in the economic problem in the form of constraints. In other words, we establish a minimum requirement level for the indicators of biodiversity which the forest manager has to meet. In this way one does not need to express biodiversity in economic terms. Instead, one can identify the difference between the timber benefits obtained by the optimal logging regime without restrictions and with restrictions. These constraints require a minimum volume of deadwood and a minimum number of large-diameter trees.

2.3 The Decision Problem of the Forest Manager

We assume that the forest is privately owned and managed over a planning horizon of T years. Using the previously defined components of the model, the decision problem of the forest owner to maximize private net benefits from timber production when biodiversity is not considered can be stated as:

$$\int_0^T \int_{l_0}^{l_m} B\big(x(t,l), u(t,l)\big)e^{-rt}\,dl\,dt - \int_0^T C\big(p(t,l_0)\big)e^{-rt}\,dt + \int_{l_0}^{l_m} S^T\big(x(T,l)\big)e^{-rT}\,dl, \tag{2}$$

subject to (1) and

$$g\big(E(t,l_0), l_0\big)x(t,l_0) = p(t,l_0), \tag{2a}$$

$$x(t_0, l) = x_0(l), \tag{2b}$$

$$p(t, l_0) \geq 0, \tag{2c}$$

$$u(t, l) \geq 0, \tag{2d}$$

where r denotes the discount rate. The function $B(x, u)e^{-rt}$ presents the discounted net benefits of timber production. It depends not only on the amount of logged trees but also on the amount of standing trees so that the maintenance costs of the forest can be taken into account. The strictly convex function $C(p)e^{-rt}$ expresses the discounted costs of upgrown trees with diameter l_0, and the function $S^T(x)$ the value of the standing trees at the final point in time of the planning horizon. The restriction $g(E(t, l_0)l_0)x(t, l_0) = p(t, l_0)$ requires that the flux of the change in diameter at diameter l_0 coincides with the total flux of trees selected for upgrowth. The term $x_0(l)$ denotes the initial diameter distribution of the trees. Finally, the control variables, $u(t, l)$ and $p(t, l_0)$, have to be non-negative.

In order to account for biodiversity we add three additional constraints to the decision problem:

$$\int_{50}^{l_m} x(t, l)dl \geq b_{\min} \quad \forall t, \tag{3}$$

$$M(t) = \int_{l_0}^{l_m} \mu\big(E(t, l), l\big)\alpha x(t, l)^{\beta} dl, \tag{4}$$

$$\int_0^t e^{-\delta(t-\tau)} M(\tau)d\tau \geq S_{\min} \quad \forall t. \tag{5}$$

Equation (3) requires that the amount of trees with a diameter higher than 50 cm be greater or equal to a minimum ecological value, $b_{\min}$. Equation (4) is an identity that specifies the volume of the trees which die at time t, where α and β are parameters that convert the amount of trees into volume. Finally, (5) establishes a minimum stock of deadwood in the forest, $S_{\min}$. Many authors assume that all the materials of dead biomass are equally decomposable, and the single-exponential model is the most frequent model used to determine decomposition constants (Zhou et al. 2007). Therefore, based on the literature, we assume a constant decomposition rate of the deadwood in the forest ecosystem, denoted by δ.

In practice, the necessary conditions of the optimization problem (2), two equations and a system of partial integrodifferential equations, can only be solved analytically under severe restrictions with respect to the specification of the mathematical problem (Muzicant 1980). Thus, one has to resort to numerical techniques in order to solve the distributed control problem. To take the analysis further we propose to employ a numerical solution technique known as the "Escalator Boxcar Train" used initially by de Roos (1988) to describe the evolution of physiologically-structured populations. The convergence of the EBT was evaluated by Brännström et al. (2013). They found that this method converges weakly to the true solution under weak conditions on the biological parameters.

Applying the EBT allows the partial integrodifferential equations of problem (2) to be transformed into a set of ordinary differential equations which are subsequently approximated by difference equations. Besides a brief presentation of the EBT method, Goetz et al. (2008) show how this approach can be extended to account for optimization problems by incorporating decision variables. The transformed model is subsequently implemented in GAMS (General Algebraic Modeling System, Brooke et al. 1992) "that is frequently" used for solving mathematical programming problems. To transform the decision problem (2), we first divide the range of diameter into equal parts, and define $X_i(t)$ as the number of trees in the cohort i, being $i = 0, 1, 2, \ldots, n$, that is, the trees whose diameter falls within the limits l_i and l_{i+1} are grouped in the cohort i. Likewise, we define $L_i(t)$ as the average diameter, $U_i(t)$ as the amount of cut trees within cohort i, and $P(t)$ as the amount of trees selected for upgrowth.

3 Numerical Analysis

The purpose of the numerical analysis is to determine the optimal logging regime of a diameter-distributed forest taking biodiversity into account. In other words we determine the logging regime that maximizes the discounted private net benefits from timber production of a stand of *Pinus sylvestris* over a time horizon of 150 years while satisfying constraints with respect to the volume of deadwood and large-diameter trees in the forest ecosystem. The election of *Pinus sylvestris* was motivated by the fact that it is the most important commercial species for timber production in Catalonia.

3.1 Economic Data

The net benefit function (in €) of the economic model, $B(x(t,l), u(t,l))$, consists of the net revenue from the sale of timber at time *t*, minus the costs of maintenance, which comprises clearing, pruning and grinding the residues. The net revenue is given by the sum of the revenue of the timber sale minus logging costs defined as: $[\sum_{i=1}^{n}(\rho(L_i(t)) - vc(L_i(t)))tv(L_i(t))mv(L_i(t))U_i(t)] - [mc(X(t))]$ where $X(t)$ denotes the total number of trees in the stand, that is, $X(t) = \sum_{i=0}^{n} X_i(t)$. The terms in the first square brackets denote the sum of the revenue of the timber sale minus the cutting costs of each cohort *i*, and the term in the second square brackets, $mc(X(t))$, accounts for the maintenance costs. The parameter $\rho(L_i)$ denotes the timber price per cubic meter of wood as a function of the diameter; $tv(L_i)$ is the total volume of a tree as a function of its diameter; $mv(L_i)$ is the part of the total volume of the tree that is marketable; $vc(L_i)$ are logging costs.

Timber price per cubic meter was taken from a study by Palahí and Pukkala (2003), who analyzed the optimal management of a *Pinus sylvestris* forest in a clear-cutting regime. They estimated a polynomial function given by $\rho(L) = \min\{-23.24 + 13.63\sqrt{L}, 86.65\}$, which is an increasing and strictly concave function, for a diameter lower than 65 cm. At $L = 65$ the price reaches its maximum value, and for $L > 65$ it is considered constant. The logging cost comprises logging, pruning, cleaning the understory, and collecting and removing residues. Based on the work by Palahí and Pukkala (2003), the logging cost per cubic meter of logged timber is given by the function $vc(L) = 6 + \exp(4.292 - 0.506\ln(L))$. Data about maintenance costs were provided by the consulting firm Tecnosylva, which elaborates forest management plans throughout Spain. According to the data supplied by Tecnosylva, the maintenance cost function is approximated by $mc(X(t))$, and is given by $mc(X(t)) = 10 + 0.0159X(t) + 0.0000186X(t)^2$. The nursing costs are linear in the amount of trees selected for upgrowth and are given by $C(P) = 0.73P$. The thinning and nursing period, Δt, is set at 10 years, which is common practice for a *Pinus sylvestris* forest (Cañellas et al. 2000).

3.2 Biological Data

The basis for the specification of the forest dynamics is sufficiently long time series for the key biological variables of a forest. This data can be obtained either from historical observation of real data or from data generated with biogeochemical process-based models. While the first approach is widely used in forestry economics, it is not suitable for the analysis of the optimal management regime under climate change, since it is based on recorded data, that is, it implicitly assumes that future climatic conditions will be similar to current conditions (Garcia-Gonzalo et al. 2007; Hynynen et al. 2002). In contrast, biogeochemical process-based models are able to incorporate changes in the climate that most likely affect the evolution of trees (Mäkelä 1997). Therefore, we opted for the latter approach and we simulated the growth of a diameter-distributed stand of *Pinus sylvestris* without thinning with the bio-physical simulation model GOTILWA (Growth Of Trees Is Limited by Water). The model generates data related to growth and mortality and thus allows the exploration of how the life cycle of an individual tree is influenced by the climate, by the characteristics of the tree itself and environmental conditions. GOTILWA is defined by 11 input files specifying more than 90 parameters related to the site, soil composition, tree species, photosynthesis, stomatal conductance, forest composition, canopy hydrology, and climate. These parameters were chosen in accordance with the location of the study, which is situated in the Alta Garrotxa (county of Girona, Spain), since it is a region with a large extension of forest stands of *Pinus sylvestris* (Ibáñez 2004).

To obtain the data to estimate the growth function, different initial diameter distributions of a forest were chosen. These distributions were specified as a transformed beta density function $\theta(l)$ since it allows a great variety of distinct shapes of the initial diameter distributions of the trees to be defined (Hunter 1990). The shape parameters are denoted by γ and φ, and we allowed these parameters to take on either a value of 0.5, 1, or 2. We generated a variety of initial diameter distributions considering all possible combinations of the three possible values of the parameters γ and φ. The density function of the diameter of trees, $\theta(l; \gamma, \varphi)$, is defined over a closed interval, and thus the integral

$$\int_{l_i}^{l_{i+1}} \theta(l; \gamma, \varphi) dl$$

gives the proportion of trees lying within the range $[l_i, l_i + 1)$. The initial forest consists of a population of trees with diameters within the interval $[0 \text{ cm} \leq l \leq 50 \text{ cm})$. This interval was divided into 10 subintervals of identical length, and the initial number of trees in any cohort, $X_i(0)$, is calculated for each combination of γ and φ with three different basal areas (15, 20 and 25 m^2/ha). These different basal areas were chosen to isolate the effects of the initial distributions and the density of the stand on the biological processes.

The simulations in GOTILWA were conducted for different climate change scenarios to determine their effect on forest growth. For this purpose, we considered three different climate scenarios. The first one does not take climate change into account, and we refer to it as the baseline (BL). Additionally, we considered two other climate change scenarios, denoted by A2 and B2 in accordance with the IPCC's Third Assessment Report (2001) on climate change. Although both scenarios predict increases in CO_2 emissions in the near future, neither of the two scenarios is extreme compared to the range of scenarios considered in the report. A2 presents a more pessimistic setting, since it forecasts a higher increase in CO_2 emissions than B2. Specifically, scenario A2 estimates a CO_2 concentration of 870 ppm by the year 2100, and a rise of 2.0–5.4 °C in temperatures, while B2 calculates a CO_2 concentration of 621 ppm, and a temperature increase within the range of 1.4–3.8 °C. For this study we used the reported evolution of the CO_2 concentration in the atmosphere, as well as the estimated variations in temperature and rainfall in the Mediterranean region (Ruosteenoja et al. 2003) to supply GOTILWA with estimated time series for CO_2, temperature and rainfall for each of the three climate change scenarios analyzed.

Based on the previously specified initial diameter distributions and the time series related to each specific scenario, we simulated the growth of the forest over 150 years. The generated data from the series of simulations allows us to estimate the function $g(E, L_i)$, which describes the change in diameter over time. The type of function was specified as a von Bertalanffy growth curve (von Bertalanffy 1957), generalized by Millar and Myers (1990) which allows the rate of growth of the diameter to vary with environmental conditions. The concrete specification of the function is given by $g(E, L_i) = (l_m - L_i)(\beta_0 - \beta_1 \cdot BA + \beta_2 \cdot BA_i)$, where the exogenous variables of this function, provided by GOTILWA, are the diameter at breast height (L_i), the basal area of the stand (BA) and the basal area of cohort i (BA_i). Forest growth is simulated within limits by an increasing concentration of CO_2 in the atmosphere due to its fertilizer effect (Heimann and Reichstein 2008). Therefore, the parameters to be estimated, β_0, β_1 and β_2, were assumed to depend on CO_2 concentrations. The estimation yielded the growth function: $g(E, L_i) = (183.74 - L_i)[(0.021 + 0.19 \cdot 10^{-4}CO_2) + (-0.24 \cdot 10^{-3} + 0.19 \cdot 10^{-7}CO_2)BA + (0.064 - 0.58 \cdot 10^{-4}CO_2)BA_i]$.

The value of the tree volume parameters, $tv(L_i)$, has also been estimated using the data generated with GOTILWA. The tree volume is based on the allometric relation $tv(L) = 0.000135L^{2.429685}$. A study by Cañellas et al. (2000) provides information that allows the marketable part of the tree volume, $mv(L_i(t))$, to be estimated as a function of the diameter. The marketable part of the timber volume of each tree is an increasing function of the diameter and is given by $mv(L) = 0.699 + 0.0004311L$.

The mortality function was designed based on the survival function of González et al. (2005), and it is given by:

$$\mu(E, L_i) = 1 - 1/\bigl(1 + \exp\bigl(-3.954 + 0.035BA - 2.297(L_i/A)\bigr)\bigr)^2.$$

Mortality depends on the diameter of the individual, the basal area, and the average age of the stand, denoted by A. In accordance with the literature, we assumed that climate change exacerbates the rate at which the woody debris is decomposed, due to higher bacterial activity (Mackensen et al. 2003; Garrett et al. 2012). According to the latter work, the decomposition rate of the deadwood is given by $\delta = 0.0429 \cdot \exp(0.093 \cdot Te)$, where Te denotes the mean temperature.

The two stands considered are characterized by the difference in the magnitude of their average diameter. Stand 1 consists of a young population of trees and Stand 2 of a mature population and are presented in Table 1. The specification of these two different initial distributions allows the sensitivity of the optimal trajectories to be evaluated.

The initial distributions were obtained from the database of the EFIC (Ecological and Forest Inventory of Catalonia), an inventory of Catalan forests set up between 1988 and 1998 by the CREAF (Center for Ecological Research and Forestry Applications). This database offers a large variety of data, such as biomass, above-ground production of wood, leaves, branches, and the diameter distribution of the inventoried stand.

3.3 Optimization Results

To find the optimal logging regime for the two stands considered, the Conopt3 solver available within GAMS was employed. The numerical solution of the problem provides the optimal values of the stock variable X_i and decision variables, U_i and P for every 10 year period. Based on these values the economic variables, such as the revenues from timber sale, cutting and maintenance costs can be determined. We initially calculated the optimal logging regime for the stands in the baseline scenario given a discount rate of 2 %. The results show that it is optimal to harvest the trees selectively according to their size.

Table 1 Initial diameter distributions of the analyzed stands

Stand 1		Stand 2	
Basal area: 23.42 m^3/ha		Basal area: 47.82 m^3/ha	
UTM X Coordinate: 456600		UTM X Coordinate: 448500	
UTM Y Coordinate: 4664800		UTM Y Coordinate: 4674900	
Diameter class (cm)	N° of trees/ha	Diameter class (cm)	N° of trees/ha
2.5	400	2.5	37
7.5	319	7.5	74
12.5	287	12.5	294
17.5	223	17.5	368
22.5	96	22.5	404
27.5	64	27.5	147
32.5	64	32.5	74
		37.5	37

Tables 2 and 3 present the main biological variables resulting from the optimization for the case where biodiversity is not considered and where it is considered for Stands 1 and 2, respectively. Jönsson and Jonsson (2007) observed an average of 19.5 m^3/ha of coarse woody debris for key habitats in Sweden, and Penttilä et al. (2004) found that threatened species are practically only found in forests where the volume of deadwood exceeds 20 m^3/ha. Therefore, we consider this value as an adequate stand-level threshold of a high-quality habitat, and consequently, $S_{\min}$ is set equal to 20. A review of the literature reveals that a general threshold for large-diameter trees cannot be easily established. Given some references in the literature we required that the optimal management regime needs to maintain at least 15 trees with a diameter of at least 50 cm, that is, $b_{\min} = 15$. Therefore, we opted to set $b_{\min} = 15$ and present a subsequent sensitivity analysis with respect to the minimum number of large-diameter trees.

For a stand of predominantly young trees (Stand 1), Table 2 shows that when biodiversity forms part of the management objective, the investment of the forest manager in young trees is advanced and higher. For example, the number of trees selected for upgrowth is positive from year 30 onwards while, when biodiversity is not considered, only from year 70. Moreover, when biodiversity is taken into account, the first thinning is delayed until year 10, and the number of logged trees in this period is considerably lower, leading to a decrease in the volume of logged trees. As a result, the benefits from timber sales are lower when biodiversity is taken into account. Thus, when the stand is managed only for timber, the discounted sum of the net benefits obtained from forest management over 150 years (NPV of the benefits for short) is about 4663 Euro/ha. However, when biodiversity considerations are incorporated in the formulation of the decision problem, the NPV of the benefits decreases to 4174 Euro/ha, since the forest owner needs to maintain a specific amount of big trees that have passed the rotation age. Thus, the incorporation of biodiversity in the economic problem leads to a decrease of 10.48 % in the NPV of benefits

Table 2 Optimal selective-logging regime of a young stand (Stand 1)

Year	Number of trees[a]	Upgrown trees	Logged trees	Logged volume (m^3/ha)	Dead wood volume	Average forest age (years)	Average diameter (cm)	Benefits (Euro/ha)
Considering only timber								
0	1845	0	10	1.46	3.41	33.21	10.55	−1016.49
10	1509	0	249	45.70	9.45	43.44	14.59	−67.97
20	1138	0	289	83.50	16.14	52.47	17.83	1370.94
30	946	0	132	75.82	23.08	60.47	20.87	1444.87
40	730	0	166	103.64	27.50	68.02	23.16	2314.56
50	640	0	56	31.70	34.05	73.93	24.89	486.35
60	528	0	79	81.86	39.08	82.71	28.02	2426.30
70	508	94	89	60.48	42.04	77.21	25.65	1309.40
80	555	127	53	46.61	43.66	68.37	22.91	1060.13
90	539	93	80	65.74	43.47	65.63	22.20	1606.76
100	565	116	61	63.87	41.93	60.36	20.33	1752.11
110	545	62	53	70.56	39.90	60.21	20.42	2259.18
120	515	41	45	72.90	37.77	61.10	20.95	2614.63
130	456	2	38	76.96	35.87	65.26	22.77	3076.39
140	405	0	34	82.13	34.10	69.05	24.50	3582.11
150	421	36	6	17.38	33.74	67.15	24.02	628.82
Considering timber and biodiversity								
0	1855	0	0	0.00	3.48	33.21	10.55	−1036.25
10	1577	0	191	22.92	10.95	43.51	14.62	−682.71
20	1160	0	324	70.15	20.00	53.18	18.14	862.15
30	912	1	178	109.41	26.58	61.88	21.68	2425.23
40	732	1	131	79.87	32.01	68.45	23.42	1684.98
50	636	28	86	68.32	36.76	72.95	24.72	1663.87
60	520	3	86	61.40	40.98	79.65	26.89	1434.15
70	584	159	68	45.09	43.50	68.78	22.93	806.76
80	584	131	98	59.30	43.79	63.49	21.35	1126.70
90	572	90	70	55.21	43.07	61.73	20.81	1262.53
100	526	48	62	62.63	41.79	63.32	21.54	1742.12
110	518	48	27	90.61	40.38	63.60	21.82	4260.43
120	465	7	35	80.00	39.17	69.00	23.63	3470.95
130	428	20	37	71.73	37.87	70.96	24.48	2807.10
140	400	8	18	51.26	37.37	73.94	25.86	2296.40
150	491	106	0	0.30	38.32	65.33	22.79	−290.26

[a]The number of trees in the forest is calculated just after the trees are planted, and before the thinning takes place

Table 3 Optimal selective-logging regime of a mature stand (Stand 2)

Year	Number of trees	Upgrown trees	Logged trees	Logged volume (m^3/ha)	Dead wood volume	Average forest age (years)	Average diameter (cm)	Benefits (Euro/ha)
Considering only timber								
0	1127	0	271	114.71	8.52	48.65	19.74	2135.85
10	929	0	142	81.31	16.92	56.88	21.86	1556.52
20	728	0	154	78.33	24.27	65.31	24.07	1460.85
30	575	0	117	79.14	30.12	73.87	26.55	1837.65
40	520	63	91	56.95	34.79	74.99	26.14	1170.45
50	521	110	83	66.41	37.09	70.53	24.15	1595.39
60	529	100	66	67.48	37.60	66.39	22.40	1869.27
70	553	96	47	47.43	37.57	62.78	20.96	1220.31
80	549	83	58	69.81	36.66	62.03	20.76	2115.32
90	522	48	48	71.34	35.35	62.37	21.06	2445.23
100	467	11	42	78.02	33.97	65.81	22.56	3003.73
110	462	35	22	45.94	33.24	65.54	22.69	1775.59
120	404	0	39	84.30	32.41	72.10	25.42	3499.88
130	497	107	0	0.00	32.75	60.99	21.59	−303.17
140	507	70	38	44.74	33.54	64.12	22.79	1249.28
150	504	67	47	55.62	33.95	64.03	22.72	1634.98
Considering timber and biodiversity								
0	1117	0	281	108.37	8.78	48.65	19.74	1903.43
10	919	1	142	81.11	17.39	57.07	22.04	1554.21
20	720	1	152	77.39	24.93	65.49	24.24	1439.02
30	597	28	114	76.72	30.79	71.72	25.81	1740.17
40	496	27	98	59.74	35.59	76.79	26.97	1247.06
50	509	118	81	64.19	37.85	70.74	24.37	1523.10
60	564	130	49	45.33	38.94	63.83	21.59	1066.40
70	556	93	69	64.97	38.70	62.80	21.15	1698.58
80	518	46	54	64.26	37.88	63.69	21.63	1946.85
90	496	50	45	66.31	36.64	63.66	21.78	2244.41
100	480	33	26	43.83	36.02	64.89	22.44	1523.09
110	448	17	26	95.24	35.48	69.54	24.36	4584.18
120	414	11	26	72.21	35.13	73.18	25.58	3313.99
130	480	82	0	0.01	36.43	66.77	23.27	−278.84
140	499	85	43	41.79	37.77	67.40	23.48	1032.60
150	483	73	63	55.92	38.10	66.71	23.17	1422.24

obtained from forest management. This is a very significant result, since it implies that forest owners are not likely to adapt their management regime to incorporate biodiversity targets unless they are required to do so. Therefore, the promotion of conservation and enhancement of forest biodiversity can only be achieved if specific policies are put in place.

The results of the optimization for the mature stand (Stand 2), presented in Table 3 show a similar time profile for the investment. When biodiversity is incorporated into the model, nursing starts at year 10, in comparison with the case without biodiversity consideration where nursing starts at year 40. As shown in Table 3, in both cases logging begins in the initial period of the planning horizon. However, in the case where biodiversity is considered the logged trees have a smaller diameter, and consequently the volume of logged trees is lower. Hence, the structure of the logging regime required to maintain a minimum number of standing large-diameter trees reduces the net benefits from timber sale during the initial time periods. However, since the forest owner obtains their first benefits during the initial years, the losses in the NPV of the benefits due to the consideration of biodiversity are less pronounced, compared to the case of a young stand. The losses of the NPV of benefits in this case amount to 4.44 %.

The effects of climate change on the selective-logging regime can be observed in Fig. 1, cases (1a) to (1d) for the scenarios BL and A2. It depicts the evolution of the number of standing trees, the average forest age, the average diameter and the deadwood volume over time when biodiversity is not considered and when it is considered. Specifically, for the management of biodiversity, the optimal selective logging regime guarantees, as a minimum, 15 standing trees with a diameter of at least 50 cm ($b_{\min} = 15$), and not less than 20 m^3/ha of deadwood ($S_{\min} = 20$). Figure 1 shows that climate change leads, at the end of the planning horizon for the management regime for "timber benefits only", to an increase in the number of trees by 123.19 % and by 83.11 % for the management regime of timber and biodiversity. This evolution is due to the fertilization effect of the increase in the CO_2 concentration in the atmosphere, which facilitates forest growth and makes investment in forest more profitable. We also observe in Fig. 1 that the average age (Fig. 1b) and diameter of the standing trees (Fig. 1c) decrease with climate change. Finally, in the presence of climate change more deadwood is accumulated in the forest ecosystem, and this effect is even more accentuated if biodiversity is taken into account (see Fig. 1d).

To analyze to what extent the losses from biodiversity conservation targets can be limited, we varied the constraint on the number of standing large-diameter trees from 0 to 30 for the case of the young stand (Stand 1). Figure 2 depicts the NPV of the different optimization scenarios. It shows that the NPV of the benefits increase with climate change, however they decrease with the number of standing large-diameter trees. This decrease is more pronounced in scenario A2, even to the extent that the benefits of A2 are less than those of B2, when 30 large-diameter trees are required. This fact can be best observed in Fig. 3, which depicts the losses in the NPV of the benefits aggregated over 150 years as a function of the standing large-diameter trees in comparison with the case where the forest is managed for timber

Fig. 1 Variation in the evolution of the structural variables over time when accounting for biodiversity

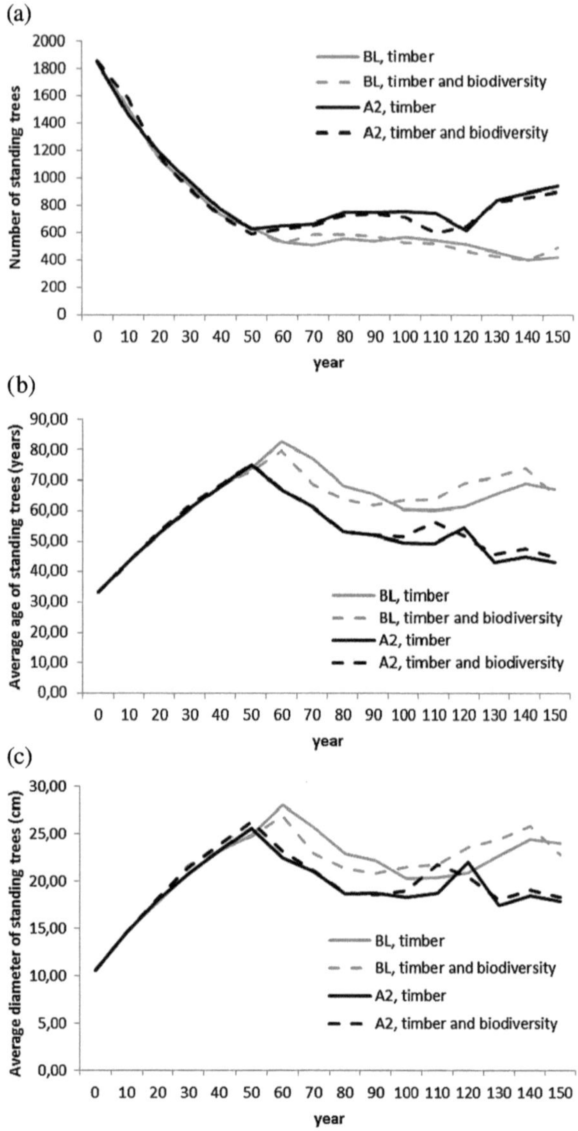

only. It shows that foregone profits can be substantial. In the most extreme case, where it is assumed that the forest owner is required to maintain a minimum stock of 30 large-diameter trees, the losses total 23.27 %, 26.15 %, and 29.12 % in the BL, B2, and A2 scenarios, respectively.

Half of key Swedish woodland habitats have more than 38.94 m^3/ha of coarse woody debris (Jönsson and Jonsson 2007). Thus, we also conducted a sensitivity analysis and evaluated the effects of a variation in the required volume of deadwood in the forest ecosystem.

Fig. 1 (Continued)

(d)

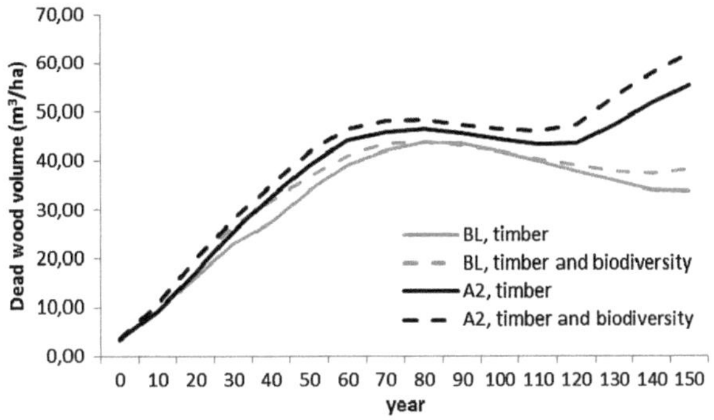

Fig. 2 Net present value of the benefits of forest management over 150 years as a function of the required minimum number of large-diameter trees

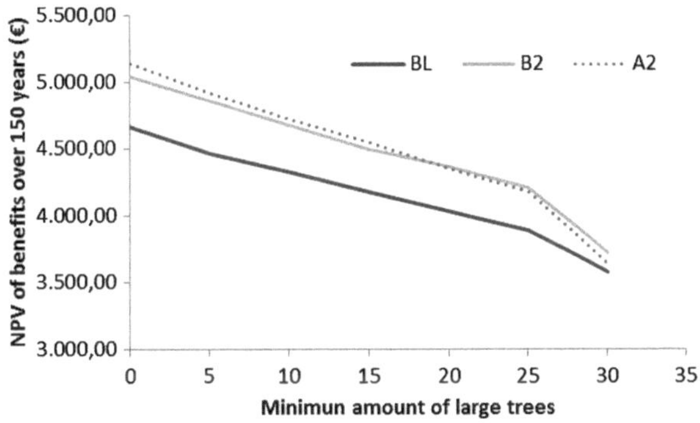

Fig. 3 Decrease in the net present value of the benefits of forest management over 150 years as a function of the required minimum number of large-diameter trees

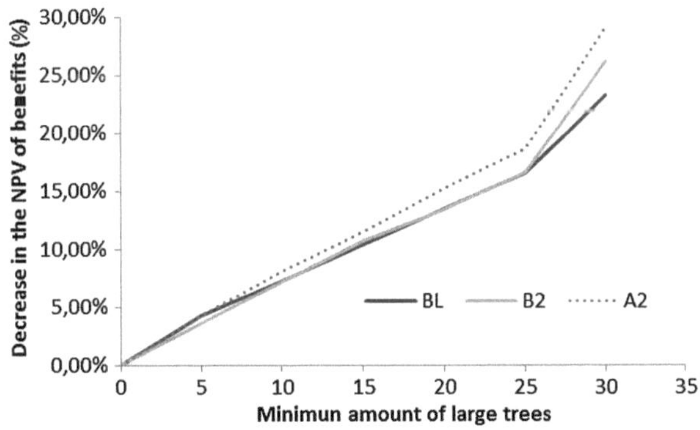

Table 4 presents the NPV benefits of the optimal selective logging regime for Stands 1 and 2 obtained by varying the number of large-diameter trees and the volume of deadwood. Table 4 shows that for the baseline scenario and scenario A2 the foregone profits of maintaining biodiversity can be very substantial and moreover, they are exacerbated by climate change. In the case of a young stand (Stand 1), for instance, the constraint of 30 large-diameter trees and a minimal volume of 38.94 m^3 of deadwood per hectare leads, for the baseline scenario, to a decrease in the NPV

Table 4 Sensitivity analysis of the NPV with respect to the large-diameter trees and the minimum amount of deadwood in the forest

Large-diameter trees (#)	Dead wood volume (m^3/ha)	NPV in the NCH scenario (Euro/ha)	NPV losses (NCH) (%)	NPV in the A2 scenario (Euro/ha)	NPV losses (A2) (%)
Stand 1					
0	0	4662.79	–	5136.35	–
10	20	4324.78	7.25	4721.64	8.07
20	20	4033.88	13.49	4353.17	15.25
30	20	3577.77	23.27	3640.53	29.12
10	38.94	2448.65	47.49	2844.80	44.61
20	38.94	2162.44	53.62	2499.40	51.34
30	38.94	1421.09	69.52	868.62	83.09
Stand 2					
0	0	9332.94	–	9821.36	–
10	20	9041.88	3.12	9408.27	4.21
20	20	8775.34	5.97	9068.49	7.67
30	20	8506.98	8.85	8653.22	11.89
10	38.94	8885.24	4.80	9323.68	5.07
20	38.94	8624.97	7.59	8924.36	9.13
30	38.94	8402.87	9.97	8556.99	12.87

of benefits by 69.52 % compared to the case where biodiversity in not considered. These losses increase to 83.09 % when climate change is taken into account.

Table 4 also shows that the foregone profits are more moderate in the case of a mature stand (Stand 2). In this case the losses are only 9.97 % for the baseline scenario and 12.87 % for scenario A2. This result shows the need to target areas to establish or conserve biodiversity, which are less sensitive to changes in the logging regime. In particular more mature forests are less affected by changes in the management regime than young forests.

4 Conclusions

This chapter presents a theoretical model that allows us to determine the optimal management regime of a diameter-distributed forest where biodiversity conservation is taken into account, that is, when the forest manager is required to meet certain requirements with respect to various indirect indicators of biodiversity in forest ecosystems. The chosen indicators are the volume of deadwood existing in the forest and the number of large-diameter trees.

The economic decision problem for determining optimal forest management is formulated as a distributed optimal control problem where the control variables and the state variable depend on both time and the diameter of the tree, where the growth and mortality processes depend not only on individual characteristics but also on the distribution of the individual characteristics over the entire population.

The problem is solved numerically by applying the Escalator Boxcar Train method and subsequently implementing the transformed model in GAMS. In the numerical analysis, the optimal selective-logging regime of two real stands of *Pinus sylvestris* in Alta Garrotxa (Catalonia) is determined, and the optimal management plans, with and without biodiversity considerations are compared, for three given climate scenarios. The results of this study indicate that the costs of biodiversity conservation in terms of reduced profitability can be significant and that they are exacerbated by climate change. They also suggest that it is especially important to assess the link between the indirect indicators and biodiversity to properly determine which type of forest can be targeted to biodiversity conservation without compromising future timber benefits.

Acknowledgements The authors gratefully acknowledge the support of the Spanish Ministry of Science and Technology grant Econ2010-17020, with partial funding from the program FEDER of the European Union, and of the Government of Catalonia grants XREPP and 2009 SGR189.

References

Alberdi, I., Saura, S., & Martínez, F. J. (2005). El estudio de la biodiversidad en el tercer inventario forestal nacional. *Cuadernos de la Sociedad Española de Ciencias Forestales*, *19*, 11–19.

Alig, R. J., Adams, D. M., & McCarl, B. A. (2002). Projecting impacts of global climate change on the US forest and agriculture sectors and carbon budgets. *Forest Ecology and Management*, *169*, 3–14.

Brännström, Å., Carlsson, L., & Simpson, D. (2013). On the convergence of the escalator boxcar train. *SIAM Journal on Numerical Analysis*, *51*, 3213–3231.

Brin, A., Brustel, H., & Jactel, H. (2009). Species variables or environmental variables as indicators of forest biodiversity: a case study using saproxylic beetles in Maritime pine plantations. *Annals of Forest Science*, *66*, 306. doi:10.1051/forest/2009009.

Brooke, A., Kendrick, D., & Meeraus, A. (1992). *GAMS: a user's guide*. San Francisco: The Scientific Press. Release 2.25.

Bütler, R., Angelstam, P., Ekelund, P., & Schlaepfer, R. (2004). Dead wood threshold values for the three-toed woodpecker presence in boreal and sub-Alpine forest. *Biological Conservation*, *119*, 305–318.

Camprodon, J., Salvanyà, J., & Soler-Zurita, J. (2008). The abundance and suitability of tree cavities and their impact on hole-nesting bird populations in beech forests of NE Iberian Peninsula. *Acta Ornithologica*, *43*(1), 17–31.

Cañellas, I., Martinez García, F., & Montero, G. (2000). *Silviculture and dynamics of Pinus sylvestris L. stands in Spain. Investigación Agraria: Sistemas y Recursos Forestales*. Fuera de Serie 1.

Conrad, J., & Clark, C. (1987). *Natural resource economics*. Cambridge: Cambridge University Press.

Croitoru, L. (2007). Valuing the non-timber forest products in the Mediterranean region. *Ecological Economics*, *63*(4), 768–775.

de Roos, A. (1988). Numerical methods for structured population models: the escalator boxcar train. *Numerical Methods for Partial Differential Equations*, *4*, 173–195.

de Roos, A. (1997). A gentle introduction to physiologically structured population models. In S. Tuljapurkar & H. Caswell (Eds.), *Structured populations models in marine, terrestrial and freshwater systems* (pp. 119–204). New York: Chapman & Hall.

Doyon, F., Gagnon, D., & Giroux, J. (2005). Effects of strip and single-tree selection cutting on birds and their habitat in a southwestern Quebec northern hardwood forest. *Forest Ecology and Management, 209*, 106–116.

EU (2006). Communication from the Commission to the Council and the European Parliament on an EU Forest Action Plan. Available in: http://europa.eu/legislation_summaries/agriculture/environment/l24277_en.htm; accessed 28.02.2013.

Garcia-Gonzalo, J., Peltola, H., Briceño-Elizondo, E., & Kellomäki, S. (2007). Effects of climate change and management on timber yield in boreal forests, with economic implications: a case study. *Ecological Modelling, 209*, 220–234.

Garrett, K. A., Dobson, A. D. M., Kroschel, J., Natarajan, B., Orlandini, S., Tonnang, H. E. Z., & Valdivia, C. (2012). The effects of climate variability and the color of weather time series on agricultural diseases and pests, and on decisions for their management. *Agricultural and Forest Meteorology, 170*, 216–227.

Goetz, R., Hritonenko, N., Xabadia, A., & Yatsenko, Y. (2008). Using the escalator boxcar train to determine the optimal management of a size-distributed forest when carbon sequestration is taken into account. *Lecture Notes in Computer Science, 4818*, 323–330.

Goetz, R., Hritonenko, N., Mur, R., Xabadia, A., & Yatsenko, Y. (2010). Forest management and carbon sequestration in size-structured forests: the case of Pinus sylvestris in Spain. *Forest Science, 56*, 242–256.

Goetz, R., Xabadia, A., & Calvo, E. (2011). Optimal forest management in the presence of intraspecific competition. *Mathematical Population Studies, 181*, 151–171.

González, J., Pukkala, T., & Palahí, M. (2005). Optimizing the management of Pinus sylvestris L. stand under risk of fire in Catalonia (north-east of Spain). *Annals of Forest Science, 62*, 493–501.

Harmon, M. E., Franklin, J. F., & Swanson, F. J. (1986). Ecology of coarse woody debris in temperate ecosystems. *Advances in Ecological Research, 15*, 133–302.

Heimann, M., & Reichstein, M. (2008). Terrestrial ecosystem carbon dynamics and climate feedbacks. *Nature, 451*, 289–292.

Hunter, M. L. (1990). *Wildlife, forest and forestry. Principles of managing forests for biological diversity* (p. 370). New Jersey: Prentice Hall.

Hynynen, J., Ojansuu, R., Hökkä, H., Siipilehto, J., Salminen, H., & Haapala, P. (2002). Models for predicting stand development in MELA system, Finnish Forest Research Institute 835.

Ibáñez, J. (2004). El bosc espcies dominants, existncies, estructura i altres caracterstiques. In J. Terradas & F. Rod (Eds.), *Els Boscos de Catalunya: Estructura, Dinàmica i Funcionament. Generalitat de Catalunya* (pp. 56–93). Barcelona: Department de Medi Ambient i Habitatge

Jönsson, M. T., & Jonsson, B. G. (2007). Assessing coarse woody debris in Swedish woodland key habitats: implications for conservation and management. *Forest Ecology and Management, 242*, 363–373.

Koskela, E., Ollikainen, M., & Pukkala, T. (2007). Biodiversity policies in commercial boreal forests: optimal design of subsidy and tax combinations. *Forest Policy and Economics, 9*, 982–995.

Mackensen, J., Bauhus, J., & Webber, E. (2003). Decomposition rates of coarse woody debris—a review with particular emphasis on Australian tree species. *Australian Journal of Botany, 51*(1), 27–37.

Mäkelä, A. (1997). A carbon balance model of growth and self-pruning in trees based on structural relationships. *Forest Science, 43*, 7–24.

Mäkelä, A., del Rio, M., Hynynen, J., Hawkins, M. J., Reyer, K., Soares, P., van Oijen, M., & Tomé, M. (2012). Using stand-scale forest models for estimating indicators of sustainable forest management. *Forest Ecology and Management, 285*, 164–178.

McComb, W., & Lindenmayer, D. (1999). Dying, dead and down trees. In M. L. Hunter (Ed.), *Managing biodiversity in forest ecosystems* (pp. 335–372). Cambridge: Cambridge University Press.

Metz, J., & Diekmann, O. (1986). *The dynamics of physiologically structured populations. Springer lecture notes in biomathematics*. Heidelberg: Springer.

Millar, R., & Myers, R. (1990). Modelling environmentally induced change in growth for Atlantic Canada cod stocks. ICES—International Council for the Exploration of the Sea C.M./G24.

Moning, C., Werth, S., Dziock, F., Bässler, C., Bradtka, J., Hothom, T., & Müller, J. (2009). Lichen diversity in temperate montane forests is influenced by forest structure more than climate. *Forest Ecology and Management*, *258*, 745–751.

Murray, B. C., Sohngen, B. L., Sommer, A. J., Depro, B. M., Jones, K. M., McCarl, B. A., Gillig, D., DeAngelo, B., & Andrasko, K. (2005). EPA-R-05-006. Greenhouse gas mitigation potential in US forestry and agriculture. Economic modeling of effects of climate change on the forest sector and mitigation options, Washington, DC: US Environmental Protection Agency, Office of Atmospheric Programs 144.

Muzicant, J. (1980). *Systeme mit verteilten Parametern in der Bioökonomie*. Wien: Technische Universitat. Dissertation.

Nilsson, S. G., Hedin, J., & Niklasson, M. (2001). Biodiversity and its assessment in boreal and nemoral forests. *Scandinavian Journal of Forest Research*, *16*(3), 10–26.

Norddahl-Kirsch, M. M., & Bradshaw, R. H. W. (2004). European forest types for biodiversity assessment—quantitative approaches. *European Forest Institute Proceedings*, *51*, 134–142.

Ódor, P., Heilmann-Clausen, J., Christensen, M., Aude, E., van Dort, K. W., Piltaver, A., Siller, I., Veerkamp, M. T., Walleyn, R., Standovár, T., van Hees, A. F. M., Kosec, J., Matočec, N., Kraigher, H., & Grebenc, T. (2006). Diversity of dead wood inhabiting fungi and bryophytes in semi-natural beech forest in Europe. *Biological Conservation*, *131*, 58–71.

Palahí, M., & Pukkala, T. (2003). Optimising the management of Scots Pine (Pinus sylvestris L.) stands in Spain based on individual-tree models. *Annals of Forest Science*, *60*, 105–114.

Penttilä, R., Siitonen, J., & Kuusinen, M. (2004). Polypore diversity in managed and old-growth boreal Picea abies forests in southern Finland. *Biological Conservation*, *117*, 271–283.

Perez-Garcia, J., Joyce, L. A., McGuire, A. D., & Xiao, X. (2002). Impacts of climate change on the global forest sector. *Climatic Change*, *54*, 439–461.

Rosenvald, R., & Lõhmus, A. (2008). For what, when and where is green-tree retention better than clearcutting? A review of the biodiversity aspects. *Forest Ecology and Management*, *255*(1), 1–15.

Ruosteenoja, K., Carter, T., Jylhä, K., & Tuomenvirta, H. (2003). Future climate in world regions: an intercomparison of model-based projections for the new IPCC emissions scenarios. The Finnish Environment. Finland, Finnish Environment Institute 644.

Sawadogo, L., Tiveau, D., & Nygard, R. (2005). Influence of selective tree cutting, livestock and prescribed fire on herbaceous biomass in the Savannah woodlands of Burkina Faso, West Africa. *Agriculture, Ecosystems & Environment*, *105*, 335–345.

Scarpa, R., Hutchinson, W., Chilton, S., & Buongiorno, J. (2000). Importance of forest attributes in the willingness to pay for recreation: a contingent valuation study of Irish forests. *Forest Policy and Economics*, *1*, 315–329.

Shono, K., Cadaweng, E. A., & Durst, P. B. (2007). Application of assisted natural regeneration to restore degraded tropical forestlands. *Restoration Ecology*, *15*, 620–626.

Shugart, H., Sedjo, R., & Sohngen, B. (2003). *Forests and global climate change: potential impacts on US forest resources* (Vol. 52). Washington: Pew Center on Global Climate Change.

Sohngen, B., Mendelsohn, R., & Sedjo, R. (2001). A global model of climate change impacts on timber markets. *Journal of Agriculture and Resource Economics*, *26*(2), 326–343.

Solberg, B., Moiseyev, A., & Kallio, A. M. I. (2003). Economic impacts of accelerating forest growth in Europe. *Forest Policy and Economics*, *5*, 157–171.

St-Laurent, M.-H., Ferron, J., Haché, S., & Gagnon, R. (2008). Planning timber harvest of residual forest stands without compromising bird and small mammal communities in boreal landscapes. *Forest Ecology and Management*, *254*, 261–275.

Steele, R. C. (1972). *Wildlife conservation in woodlands*. London: HMSO.
Summerville, K. S., & Crist, T. O. (2002). Effects of timber harvest on forest Lepidoptera: community, guild, and species responses. *Ecological Applications*, *12*(3), 820–835.
Tahvonen, O., Pukkala, T., Laiho, O., Lahde, E., & Niinimäki, S. (2010). Optimal management of uneven-aged Norway spruce stands. *Forest Ecology and Management*, *260*, 106–115.
Torras, O., & Saura, S. (2008). Effects of silvicultural treatments on forest biodiversity indicators in the Mediterranean. *Forest Ecology and Management*, *255*, 3322–3330.
UNFCCC (1997). Kyoto protocol to the United Nations framework convention on climate change.
Vaillancourt, M. A., Drapeau, P., Gauthier, S., & Robert, M. (2008). Availability of standing trees for large cavity-nesting birds in the eastern boreal forest of Quebéc, Canada. *Forest Ecology and Management*, *255*, 2272–2285.
van Kooten, G. C., & Sohngen, B. (2007). Economics of forest ecosystem carbon sinks: a review. *International Review of Environmental and Resource Economics*, *1*, 237–269.
von Bertalanffy, L. (1957). Quantitative laws in metabolism and growth. *The Quarterly Review of Biology*, *32*, 217–231.
White, E., Alig, R., & Haight, R. (2010). The forest sector in a climate-changed environment. Economic modeling of effects of climate change on the forest sector and mitigation options: a compendium of briefing papers. General Technical Report PNW-GTR-833, US Department of Agriculture, Forest Service, Pacific Northwest Research Station.
Whittam, R., McCracken, J., Francis, C., & Gartshore, M. (2002). The effects of selective logging on nest-site selection and productivity of hooded warblers (Wilsonia citrina) in Canada. *Canadian Journal of Zoology*, *80*, 644–654.
Zhou, L., Dai, L., Gu, H., & Zhong, L. (2007). Review on the decomposition and influence factors of coarse woody debris in forest ecosystem. *Journal of Forestry Research*, *18*, 48–54.

Carbon Taxes and Comparison of Trading Regimes in Fossil Fuels

Seiichi Katayama, Ngo Van Long, and Hiroshi Ohta

Abstract We study a dynamic game involving a fossil fuel exporting cartel and a coalition of importing countries that imposes carbon taxes. We show that there exists a unique Nash equilibrium, where all countries use feedback strategies. We also obtain two Stackelberg equilibria, one where the exporting cartel is the leader, and one where the coalition of importing countries is the leader. Not surprisingly, the world welfare under the Nash equilibrium is lower than that under the social planning, even though both solutions have the same steady state. Comparison of the Stackelberg equilibria with the Nash equilibrium is performed numerically. All our numerical examples reveal that world welfare under the Nash equilibrium is higher than that under the Stackelberg game where the exporting cartel is the leader. The worst outcome for world welfare occurs when the importing coalition is the Stackelberg leader.

1 Introduction

The publication of the Stern Review of the Economics of Climate Change (Stern 2007) has provided impetus to economics analysis of climate change. Much progress has been achieved in climate change studies over the last decade. A large literature has appeared, bringing new insights to the field of climate change research. Some of the pressing economic issues are discussed in Heal (2009) and Haurie et al. (2012), among others.

S. Katayama (✉)
Department of Economics, Aichi Gakuin University, Aichi, Japan
e-mail: skataya@dpc.agu.ac.jp

N. Van Long
Department of Economics, McGill University, Montreal H3A 2T7, Canada
e-mail: ngo.long@mcgill.ca

H. Ohta
GSICS, Kobe University, Kobe, Japan
e-mail: ohta@kobe-u.ac.jp

E. Moser et al. (eds.), *Dynamic Optimization in Environmental Economics*, Dynamic Modeling and Econometrics in Economics and Finance 15,
DOI 10.1007/978-3-642-54086-8_13,

Issues of climate change are broad and can be analyzed from multiple perspectives. In this paper we adopt a dynamic game approach, because there are strategic considerations that extend into the far future. We hope that the model will contribute toward a formulation of a useful framework for thinking about policies to combat climate change. We focus on international aspects of the exploitation of fossil fuels under the threat of global warming, where carbon taxes are used as policy instruments for mitigating its adverse effects.

We study a dynamic game involving a fossil-fuel exporting cartel and a coalition of fuel importing countries that impose carbon taxes. The fossil fuel is a nonrenewable resource, and its consumption leads to stock externality in the form of carbon dioxide concentration which is largely responsible for global warming. We will focus on the case where the importing countries form a coalition and agree on their carbon policies. We show that there exists a unique Nash equilibrium in a game by exporting and importing countries, where they use feedback strategies to set fuel price and carbon tax. Further we compare the Nash outcome with the Stackelberg equilibria in which either Stackelberg leadership rests with exporting or importing countries.

Our model borrows some features from Wirl (1995) and Fujiwara and Long (2011). The main differences are that Wirl (1995) derives a Nash equilibrium but does not deal with the Stackelberg leader-follower relationship and Fujiwara and Long (2011) do not consider the externalities of fossil fuel consumption.

After deriving the solutions, we compare welfare levels of participants under Nash equilibrium with the efficient outcome, which is a benchmark scenario where a single world social planner maximizes world welfare. Furthermore, we take two Stackelberg leadership scenarios, one where the importing coalition is the leader and the other with the leadership by exporters. After showing analytical results, we provide numerical comparisons among alternative regimes under a range of possible parameter values.

The paper is organized as follows. Section 2 presents the basic model. In Sect. 3 a benchmark scenario is analyzed by assuming the existence of a world social planner. In Sect. 4 we consider the optimal behavior of the oil cartel facing an arbitrary carbon-tax rule set by oil importing countries and, in turn, in Sect. 5 the behavior of oil importing countries against an arbitrary price-setting rule of the oil cartel. Section 6 derives the feedback Nash equilibrium. Section 7 compares the Nash equilibrium with the outcome under the social planner, both in terms of welfare and in terms of speed of accumulation of the pollution stock. Extending the analyses in the previous sections, Sect. 8 derives the global Stackelberg equilibria in linear strategies of the importing and exporting countries as leader. After pinning down the analytical conditions to solve, numerical examples are presented to shed light on the comparison of welfare under four different regimes.

2 Model

There are three countries, denoted by 1, 2, and 3. Countries 1 and 2 import fossil fuels from country 3. The consumption of fossil fuels generates CO_2 emissions, which contributes to greenhouse gas concentration, causing climate change damages. We assume that climate change damages to country 3 are negligible.

For simplicity, assume that country 3 consists of N identical oil producers. (In what follows, "oil" stands for "fossil fuels".) Each producer takes the price path of oil as beyond its control. Its sole objective is to maximize the present value of its stream of revenue. Extraction is assumed to be costless. Each producer j is endowed at time $t = 0$ with a deposit of size R_{j0}. Let $R_0 = \sum_{j=1}^{N} R_{j0}$. Let $q(t) \geq 0$ denote their aggregate extraction at time t. Let $Y(t)$ denote their cumulative extraction. Then

$$\dot{Y}(t) = q(t), \quad Y(0) = Y_0.$$

It is required that total cumulative extraction from time zero to time t cannot exceed the available stock at time zero, R_0:

$$Y(t) - Y_0 \leq R_0 \quad \text{for all } t \geq 0.$$

The importing country i (where $i = 1, 2$) consists of M_i identical consumers. Let $M = M_1 + M_2$. Each consumer k has a utility function $U(c_k, x_k, g_k)$ where c_k is the consumption of oil, x_k is the consumption of a numeraire good, and g_k is the damage caused by global warming. Assume that $U(c_k, x_k, g_k)$ is of the form

$$U(c_k, x_k, g_k) = Ac_k - \frac{1}{2}c_k^2 + x_k - g_k = u(c_k) + x_k - g_k$$

where $u'(c_k) = A - c_k$ is the consumer's marginal utility of oil consumption.

For simplicity, assume that the damage is quadratic in cumulative extraction:

$$g_k(t) = \frac{\gamma}{2}Y(t)^2, \quad \gamma \geq 0.$$

Note that this view (relating damages to cumulative extraction, rather than GHG concentration level) is based on the scientific work of Allen et al. (2009).

At each point in time, each consumer is endowed with $\overline{x}$ units of the numeraire good. It is assumed that $\overline{x}$ is sufficiently large, so that the consumers after paying for the oil they purchase still have some positive amount of the numeraire good to consume.

3 A Benchmark Scenario: World Social Planner

As a benchmark, suppose there is a world social planner who wants to maximize the welfare of all consumers and producers. The planner treats all consumers identically.

Then, if the aggregate oil extraction at t is $q(t)$, the planner would let each individual consume $c(t) = q(t)/M$ units of oil. Each individual is asked to pay $p(t)$ for each unit of oil consumed. The revenue to the producers is then $p(t)q(t)$. The utility at time t of the representative consumer k is then

$$U(t) = A\frac{q(t)}{M} - \frac{1}{2}\left(\frac{q(t)}{M}\right)^2 + \left(\overline{x} - p(t)\frac{q(t)}{M}\right) - \frac{\gamma}{2}Y(t)^2$$

and the revenue of the collection of producers is $\Pi(t) = p(t)q(t)$. The world's welfare is the weighted sum of producers' welfare and consumers' welfare, where ω is the weight given to producers:

$$W = \int_0^{\infty} e^{-rt}\big(\omega\Pi(t) + MU(t)\big)dt. \tag{1}$$

The rate of discount $r > 0$ is exogenously given.

Considering the standard case where $\omega = 1$, i.e., consumers and producers receive the same weight, the social welfare function (1) reduces to

$$W = \int_0^{\infty} e^{-rt}\left(Aq(t) - \frac{1}{2M}q(t)^2 + M\overline{x} - \frac{M\gamma}{2}Y(t)^2\right)dt. \tag{2}$$

The social planner chooses $q(t)$ to maximize (2) subject to

$$\dot{Y} = q,$$
$$\text{given } Y(0) = Y_0, \quad \lim_{t\to\infty}\big(Y(t) - Y_0\big) \le R_0.$$

Before solving this problem, consider some extreme cases that will provide us some useful intuition.

First, the case where $\gamma = 0$ (i.e. no climate change damages). Then the problem (2) reduces to a standard resource-extraction problem with a quadratic utility function. The marginal benefit of extracting q is

$$A - \frac{1}{M}q.$$

In this case, it is optimal to exhaust the resource at some finite time T. The extraction rate $q(t)$ will fall over time, with $q(T) = 0$. At time T, the price of the resource reaches its "choke price" level A, and extraction stops.

Second, consider the case where γ and Y_0 are so large that at time zero the present value of the stream of marginal damage cost of adding to the cumulative extraction, $\frac{\gamma M Y_0}{r}$, is greater than the marginal utility of consuming oil, A. Then clearly it is optimal not to extract the resource, i.e. $q^*(t) = 0$ for all $t \ge 0$.

Armed with the above intuition, we now consider the case where $0 < \gamma M Y_0/r < A$.

It is easy to see that in this case, the following result holds:

Proposition 1 *Assume that* $0 < \gamma M Y_0/r < A$. *Define* Y_∞ *by*

$$\frac{M\gamma Y_\infty}{r} = A. \tag{3}$$

Then,

(i) *it is optimal to extract the resource during some time interval, and*

(ii-a) *if* $(Y_0 + R_0) \geq Y_\infty$, *then exhaustion will not take place, and the remaining resource stock* $R(t)$ *will asymptotically approach a critical level* R_L *defined by*

$$Y_0 + R_L = Y_\infty.$$

In this case the steady state pollution is Y_∞. *If* $Y_0 = 0$, *the social welfare is given by*

$$\alpha = \frac{M}{2r}\left[(A+\beta)^2 + 2\overline{x}\right] = \frac{M}{2r}\left[A + \frac{A\mu M}{(r - M\mu)}\right]^2 + \frac{M\overline{x}}{r}$$

where

$$\mu = \frac{r - \sqrt{r^2 + 4\gamma M^2}}{2M} < 0.$$

(ii-b) *If* $Y_0 + R_0 < Y_\infty$, *then extraction should proceed until the remaining resource stock falls to zero (in finite time).*

All the detailed derivation of equations in the paper is found in Katayama et al. (2013).

In what follows, we focus on the case where

$$Y_0 + R_0 > \frac{rA}{\gamma M}.$$

Then, as shown in Katayama et al. (2013), the social planner will not exhaust the stock of the resource. The optimal extraction path is positive, with $q(t)$ approaching zero asymptotically, as $t \to \infty$. The optimal consumer price for oil is

$$p^c(t) = A - c(t) = A - \frac{q(t)}{M} = A - \frac{(\frac{rA}{\gamma M} - Y_0)\gamma M}{(r - \lambda_1)}e^{\lambda_1 t} \tag{4}$$

where

$$\lambda_1 = \frac{1}{2}\left(r - \sqrt{r^2 + 4\gamma M^2}\right) < 0.$$

Remark 1 In case (i), the resource will never be exhausted. Therefore its scarcity value is zero. This implies that the producer price is zero, while the consumer price is $A - (q/M)$. The difference between the consumer price and the producer price is the carbon tax. We see that the carbon tax rises over time.

Remark 2 It is easy to introduce a constant extraction cost b, where $A > b > 0$. In this case, we can define $\widetilde{A} = A - b$. Then Eqs. (3) and (4) apply, with $\widetilde{A}$ replacing A. The carbon tax per unit is then $p^c - b$. The ad valorem carbon tax is τ where $(1 + \tau)b = p^c - b$.

4 Behaviour of the Oil Cartel Facing an Arbitrary Carbon-Tax Rule by Oil Importing Countries

In this section, we assume that the coalition of importing countries set a carbon tax rate $\theta(t)$ per barrel of oil at time t. Assume $\theta(t)$ is linked to $Y(t)$ by the following rule

$$\theta = \sigma + \eta Y$$

where $\sigma \geq 0$ and $\eta > 0$ are some constants. Assume $\sigma < A$. Then the tax θ will approach value A when Y approaches the value $\overline{Y}$ defined by

$$\overline{Y} = \frac{A - \sigma}{\eta}.$$

When Y reaches this level, the carbon tax is so high that even if the producer price p is zero, the consumer will not buy oil.

The cartel of oil producers takes the linear Markovian tax rule $\theta = \sigma + \eta Y$ as given. It knows that if it charges a price $p(t) \geq 0$ per barrel at time t, the representative consumer will demand the quantity $c(t)$ such that

$$u'(c) = p(t) + \theta(t) = p(t) + \sigma + \eta Y(t)$$

i.e.

$$A - c(t) = p(t) + \sigma + \eta Y(t)$$

i.e. the demand function from each consumer is

$$c(t) = A - p(t) - \sigma - \eta Y(t).$$

Since there are M consumers, the market demand is

$$q(t) = Mc(t) = M\big(A - p(t) - \sigma - \eta Y(t)\big) \equiv q(p, Y).$$

Since extraction cost is zero, the profit of the cartel at time t is

$$\pi(t) = p(t)q(t) = M\big(A - p(t) - \sigma - \eta Y(t)\big)p(t).$$

The cartel seeks to maximize

$$\int_0^\infty e^{-rt}\big[M\big(A - p(t) - \sigma - \eta Y(t)\big)p(t)\big]dt$$

subject to

$$\begin{aligned} &\dot{Y}(t) = M\big(A - p(t) - \sigma - \eta Y(t)\big) \\ &Y(0) = Y_0 \\ &Y(t) - Y(0) \le R_0 \quad \text{for all } t. \end{aligned} \tag{5}$$

Let us solve the cartel's optimal extraction path, and show how it depends on the tax parameters σ and η.

To proceed with the analysis, we make the following assumption, which implies that the cartel will never exhaust the stock of oil:

$$R_0 > \overline{Y} - Y_0 = \frac{A - \sigma}{\eta} - Y_0.$$

To solve the cartel's optimization problem, we use the Hamilton-Jacobi-Bellman (HJB) equation. Let $V_X(Y)$ be the value function of the cartel of oil exporters. Its HJB equation is

$$rV_X(Y) = \max_p \big\{M(A - p - \sigma - \eta Y)p + V_X'(Y)M(A - p - \sigma - \eta Y)\big\}. \tag{6}$$

Maximizing the right-hand side (RHS) of the HJB equation with respect to p yields the FOC

$$-2p + A - \sigma - \eta Y - V_X'(Y) = 0.$$

Therefore the cartel's producer price rule satisfies

$$p = \frac{1}{2}\big(A - \sigma - \eta Y - V_X'(Y)\big) \equiv p(Y). \tag{7}$$

Then the RHS of the HJB equation can be written as

$$\begin{aligned} M\big(p(Y) + V_X'(Y)\big)\big(A - \sigma - p(Y) - \eta Y\big) &= M\big(A - \sigma - p(Y) - \eta Y\big)^2 \\ &= \frac{M}{4}\big(A - \sigma - \eta Y + V_X'\big)^2. \end{aligned}$$

Let us conjecture that the value function is quadratic:

$$V_X(Y) = \alpha_X + \beta_X Y + \frac{1}{2}\mu_X Y^2,$$

where α_X, β_X and μ_X are to be determined. Then

$$V_X'(Y) = \beta_X + \mu_X Y \tag{8}$$

and (6) becomes

$$r\left(\alpha_X + \beta_X Y + \frac{1}{2}\mu_X Y^2\right) = \frac{M}{4}(A - \sigma - \eta Y + \beta_X + \mu_X Y)^2$$

i.e.

$$\begin{aligned} &\frac{4r}{M}\left(\alpha_X + \beta_X Y + \frac{1}{2}\mu_X Y^2\right) \\ &= \left[(A - \sigma + \beta_X) + (\mu_X - \eta)Y\right]^2 \\ &= (A - \sigma + \beta_X)^2 + 2(A - \sigma + \beta_X)(\mu_X - \eta)Y + (\mu_X - \eta)^2 Y^2. \end{aligned}$$

This equation must hold for all feasible values of Y. Therefore the coefficient of the Y^2 term on the left-hand side must equal the coefficient of the Y^2 term on the right-hand side:

$$\frac{2r\mu_X}{M} = (\mu_X - \eta)^2. \tag{9}$$

Similarly, the coefficient of the Y term on the left-hand side must equal the coefficient of the Y term on the right-hand side:

$$\frac{4r\beta_X}{M} = 2(A - \sigma + \beta_X)(\mu_X - \eta) \tag{10}$$

and, likewise for the constant term:

$$\frac{4r\alpha_X}{M} = (A - \sigma + \beta_X)^2. \tag{11}$$

The three equations (9), (10), and (11) determine the three coefficients α_X, β_X, μ_X of the quadratic value function $V_X(Y)$. We first determine μ_X from (9):

$$\frac{2r\mu_X}{M} = \mu_X^2 + \eta^2 - 2\eta\mu_X$$

i.e.

$$\mu_X^2 - 2\left(\eta + \frac{r}{M}\right)\mu_X + \eta^2 = 0.$$

This quadratic equation in μ_X has two positive real roots, μ_{X1} and μ_{X2} where $\mu_{X1} > \mu_{X2} > 0$,

$$\mu_{X1} = \frac{1}{2}\left[2\left(\eta + \frac{r}{M}\right) + \sqrt{2^2\left(\eta + \frac{r}{M}\right)^2 - 4\eta^2}\right] = \eta + \frac{r}{M} + \sqrt{\left(\frac{r}{M}\right)^2 + \frac{2}{M}\eta r}$$

and

$$\mu_{X2} = \frac{1}{2}\left[2\left(\eta + \frac{r}{M}\right) - \sqrt{2^2\left(\eta + \frac{r}{M}\right)^2 - 4\eta^2}\right] = \eta + \frac{r}{M} - \sqrt{\left(\frac{r}{M}\right)^2 + \frac{2}{M}\eta r}.$$

Which root should we select? As usual, we should choose the root such that the differential equation for Y has a solution that converges to a steady state. The differential equation is, from (5), (7) and (8),

$$\begin{aligned}\dot{Y} &= M(A - p - \sigma - \eta Y)\\ &= M\left[A - \frac{1}{2}\left(A - \sigma - \eta Y - V_X'(Y)\right) - \sigma - \eta Y\right]\\ &= \frac{M}{2}\left(A - \sigma + \beta_X - (\eta - \mu_X)Y\right). \qquad (12)\end{aligned}$$

This equation gives a converging solution to a steady state if and only if $(\eta - \mu_X) > 0$. This requires that the smaller root μ_{X2} be chosen.

Therefore

$$\mu_X^* = \mu_{X2} = \eta + \frac{r}{M} - \sqrt{\left(\frac{r}{M}\right)^2 + \frac{2}{M}\eta r}.$$

Notice that

$$\eta - \mu_X^* = \sqrt{\left(\frac{r}{M}\right)^2 + \frac{2}{M}\eta r} - \frac{r}{M} > 0.$$

Having solved for μ_X, we now turn to (10) to solve for β_X:

$$\frac{4r\beta_X}{M} = 2(A - \sigma + \beta_X)\left(\mu_X^* - \eta\right).$$

Then

$$\beta_X\left[\frac{4r}{M} + 2\left(\eta - \mu_X^*\right)\right] - 2(A - \sigma)\left(\mu_X^* - \eta\right) < 0.$$

Thus

$$\beta_X^* = -\frac{(A - \sigma)(\eta - \mu_X^*)}{(\eta - \mu_X^*) + \frac{2r}{M}} < 0$$

since $\eta - \mu_X^* > 0$. And thus

$$\begin{aligned}A - \sigma + \beta_X^* &= (A - \sigma)\left[1 + \frac{(\mu_X^* - \eta)}{\frac{2r}{M} + (\eta - \mu_X^*)}\right]\\ &= (A - \sigma)\left[\frac{\frac{2r}{M}}{\frac{2r}{M} + (\eta - \mu_X^*)}\right] > 0. \qquad (13)\end{aligned}$$

Finally, from (11),

$$\frac{4r\alpha_X}{M} = (A - \sigma + \beta_X)^2,$$

we obtain

$$\alpha_X^* = \frac{M}{4r}\left(A - \sigma + \beta_X^*\right)^2 > 0.$$

Substituting (13) into (12) we get

$$\dot{Y} = \frac{M}{2}\left\{(A-\sigma)\left[\frac{\frac{2r}{M}}{\frac{2r}{M} + (\eta - \mu_X^*)}\right] - \left(\eta - \mu_X^*\right)Y\right\}.$$

This equation has a stable steady state $\widehat{Y}$ defined by

$$\widehat{Y} = \frac{(A-\sigma)[\frac{\frac{2r}{M}}{\frac{2r}{M}+(\eta-\mu_X^*)}]}{(\eta - \mu_X^*)}$$

$$= (A-\sigma)\left[\frac{\frac{2r}{M}}{\frac{2r}{M}(\eta - \mu_X^*) + (\eta - \mu_X^*)^2}\right].$$

Now we use (9) to simplify $\widehat{Y}$:

$$\widehat{Y} = (A-\sigma)\left[\frac{\frac{r}{M}}{\frac{r}{M}(\eta - \mu_X^*) + \frac{r\mu_X^*}{M}}\right] = \frac{A-\sigma}{\eta}.$$

Thus

$$\widehat{Y} = \overline{Y}.$$

The following Proposition summarizes the result of this section:

Proposition 2 *When the oil cartel faces a carbon tax rule of the form $\theta = \sigma + \eta Y$, where $\sigma < A$, and $\eta > 0$, its optimal response is to set the producer price according to the rule*

$$p = \frac{1}{2}\left[\left(A - \sigma - \beta_X^*\right) - \left(\eta + \mu_X^*\right)Y\right] \quad \textit{for all } Y \le \overline{Y}$$

where

$$\mu_X^* = \eta + \frac{r}{M} - \sqrt{\left(\frac{r}{M}\right)^2 + \frac{2}{M}\eta r} > 0$$

and

$$\beta_X^* = -\frac{(A-\sigma)(\eta - \mu_X^*)}{(\eta - \mu_X^*) + \frac{2r}{M}} < 0$$

with $\eta + \mu_X^* > 0$ and $A - \sigma - \beta_X^* > 0$. Thus the producer's price will fall over time, and the quantity demanded, q, will also fall over time. As Y approaches $\overline{Y} = (A - \sigma)/\eta$, the producer's price approaches zero, while the consumer's price, $p + \theta$, approaches A.

Proof It remains to show that $\dot{q}(t) < 0$. Now

$$\begin{aligned} q &= M(A - \sigma - p - \eta Y) \\ &= M(A - \sigma) - M(p + \eta Y). \end{aligned}$$

So

$$\begin{aligned} \dot{q} &= -M(\dot{p} + \eta \dot{Y}) \\ &= -M\left[-\frac{1}{2}(\eta + \mu_X^*) + \eta\right]\dot{Y} \\ &= \frac{1}{2}M(\mu_X^* - \eta)\dot{Y} \\ &= \frac{1}{2}M\left[\frac{r}{M} - \sqrt{\left(\frac{r}{M}\right)^2 + \frac{2}{M}\eta r}\right]\dot{Y} < 0. \end{aligned}$$

□

5 Behavior of Oil Importing Countries Facing an Arbitrary Price-Setting Rule of the Oil Cartel

Now suppose that the oil cartel uses a price-setting rule which relates the price at time t to the state variable $Y(t)$, where $Y(t) \le Y_0 + R_0$,

$$p = \delta - \lambda Y \tag{14}$$

with $\delta < A$ and $\lambda \gtrless 0$.

Suppose the governments of the oil importing countries take δ and λ as given, and agree on coordinating their carbon-tax policy to maximize the welfare of the representative consumer.

Let $\theta(t)$ be the carbon tax that consumers must pay to their governments per barrel of oil consumed. Let $c(t)$ be the quantity of oil demanded per person, and $q(t) = Mc(t)$ be the aggregate demand for oil. The aggregate consumer surplus at time t is

$$Aq(t) - \frac{1}{2M}q(t)^2 - \big(p(t) + \theta(t)\big)q(t).$$

The quantity demanded is

$$q(t) = M\big(A - p(t) - \theta(t)\big) \tag{15}$$

and the carbon-tax revenue is

$$R(t) \equiv \theta(t)q(t). \tag{16}$$

Assume that the carbon-tax revenue is redistributed in a lump-sum fashion to consumers. Let $L(t)$ be the lump-sum transfer to the consumers. The instantaneous welfare flow of the consumers at time t is

$$W(t) = Aq(t) - \frac{1}{2M}q(t)^2 - \big(p(t) + \theta(t)\big)q(t) + L(t) + M\overline{x} - M\frac{\gamma Y(t)^2}{2}, \quad (17)$$

where $q(t)$ is given by (15) and $p(t) = \delta - \lambda Y(t)$. The coalition of the two governments chooses $\theta(t)$ and $L(t)$ to maximize the integral of the discounted flow of welfare:

$$\max \int_0^{\infty} e^{-rt} W(t) dt$$

subject to the government's budget constraint

$$L(t) = R(t) \quad (18)$$

and the dynamic equation

$$\dot{Y}(t) = M\big(A - p(t) - \theta(t)\big)$$

where $Y(0) = Y_0$ and $Y(t) \le \widetilde{Y}$.

Using (15), (16), (17) and (18), the instantaneous welfare flow $W(t)$ becomes

$$\begin{aligned} W &= \left(A - p - \frac{1}{2M}q\right)q + M\overline{x} - M\frac{\gamma Y^2}{2} \\ &= \left[A - p - \frac{1}{2}(A - p - \theta)\right]M(A - p - \theta) + M\overline{x} - M\frac{\gamma Y^2}{2} \\ &= \frac{M}{2}\left[(A - p + \theta)(A - p - \theta) + 2\overline{x} - \gamma Y^2\right] \\ &= \frac{M}{2}\left[(A - p)^2 - \theta^2 + 2\overline{x} - \gamma Y^2\right]. \end{aligned} \quad (19)$$

Let $V_I(Y)$ denote the value function for the coalition of the two oil importing countries. The HJB equation is

$$rV_I(Y) = \max_{\theta}\left\{\frac{M}{2}\left[(A - p)^2 - \theta^2 + 2\overline{x} - \gamma Y^2\right] + V_I'(Y)M(A - p - \theta)\right\}. \quad (20)$$

Maximizing the right-hand side of (20) with respect to θ gives the first order condition (FOC)

$$-\theta - V_I'(Y) = 0.$$

Substitute this FOC into (20) to get

$$\begin{aligned} rV_I(Y) &= \frac{M}{2}\left[(A - p - V_I')(A - p + V_I') + 2\overline{x} - \gamma Y^2 + 2V_I'(A - p + V_I')\right] \\ &= \frac{M}{2}\left[(A - p + V_I')^2 + 2\overline{x} - \gamma Y^2\right]. \end{aligned} \quad (21)$$

Let us conjecture that

$$V_I(Y) = \alpha_I + \beta_I Y + \frac{\mu_I}{2} Y^2.$$

Then

$$V_I' = \beta_I + \mu_I Y. \tag{22}$$

Substituting (22) and (14) into (21) yields

$$r\left(\alpha_I + \beta_I Y + \frac{\mu_I}{2} Y^2\right)$$
$$= \frac{M}{2}\{2\overline{x} + (A - \delta + \beta_I)^2 + 2(\mu_I + \lambda)(A - \delta + \beta_I)Y + \left[(\mu_I + \lambda)^2 - \gamma\right]Y^2\}.$$

It follows, by comparison, that

$$\begin{aligned} r\mu_I &= M\left[(\mu_I + \lambda)^2 - \gamma\right], \\ r\beta_I &= M(\mu_I + \lambda)(A - \delta + \beta_I), \\ r\alpha_I &= \frac{M}{2}\left[2\overline{x} + (A - \delta + \beta_I)^2\right]. \end{aligned} \tag{23}$$

Equation (23) gives the quadratic equation

$$\mu_I^2 + \left(2\lambda - \frac{r}{M}\right)\mu_I - (\gamma - \lambda^2) = 0.$$

To avoid complex roots and repeated roots, let us assume that the discriminant is positive:

$$\Delta \equiv \left(\frac{r}{M} - 2\lambda\right)^2 + 4(\gamma - \lambda^2) > 0.$$

For this to hold, it is *necessary and sufficient* that

$$\gamma > \frac{r}{M}\left(\lambda - \frac{r}{4M}\right).$$

Note: Either of the following conditions is *sufficient* for $\Delta > 0$:

$$\begin{aligned} \gamma &> \frac{r\lambda}{M}, \\ \gamma &> \lambda^2. \end{aligned} \tag{24}$$

With $\Delta > 0$, we have two roots, $\mu_{I1} > \mu_{I2}$,

$$\mu_{I1} = \frac{1}{2}\left[\left(\frac{r}{M} - 2\lambda\right) + \sqrt{\left(\frac{r}{M} - 2\lambda\right)^2 + 4(\gamma - \lambda^2)}\right],$$

$$\mu_{I2} = \frac{1}{2}\left[\left(\frac{r}{M} - 2\lambda\right) - \sqrt{\left(\frac{r}{M} - 2\lambda\right)^2 + 4(\gamma - \lambda^2)}\right].$$

As before, we should choose the root such that the differential equation for Y has a solution that converges to a steady state. The differential equation is

$$\begin{aligned}\dot{Y} &= M(A - p - \theta)\\ &= M\big(A - \delta + \lambda Y + V_I'\big)\\ &= M\big[(A - \delta + \beta_I) + (\mu_I + \lambda)Y\big].\end{aligned}$$

This equation gives a converging solution to a steady state if and only if

$$(\mu_I + \lambda) < 0. \tag{25}$$

We must choose μ_I that satisfies the convergence condition (25). Since the bigger root μ_{I1} gives

$$\mu_{I1} + \lambda = \lambda - \lambda + \frac{1}{2}\left[\frac{r}{M} + \sqrt{\left(\frac{r}{M} - 2\lambda\right)^2 + 4(\gamma - \lambda^2)}\right] > 0$$

we reject μ_{I1}. Turning to the smaller root, λ_{I2}, we find that

$$\mu_{I2} + \lambda = \frac{1}{2}\left[\frac{r}{M} - \sqrt{\left(\frac{r}{M} - 2\lambda\right)^2 + 4(\gamma - \lambda^2)}\right] \tag{26}$$

is negative if and only if

$$\left(\frac{r}{M}\right)^2 < \left(\frac{r}{M} - 2\lambda\right)^2 + 4(\gamma - \lambda^2)$$

i.e. iff

$$\gamma > \frac{r}{M}\lambda. \tag{27}$$

In what follows, we assume that condition (27) is satisfied. This condition is satisfied if $\lambda < 0$ or $\lambda > 0$ but sufficiently small.

Under Assumption (24), we select the *smaller root* μ_{I2} and denote it by μ_I^*:

$$\mu_I^* = \frac{1}{2}\left[\frac{r}{M} - 2\lambda - \sqrt{\left(\frac{r}{M} - 2\lambda\right)^2 + 4(\gamma - \lambda^2)}\right].$$

Next, we solve for β_I.

$$\beta_I\big[M(\mu_I^* + \lambda) - r\big] = -M(A - \delta)(\mu_I^* + \lambda) > 0.$$

Then $\beta_I < 0$ because $(\mu_I^* + \lambda) < \frac{r}{M}$ by (26).

$$\beta_I^* = \frac{(A - \delta)(\mu_I^* + \lambda)}{(r/M) - (\mu_I^* + \lambda)} < 0$$

i.e.

$$\beta_I^* = \frac{(A-\delta)[\frac{r}{M} - \sqrt{(\frac{r}{M}-2\lambda)^2 + 4(\gamma-\lambda^2)}]}{\frac{r}{M} + \sqrt{(\frac{r}{M}-2\lambda)^2 + 4(\gamma-\lambda^2)}} < 0$$

given that condition (27) is satisfied. Finally,

$$\alpha_I^* = \frac{M}{2r}\left[2\overline{x} + \left(A - \delta + \beta_I^*\right)^2\right] > 0.$$

The steady state is

$$\begin{aligned}
\widetilde{Y} &= \frac{A-\delta+\beta_I}{-(\mu_I+\lambda)} \\
&= \frac{1}{-(\mu_I+\lambda)}\left[\frac{(A-\delta)[(r/M)-(\mu_I^*+\lambda)] + (A-\delta)(\mu_I^*+\lambda)}{(r/M)-(\mu_I^*+\lambda)}\right] \\
&= \frac{(A-\delta)(r/M)}{(\mu_I^*+\lambda)^2 - (\mu_I^*+\lambda)(r/M)} \\
&= \frac{(A-\delta)(r/M)}{(\frac{r\mu_I^*}{M}+\gamma) - \frac{r\mu_I^*}{M} - \frac{r\lambda}{M}} = \frac{(A-\delta)(r/M)}{\gamma - \frac{r\lambda}{M}} > 0.
\end{aligned}$$

The following Proposition summarizes the result of this section.

Proposition 3 *Suppose that the coalition of oil importing countries faces an arbitrary producer's price rule of the form* $p = \delta - \lambda Y$, *where* $\delta < A$, $\lambda \gtrless 0$, *and* $\gamma > \frac{r}{M}\lambda$.

Assume that

$$Y_0 + R_0 \geq \widetilde{Y} \equiv \frac{(A-\delta)(r/M)}{\gamma - \frac{r\lambda}{M}}.$$

The intertemporal welfare maximizing behaviour of the coalition of importing countries will result in setting the carbon tax according to the rule

$$\theta = -\beta_I^* - \mu_I^* Y$$

where

$$\mu_I^* = \frac{1}{2}\left[\frac{r}{M} - 2\lambda - \sqrt{\left(\frac{r}{M} - 2\lambda\right)^2 + 4(\gamma-\lambda^2)}\right]$$

and

$$\beta_I^* = \frac{(A-\delta)[\frac{r}{M} - \sqrt{(\frac{r}{M}-2\lambda)^2 + 4(\gamma-\lambda^2)}]}{\frac{r}{M} + \sqrt{(\frac{r}{M}-2\lambda)^2 + 4(\gamma-\lambda^2)}} < 0.$$

Thus the consumer's price satisfies

$$p^c = p + \theta = (\delta - \beta_I^*) - (\mu_I^* + \lambda)Y.$$

As Y rises, the consumer's price rises (recall $\mu_I^* + \lambda < 0$). The quantity demanded, q, will fall over time.[1] *As Y approaches $\widetilde{Y}$, the carbon tax approaches A, and the consumer's price, $p + \theta$, approaches A.*[2]

Note that since we do not make any assumption about the sign of λ, it is possible that the carbon tax falls as Y rises, provided that $\lambda < 0$, so that the producer's price rises with Y. We will see later that this cannot happen in a Nash equilibrium.

Finally, to prove that q falls over time, we write

$$q = M(A - p - \theta) = M(A - \delta + \lambda Y + \beta_I^* + \mu_I^* Y).$$

Then

$$\dot{q} = (\lambda + \mu_I^*)\dot{Y} < 0$$

because $\lambda + \mu_I^* < 0$.

6 Nash Equilibrium

In the two preceding sections, we looked at the reaction of one player (either the cartel, or the coalition of importing countries) to a given linear Markovian strategy (either a carbon-tax rule, or a producer-price setting rule) of the other player. It is now time to put our pieces together to find the Nash equilibrium of the games between the two players.

Given any linear Markovian tax rule $\theta = \sigma + \eta Y$, we found that the cartel's reaction function (or best reply) is the following pricing rule

$$p = \frac{1}{2}(A - \sigma - \beta_X^*) - \frac{1}{2}(\eta + \mu_X^*)Y$$

where

$$\mu_X^* = \eta + \frac{r}{M} - \sqrt{\left(\frac{r}{M}\right)^2 + \frac{2}{M}\eta r} \equiv \mu_X^*(\eta)$$

and

$$\beta_X^* = -\frac{(A - \sigma)(\eta - \mu_X^*(\eta))}{(\eta - \mu_X^*(\eta)) + \frac{2r}{M}} \equiv \beta_X^*(\sigma, \eta).$$

---

[1] See the proof below.

[2] This follows from $(A - \delta + \beta_I^*) + (\mu_I^* + \lambda)Y \to 0$ as $Y \to \widetilde{Y}$.

Conversely, given any linear Markovian producer-price setting rule $p = \delta - \lambda Y$ (where $\lambda \gtrless 0$), we found that the coalition's reaction function (or best reply) is the following carbon-tax rule

$$\theta = -\beta_I^* - \mu_I^* Y$$

where

$$\mu_I^* = \frac{1}{2}\left[\frac{r}{M} - 2\lambda - \sqrt{\left(\frac{r}{M} - 2\lambda\right)^2 + 4(\gamma - \lambda^2)}\right] \equiv \mu_I^*(\lambda)$$

and

$$\beta_I^* = \frac{(A-\delta)[\frac{r}{M} - \sqrt{(\frac{r}{M} - 2\lambda)^2 + 4(\gamma - \lambda^2)}]}{\frac{r}{M} + \sqrt{(\frac{r}{M} - 2\lambda)^2 + 4(\gamma - \lambda^2)}} \equiv \beta_I^*(\delta, \lambda).$$

In a Nash equilibrium, it must hold that, for all Y,

$$\sigma + \eta Y = -\beta_I^* - \mu_I^* Y$$

and

$$\delta - \lambda Y \equiv \frac{1}{2}(A - \sigma - \beta_X^*) - \frac{1}{2}(\eta + \mu_X^*)Y.$$

These two conditions are satisfied if and only if the following four equalities are met:

$$\sigma = -\beta_I^*(\delta, \lambda), \tag{28}$$

$$\delta = \frac{1}{2}\big(A - \sigma - \beta_X^*(\sigma, \eta)\big), \tag{29}$$

$$\eta = -\mu_I^*(\lambda), \tag{30}$$

$$\lambda = \frac{1}{2}\big(\eta + \mu_X^*(\eta)\big). \tag{31}$$

Note that the right-hand side of (28) is positive; and the right-hand side of (29) is positive if $A - \sigma > 0$. We will verify that in a Nash equilibrium, $A - \sigma > 0$. The four equations (28) to (31) determine the Nash equilibrium tuple $(\sigma, \delta, \eta, \lambda)$.

We are able to show that a solution $(\sigma, \delta, \eta, \lambda)$ exists and is unique.

We find that (see Katayama et al. 2013) under assumption (27), there are two possible values of λ:

$$\lambda_1^* = \frac{2}{3}\frac{M\gamma}{r} + \frac{1}{9}\left[\frac{r}{M} - \sqrt{3\gamma + \left(\frac{r}{M}\right)^2}\right] > 0,$$

$$\lambda_2^* = \frac{2}{3}\frac{M\gamma}{r} + \frac{1}{9}\left[\frac{r}{M} + \sqrt{3\gamma + \left(\frac{r}{M}\right)^2}\right] > 0.$$

However, the bigger root is not admissible (see Katayama et al. 2013 for a proof). So in what follows, we define

$$\lambda^* = \lambda_1^* = \frac{2}{3}\frac{M\gamma}{r} + \frac{1}{9}\left[\frac{r}{M} - \sqrt{3\gamma + \left(\frac{r}{M}\right)^2}\right] > 0.$$

After some simple manipulations, we obtain the solution

$$\eta^* = 2\left(\frac{M\gamma}{r} - \lambda^*\right) > 0.$$

Finally, we can solve for σ^* and δ^*. We can show that

$$\delta^* = \frac{Ar\lambda^*}{\gamma M} > 0,$$

$$\sigma^* = 2\delta^* - A > 0.$$

Proposition 4 *There exists a unique Nash equilibrium. At the equilibrium, the coalition of importing countries imposes a carbon tax rule of the form* $\theta(t) = \sigma^* + \eta^* Y(t)$ *where* $\eta^* > 0$ *and* $0 < \sigma^* < A$*, and the oil cartel sets producer's price according to the pricing rule of the form* $p(t) = \delta^* - \lambda^* Y(t)$ *where* $\delta^* > 0$ *and* $0 < \lambda^* < M\gamma/r$*. The quantity demanded will fall over time, and the consumer price will approach A as the stock of pollution Y approaches* $\overline{Y}$ *where*

$$\overline{Y} \equiv \frac{A - \sigma^*}{\eta^*} = \frac{A - \sigma^*}{2[(M\gamma/r) - \lambda^*]} = \widetilde{Y} \equiv \frac{A - \delta^*}{(M\gamma/r) - \lambda^*} = Y_\infty \equiv \frac{Ar}{\gamma M}.$$

In the Nash equilibrium, the importing countries use the carbon tax strategy

$$\theta = \sigma^* + \eta^* Y$$

while the oil cartel uses the price setting strategy

$$p = \delta^* - \lambda^* Y = \lambda^*\left(\frac{Ar}{\gamma M} - Y\right).$$

Therefore the tax increases and the producer price falls as the stock of pollution increases. The consumer price is

$$p^c = p + \theta = (\sigma^* + \delta^*) + (\eta^* - \lambda^*)Y$$

where $\eta^* - \lambda^* > 0$*. As Y increases toward its steady state value* $Y_\infty = Ar/(\gamma M)$*, the carbon tax tends to A and the cartel's producer price tends to zero.*

The rate of increase in pollution is

$$\dot{Y} = Mc = q = M(A - p - \theta) = M\left[A - (\sigma^* + \delta^*) - (\eta^* - \lambda^*)Y\right].$$

Thus, as Y rises, $\dot{Y}$ (the rate of increase in pollution) becomes smaller and smaller:

$$\frac{d\dot{Y}}{dY} = -M(\eta^* - \lambda^*) = \frac{M}{3}\left[\frac{r}{M} - \sqrt{3\gamma + \left(\frac{r}{M}\right)^2}\right] < 0.$$

Remark 3 The socially optimal steady state stock is

$$Y_\infty = \frac{Ar}{M\gamma}.$$

Then $Y_\infty = \widetilde{Y}$ is true if and only if

$$\delta^* = \frac{\lambda^* A r}{\gamma M}.$$

See Katayama et al. (2013).

Remark 4 Under the social planner, the pollution stock also tends to the steady state $Y_\infty = Ar/(\gamma M)$. However, the rate of change in Y is not the same in the two regimes. In fact,

$$\frac{d\dot{Y}}{dY} = -\frac{M}{3}\left[\sqrt{3\gamma + \left(\frac{r}{M}\right)^2} - \frac{r}{M}\right] \quad \text{for Nash equilibrium}$$

while

$$\frac{d\dot{Y}}{dY} = -\frac{M}{2}\left[\sqrt{4\gamma + \left(\frac{r}{M}\right)^2} - \frac{r}{M}\right] \quad \text{for social planning.}$$

It is clear that the former takes a smaller negative value than the latter. Thus, compared with the social planner case, the Nash equilibrium results in lower consumption earlier on. This is because cartel conserves the resource stock. This is another confirmation of Solow's claim that the resource monopolist is the conservationist's best friend.

7 Welfare Comparison

Since the social planner maximizes world welfare, it is clear that, in terms of world welfare, the Nash equilibrium outcome cannot dominate the outcome under the social planner.

For the sake of illustration, we provide a numerical example. Assume $M = 1$, $Y_0 = 0$, $r = 0.05$, $A = 5$, and $\gamma = 0.02$.

Then the social planner's optimal pollution stock is

$$Y_\infty = \frac{rA}{\gamma M} = \frac{(0.05)5}{0.02} = 12.5$$

and welfare under the social planner is[3]

$$V(0)=\alpha=\frac{M}{2r}\left[(A+\beta)^2+2\overline{x}\right]=\frac{M}{2r}\left(A+\frac{A\xi M}{r-M\xi}\right)^2+\frac{M\overline{x}}{r}$$

where

$$\xi=\frac{r-\sqrt{r^2+4\gamma M^2}}{2M},$$
$$\frac{M}{2r}\left(A+\frac{A\mu M}{r-M\mu}\right)^2=21.983.$$

In the case of Nash equilibrium,

$$\gamma=\lambda_1=\frac{1}{9}\left[6\frac{M\gamma}{r}+\frac{r}{M}-\sqrt{3\gamma+\left(\frac{r}{M}\right)^2}\right]=0.244.$$

We keep only λ_1 and call it λ^*. Next, compute η^*:

$$\eta^*=2\left(\frac{M\gamma}{r}-\lambda^*\right)=0.311.$$

Then

$$\mu_X(\eta^*)=\eta^*+\frac{r}{M}-\sqrt{\left(\frac{r}{M}\right)^2+\frac{2}{M}\eta^* r}=0.177,$$
$$z(\eta^*)=\frac{\eta^*-\mu_X^*}{\eta^*-\mu_X^*+\frac{2r}{M}}=0.571,$$
$$\delta^*=\frac{A(1-z(\eta^*))(1+z(\eta^*))}{2-z(\eta^*)(1+z(\eta^*))}=3.055,$$
$$\sigma^*=\frac{Az(\eta^*)(1-z(\eta^*))}{2-z(\eta^*)(1+z(\eta^*))}=1.111,$$
$$2\delta^*-A=1.111,$$
$$\overline{Y}=\frac{A-\sigma^*}{\eta^*}=12.499,$$
$$Y_\infty=\frac{A-\delta^*}{(M\gamma/r)-\lambda^*}=12.499.$$

Therefore, we confirm that $\overline{Y}=Y_\infty$.

---

[3]Since the term $\frac{M\overline{x}}{r}$ is a constant, we can omit it in all welfare expressions.

Notice that the steady state pollution stock in the Nash equilibrium is the same as under the social planner. However, the rates at which the pollution stock grows toward the steady state are different under the two regimes.

Concerning welfare in the Nash equilibrium, for simplicity, we set $Y_0 = 0$. The welfare of the importing coalition, as seen from time $t = 0$, is

$$V_I(0) = \alpha_I^* = \frac{M}{2r}\left[2\overline{x} + \left(A - \delta^* + \beta_I^*\right)^2\right] = \frac{M\overline{x}}{r} + \frac{M}{2r}\left(A - \delta^* + \beta_I^*\right)^2.$$

The welfare of the cartel of oil exporters is

$$\begin{aligned} V_X(0) = \alpha_X^* &= \frac{M}{4r}\left(A - \sigma^* + \beta_X^*\right)^2 \\ &= \frac{M}{4r}\left[A - \sigma^* - \frac{(A - \sigma^*)(\eta - \mu_X^*)}{(\eta - \mu_X^*) + \frac{2r}{M}}\right]^2 \\ &= \frac{M}{4r}\left[\left(A - \sigma^*\right)\left(1 - z(\eta^*)\right)\right]^2. \end{aligned}$$

In the Nash equilibrium, $\lambda = \lambda_1 = 0.244\,44$. So the welfare of the coalition of importers is $\frac{M\overline{x}}{r}$ plus

$$\frac{1}{2r}\left(A - \delta^* + \beta_I^*\right)^2 = \frac{1}{2r}\left(A - \delta^* - \sigma^*\right)^2 = 6.944.$$

The welfare of the cartel of exporters is

$$\alpha_X^* = 13.888.$$

The sum of their welfare levels is

$$\frac{M\overline{x}}{r} + 6.944 + 13.888 = \frac{M\overline{x}}{r} + 20.832.$$

Recall the welfare under the social planner, which is $\frac{M\overline{x}}{r} + 21.983$. This implies that welfare in the social planner regime is greater than that in the Nash equilibrium.

8 Stackelberg Solutions

In Sect. 4, we have shown how the cartel determines its pricing strategy facing a given tax rule $\theta = \sigma + \eta Y$ by the importing coalition, i.e., given the parameters $\sigma < A$ and $\eta > 0$. Suppose the importing coalition knows this "reaction function" of the cartel. Then it seems tempting for the coalition to choose the "best" parameters σ and η to maximize its welfare.

Let us formulate this problem. We have found that given (σ, η) the cartel's best reply takes the form

$$p = \frac{1}{2}\big(A - \sigma - \beta_X^*(\sigma, \eta)\big) - \frac{1}{2}\big(\eta + \mu_X^*(\eta)\big)Y.$$

For simplicity, define

$$G(\eta) = \left(\frac{r}{M}\right)^2 + \frac{2r\eta}{M}$$

and define the cartel's reaction functions

$$\delta^R(\sigma, \eta) \equiv \frac{1}{2}\big(A - \sigma - \beta_X^*(\sigma, \eta)\big) = \frac{(A-\sigma)(\frac{r}{M} + 2\eta - \sqrt{G(\eta)})}{2\eta}$$

and

$$\lambda^R(\eta) \equiv \frac{1}{2}\big(\eta + \mu_X^*(\eta)\big) = \frac{1}{2}\left(\frac{r}{M} + 2\eta - \sqrt{G(\eta)}\right)$$

then the cartel's price setting reaction is

$$p^R = \delta^R(\sigma, \eta) - \lambda^R(\eta)Y.$$

The consumer price is

$$p = p^R + \theta.$$

Then the transition equation becomes

$$\dot{Y} = M\big(A - p^R - \theta\big) = M\big[A - \big(\sigma + \delta^R\big)\big] - M\big(\eta - \lambda^R\big)Y.$$

Then $Y(t)$ converges to the steady state $\overline{Y} = (A - \sigma)/\eta$ and

$$Y(t) = \overline{Y} + (Y_0 - \overline{Y})\exp\left(\frac{(r - M\sqrt{G(\eta)})t}{2}\right). \tag{32}$$

From (19), the instantaneous welfare of the importing country is

$$W = M\overline{x} + M\left[\frac{(A - p^R)^2 - \theta^2}{2} - \frac{\gamma Y^2}{2}\right],$$

which can be expressed as

$$W = \kappa Y^2 + \rho Y + \psi + M\overline{x}$$

where

$$\kappa(\eta) \equiv -\frac{M}{2}\big[\big(\eta - \lambda^R(\eta)\big)\big(\eta + \lambda^R(\eta)\big) + \gamma\big],$$

$$\rho(\sigma,\eta) \equiv M\big[-\sigma\eta + \big(A - \delta^R(\sigma,\eta)\big)\lambda^R(\eta)\big],$$

$$\psi(\sigma,\eta) \equiv \frac{M}{2}\big(A - \sigma - \delta^R(\sigma,\eta)\big)\big(A + \sigma - \delta^R(\sigma,\eta)\big).$$

It follows that, after substituting for $Y(t)$ using (32), instantaneous welfare at t is

$$W(t) = \kappa(Y_0 - \overline{Y})^2 e^{(r-MG^{1/2})t} + (2\kappa\overline{Y} + \rho)(Y_0 - \overline{Y})e^{(r-MG^{1/2})t/2} + \kappa\overline{Y}^2 + \rho\overline{Y} + \psi + M\overline{x}.$$

Thus

$$\int_0^\infty e^{-rt}W(t)dt = \frac{\kappa}{MG^{1/2}}(Y_0 - \overline{Y})^2 + \frac{2(2\kappa\overline{Y} + \rho)}{r + MG^{1/2}}(Y_0 - \overline{Y}) + \frac{\kappa\overline{Y}^2 + \rho\overline{Y} + \psi}{r} + \frac{M\overline{x}}{r}$$

which can be simplified as

$$\int_0^\infty e^{-rt}W(t)dt = \frac{\kappa}{MG^{1/2}}(Y_0 - \overline{Y})^2 + \frac{2(2\kappa\overline{Y} + \rho)}{r + MG^{1/2}}(Y_0 - \overline{Y}) + \frac{M\overline{x}}{r} - \frac{M\gamma\overline{Y}^2}{2r}. \tag{33}$$

The task of the importing coalition, acting as leader, is to choose η and σ to maximize the right-hand side of (33). Note that $\overline{Y} = \overline{Y}(\sigma,\eta)$. The first order conditions that determine the optimal pair (σ,η) would involve the term Y_0. Suppose that at some future time τ the leader can replan, by choosing η and σ again to maximize the integral of instantaneous welfare flow starting from time τ, where Y_τ is the current pollution stock:

$$V_\tau(Y_\tau) = \int_\tau^\infty e^{-r(t-\tau)}W(t)dt = \frac{\kappa}{MG^{1/2}}(Y_\tau - \overline{Y})^2 + \frac{2(2\kappa\overline{Y} + \rho)}{r + MG^{1/2}}(Y_\tau - \overline{Y}) + \frac{M\overline{x}}{r} - \frac{M\gamma\overline{Y}^2}{2r}. \tag{34}$$

Then the new first order conditions that determine the optimal pair (σ,η) would involve the term Y_τ, which is different from Y_0. This observation leads us to conclude that the optimal policy of the leader is time-inconsistent. This time inconsistency in dynamic games with Stackelberg leadership is a well-known result, see e.g. Kemp and Long (1980).

To resolve the problem of time inconsistency, several authors have imposed time-consistent conditions that would constrain the choice set available to the Stackelberg leader, see for example Karp (1984), Fujiwara and Long (2011). In what follows we use the approach advocated by Fujiwara and Long (2011). They propose that the leader's choice of the parameters of the tax function should lead to the socially optimal steady state. The rationale for this requirement is that if a policy leads to a steady state that is not efficient, there will be the incentive for the leader to deviate from it to achieve gains. In terms of our model, this requirement is

$$\overline{Y} = \frac{A - \sigma}{\eta} = \frac{Ar}{\gamma M} = Y_\infty$$

i.e.

$$\sigma = \frac{A(M\gamma - \eta r)}{M\gamma}.$$

This requirement allows us to simplify the coefficients of the terms $(Y_\tau - \overline{Y})$ and $(Y_\tau - \overline{Y})^2$ as follows:

$$\frac{2(2\kappa\overline{Y} + \rho)}{r + MG^{1/2}} = -A,$$

$$\frac{\kappa}{MG^{1/2}} = \frac{r(3\eta + r/M) - (2\eta + r/M)MG^{1/2} - 2\gamma M}{4MG^{1/2}}.$$

Therefore the right-hand side of (34) becomes

$$V_\tau(Y_\tau) = \left[\frac{r(3\eta + r/M) - (2\eta + r/M)MG^{1/2} - 2\gamma M}{4MG^{1/2}}\right]\left(Y_\tau - \frac{Ar}{\gamma M}\right)^2$$
$$- A\left(Y_\tau - \frac{Ar}{\gamma M}\right) + \frac{M\overline{x}}{r} - \frac{M\gamma}{2r}\left(\frac{Ar}{\gamma M}\right)^2.$$

It follows that the time-consistent leader's optimization problem amounts to choosing η to maximize the term inside the square brackets. Set $M = 1$ for simplicity. The first order condition for this optimization problem is

$$3rG - \left[r(3\eta + r) - 2\gamma\right]r - 2G^{3/2} = 0.$$

We can show (see Katayama et al. 2013) that the above FOC has a unique positive root η. Unfortunately, it is not possible to express the leader's optimal choice of η as an explicit function of the parameter values r and γ. We must therefore resort to numerical computations.

Given the numerical values of r and γ, we must solve for η. The solution proceeds as follows. Define the new variable $s = 2\eta + r$. First, we must find the unique positive root of the following cubic equation in s:

$$4s^3 - \frac{9r}{4}s^2 - 3r\left(\frac{r}{2} + \frac{2\gamma}{r}\right)s - r\left(\frac{r}{2} + \frac{2\gamma}{r}\right)^2 = 0.$$

Next, we find η from $s = 2\eta + r$, and compute

$$G^{1/2} = \sqrt{\left(\frac{r}{M}\right)^2 + \frac{2r\eta}{M}}.$$

After that, we find the welfare $V_\tau(Y_\tau)$.

Numerical Example Assume $M = 1$, $r = 0.05$, $\gamma = 0.02$, $A = 5$. Solving the cubic equation in s, we obtain the real root $s = 0.265$. Then

$$\eta = 0.107.$$

Turning to $G(\eta)^{1/2} = \sqrt{(\frac{r}{M})^2 + \frac{2r\eta}{M}}$,

$$G(\eta)^{1/2} = 0.115.$$

Moreover,

$$\sigma = \frac{A(M\gamma - \eta r)}{M\gamma} = 3.653.$$

Then, assuming $Y_0 = 0$, the leader's payoff is $13.644 + \frac{\overline{x}}{r}$. This is an improvement over the Nash equilibrium welfare (which was $6.944 + \frac{\overline{x}}{r}$).

What about the follower's welfare?

Recall that in Sect. 4, for any arbitrary tax function $\theta = \sigma + \eta$, the payoff of the cartel, when $Y_0 = 0$, is

$$\alpha_{X^*} = \frac{M}{4r}\left(A - \sigma + \beta_X^*\right)^2$$

where

$$\beta_X^* = -\left[\frac{(A - \sigma)(\eta - \mu_X^*)}{(\eta - \mu_X^*) + \frac{2r}{M}}\right]$$

and

$$\eta - \mu_X^* = \sqrt{\left(\frac{r}{M}\right)^2 + \frac{2}{M}\eta r} - \frac{r}{M} = \sqrt{G(\eta)} - \frac{r}{M} = 0.065.$$

Then $\beta_X^* = -0.531$ and $\alpha_X^* = 3.321$. Therefore the payoff for the follower (the cartel) is much smaller than under the Nash equilibrium (which was 13.888).

World welfare under the leadership of the importing coalition is $16.966 + \frac{\overline{x}}{r}$, which is smaller than under the Nash equilibrium, which is in turn smaller than under the social optimum.

Next, we turn to the Stackelberg regime with the exporter cartel as the leader. However, the analysis and derivation under this regime are quite similar to the case of Stackelberg under importer coalition leadership that we have just developed. Therefore, let us avoid cumbersome repetition of equations but present only the result of the numerical example. (The full exposition is available in Katayama et al. 2013). Assuming as before $M = 1$, $r = 0.05$, $\gamma = 0.02$, $A = 5$, we obtain that the payoff of the cartel (the leader) is 15.810 and that of the importer coalition is $2.901 + \frac{\overline{x}}{r}$. Thus the world welfare is $18.711 + \frac{\overline{x}}{r}$, which is greater than that under importer's leadership, $16.966 + \frac{\overline{x}}{r}$.

To check the robustness of our results, we have computed a number of numerical examples to make the welfare comparisons for three entities (world welfare, importer and exporter welfare) across four regimes (social planner, Nash, leadership by exporters, and leadership by importers). They are conducted with different values of r, keeping other parameters as fixed at $M = 1$, $\gamma = 0.02$, $A = 5$. The rate of discount is varied from 0.001 to 0.10. For all the different values of r, we find that world welfare is highest under social planning, which dominates world welfare under Nash, which is in turn superior to world welfare when the exporters are the Stackelberg leader. World welfare is always lowest when the coalition of importers is the Stackelberg leader. The last inequality is interesting. This direction of inequality seems to hold in general (or at least over all numerical examples of ours). This could be model-specific, since the exporters manage to control the resource and to act as the conservationist's friend and by that reason, atmospheric carbon accumulates at a slower rate.

Thus, as the conclusion of the present paper, we interpret the order of ranking in welfare comparison among social planning, Nash equilibrium, Stackelberg solution under exporter leadership and Stackelberg solution under importer leadership as follows. Naturally, we would expect that social planning gives the highest welfare for the world as a whole and Nash and Stackelberg would not surpass the socially optimal solution. Then, the ranking between Nash and Stackelberg is more of our interest and our conjecture is that the degree of competitiveness is greater under Nash type interaction between two players than under Stackelberg interactions since neither of Nash players exerts any power to infer the other party's reaction function while in the Stackelberg case, one of them does. In other words, the Nash equilibrium is established on equal footing among players, while Stackelberg solutions require stronger imperfectness in competition implying, in turn, greater distortion from social optimality.

Finally, between two Stackelberg welfare results, the one with exporter leadership is better than the other because the exporter/producer of natural resource fuels cares for its profit by directly controlling export price while the importer's main interest, in our model, is not in directly controlling resource reserves but in implementing environmental tax to curb the cumulative effect of fossil fuel consumption, while at the same time seeking more favorable terms of trade. It has been known from the literature that the resource monopolist is in favor of a slow mode of exploitation. Therefore, in the context of climate change driven largely by emissions from the burning of fossil fuels, we can expect to have higher welfare when the sole exporter of resource in our model leads the world rather than when the importing coalition targeting on environmental damages takes the lead.

Acknowledgements Katayama and Ohta gratefully acknowledge the financial support from Grant-in-Aid for Scientific Research of Japan Society for the Promotion of Science (JSPS). An earlier version of this paper was presented at the 73rd International Atlantic Economic Conference, the 12th Viennese Workshop on Optimal Control, Dynamic Games and Nonlinear Dynamics and the 10th Biennial Pacific Rim Conference. We thank the participants for many useful comments; however the usual disclaimer applies.

References

Allen, M. R., Frame, D. J., Huntingford, C., Jones, C. D., Lowe, J. A., Meinhausen, M., & Meinhausen, N. (2009). Warming caused by cumulative carbon emissions towards the trillionth tonne. *Nature*, *458*(7242), 1163–1166. doi:10.1038/nature08019.

Fujiwara, K., & Long, N. V. (2011). Welfare implications of leadership in a resource market under bilateral monopoly. *Dynamic Games and Applications*, *1*(4), 479–497.

Haurie, A., Tavoni, M., & van der Zwaan, B. C. C. (2012). Modeling uncertainty and the economics of climate change: recommendations for robust energy policy. *Environmental Modeling & Assessment*, *17*, 1–5.

Heal, G. (2009). The economics of climate change: a post-Stern perspective. *Climate Change*, *96*, 275–297.

Karp, L. (1984). Optimality and consistency in a differential game with non-renewable resources. *Journal of Economic Dynamics & Control*, *8*, 73–97.

Katayama, S., Long, N. V., & Ohta, H. (2013). Carbon taxes in a trading world. GISCS Working paper Series, No. 26. http://www.research.kobe-u.ac.jp/gsics-publication/gwps/.

Kemp, M. C., & Long, N. V. (1980). Optimal tariffs on exhaustible resources. In Kemp & Long (Eds.), *Exhaustible resources, optimality, and trade* (pp. 183–186). Amsterdam: North-Holland.

Stern, N. (2007). *The economics of climate change: the Stern review*. Cambridge: Cambridge University Press.

Wirl, F. (1995). The exploitation of fossil fuels under the threat of global warming and carbon taxes: a dynamic game approach. *Environmental & Resource Economics*, *39*, 1125–1136.

Landowning, Status and Population Growth

Ulla Lehmijoki and Tapio Palokangas

Abstract This paper considers the effects of the landowning and land reforms on economic and demographic growth by a family-optimization model with endogenous fertility and status-seeking. A land reform provides the peasants with strong incentives to limit their family size and to improve the productivity of land. Even though the income effect due to the land reform tends to raise fertility, a strong enough status-effect outweighs it, thus generating a decrease in population growth. The European demographic history provides supporting anecdotal evidence for this theoretical result.

1 Introduction

The core of the Malthusian thinking is the inescapable relationship between population and land: as land is fixed but population growing, a contradiction cannot be avoided (Malthus 1798). This document shows that it is essential to know who owns the land. Land ownership creates incentives to increase the productivity of land and to limit the family size. Therefore, land reforms have often diminished population growth, in particular where land ownership generates social status and appreciation.

Lucas (2002) characterizes land-population relationship by models of human history as follows. In primitive economies, the land is commonly owned so that even altruistic parents cannot improve the lot of their descendants. Nevertheless, once land property rights are established, parents decide on the optimal number of children and hand their farm over to their children. With private ownership, a newcomer decreases income per capita so that the steady state population growth rate falls.

U. Lehmijoki (✉)
University of Helsinki and HECER, Economicum Building, Arkadiankatu 7, 00014 Helsinki, Finland
e-mail: ulla.lehmijoki@helsinki.fi

T. Palokangas
University of Helsinki, HECER and IIASA, Economicum Building, Arkadiankatu 7, 00014 Helsinki, Finland
e-mail: tapio.palokangas@helsinki.fi

E. Moser et al. (eds.), *Dynamic Optimization in Environmental Economics*, Dynamic Modeling and Econometrics in Economics and Finance 15, DOI 10.1007/978-3-642-54086-8_14,

Parents can also educate their children which increases the cost of the newcomers, decreasing the steady state population growth even further. However, the transition from high to low fertility occurs only if there is a mechanism through which modern technology can gradually replace agricultural technology (Lucas 2002). A mechanism of this kind was postulated by Galor et al. (2009): they argue that unequal land ownership discourages human capital, thus preventing the decline in population growth. Benefiting from cheap labor, the landed aristocracy retards education by its political and social status. For this reason, land reforms have triggered both modernization and demographic change.

The essential difference between Lucas (2002) and Galor et al. (2009) is that the former focuses on the productive role of land while the latter consider the land also as a source of social status and political power. We extend the concept of status from landowning in two ways. First, we assume that status-seeking is important not only for the landed aristocracy but also for peasants. Where the status of the peasant depends on land per capita, farming families have a strong incentive to limit their family size. Second, we show that land reforms generate modernization, i.e., a shift from high fertility and low income to low fertility and high income.

Land reforms redistribute land from the landed aristocracy to tenants. We model a channel from land reforms to population growth through the social status, which is characterized by land per capita in the family relative to that elsewhere in the economy. A land reform decreases population growth the more, the stronger is the desire of status. The importance of status has already been recognized by Smith (1776), who denoted the appreciation of productive assets as the "Spirit of Capitalism". Kurz (1968), Corneo and Jeanne (2001) and Fisher and Hof (2005) used status to explain economic growth in advanced economies. Later, Lehmijoki and Palokangas (2009, 2010) applied status-seeking to explain economic and demographic growth in developing countries.

This document is organized as follows: Sect. 2 considers the optimal behavior of peasant families. Section 3 examines the dynamics of the economy. Sections 4 and 5 consider the long-run and short-run effects of land reforms, illustrating the transition from high fertility and low income to low fertility and high income. Section 6 provides supporting evidence from Europe. Section 7 summarizes the results.

2 Peasant Families

We consider an economy containing a large number of similar *peasant families*, who derive welfare from their children and status, and invest in agricultural technology to improve the productivity of land, and *landowners*, who do not farm but consume all their income. The analysis is based on a tradeoff between investment in children and investment in land productivity. We construct a model of a representative peasant family for this purpose.

2.1 Fertility, Production and Investment

The representative peasant family (hereafter *the family*) has $L(t)$ members at time t.[1] Its fertility rate n is

$$n \doteq \frac{\dot{L}}{L} \doteq \frac{1}{L}\frac{dL}{dt}, \tag{1}$$

where $(\dot{\cdot})$ is the time derivative.

We normalize the area of land at unity. The family owns a fixed proportion $\beta \in [0, 1]$ of the land it farms, but rents the remainder $1 - \beta$ from the representative landowner (hereafter *the landowner*). The model contains two important special cases: an independent farmer for $\beta = 1$ and a landless tenant farmer for $\beta = 0$.

The landowner requires the family to obey "good farming practises" when cultivating the rented land. To simplify the dynamics of the model, we specify this requirement as follows: the family must keep its rented land as productive as its own land. Let A be the productivity of land and simultaneously the supply of efficient land. The family improves the productivity A by its investment I in agricultural technology:

$$\dot{A} \doteq \frac{dA}{dt} = I. \tag{2}$$

There is only one good which is used in consumption and investment, and which we choose as the numeraire. The number of family members employed in child rearing, qnL, is in fixed proportion q to total fertility nL at any time. The rest of the family, $L - qnL = (1 - qn)L$, works in the family farm. The output Y of the good is produced from labor $(1 - qn)L$ and efficient land A according to the neoclassical technology with constant returns to scale:

$$Y = F\big((1-qn)L, A\big), \qquad F_1 \doteq \frac{\partial F}{\partial[(1-qn)L]} > 0, \qquad F_2 \doteq \frac{\partial F}{\partial A} > 0,$$

$$F_{11} \doteq \frac{\partial^2 F}{\partial[(1-qn)L]^2} < 0, \qquad F_{22} \doteq \frac{\partial F}{\partial A^2} < 0, \qquad F_{12} \doteq \frac{\partial^2 F}{\partial[(1-qn)L]\partial A} > 0,$$

F linearly homogeneous.

$$\tag{3}$$

We denote per capita efficient land by $a \doteq A/L$ for the family and by $\underline{a}$ for the entire economy. The family takes the macroeconomic variable $\underline{a}$ as given, but in equilibrium $a = \underline{a}$ holds true.

Because both the family and the landowner observe the productivity of land, rent r is set on efficient land. The family's investment I is thus equal to output Y minus

[1] We ignore mortality in this document, for convenience. The introduction of a constant mortality rate would complicate the analysis, without qualitatively changing the results.

rents for efficient land $r(1-\beta)A$ minus consumption C. Noting $c = C/L, a = A/L$, (2) and (3), this implies

$$\dot{A} = I = Y - r(1-\beta)A - C = F\big((1-qn)L, A\big) - (1-\beta)rA - C$$
$$= \big[F(1-qn, a) - (1-\beta)ra - c\big]L. \tag{4}$$

Noting $a = A/L$, (1) and (4), we obtain the per capita budget constraint

$$\dot{a} = \frac{\dot{A}}{L} - \frac{\dot{L}}{L}\frac{A}{L} = \frac{\dot{A}}{L} - na = F(1-qn, a) - (1-\beta)ra - c - na. \tag{5}$$

2.2 Utility

Following Razin and Ben-Zion (1975) and Becker (1991), the family derives temporary utility from the (logarithm of) per capita consumption and the proportion of new people in population, n (= the fertility rate). In addition, the peasant family benefits from its status in the society. This is proxied by the per capita efficient land the family cultivates, a, relative to that for the economy as a whole, $\underline{a}$. Thus, we augment the temporary utility by an increasing and concave function $v(a - \underline{a})$ of the status $a - \underline{a}$:[2]

$$u(t) = \log c + \theta \log n(t) + \varepsilon v\big(a(t) - \underline{a}(t)\big), \quad \theta > 0,\ v' > 0,\ v'' < 0,\ v'(0) = 1, \tag{6}$$

where $\theta > 0$ and $\varepsilon > 0$ are the constant weights for children and status. The bigger ε, the higher desire for status due to land. The bigger θ, the more children the families should like to have.

Let the constant ρ be a family's rate of time preference. Noting (1) and (6), the representative peasant family's expected utility at time $t = 0$ is then

$$U = \int_0^\infty u(t)e^{-\rho t}dt = \int_0^\infty \big[\log c + \theta \log n + \varepsilon v(a - \underline{a})\big]e^{-\rho t}dt,$$
$$v' > 0,\ v'' < 0,\ v'(0) = 1,\ \rho > 0,\ \theta > 0. \tag{7}$$

2.3 The Family's Optimal Behavior

The representative family maximizes its utility (7) by choosing its fertility n and consumption per capita, c, subject to its budget constraint (5), given the rent r. The

[2]If the measure for status, v, were a linearly homogeneous function of a and $\underline{a}$, we would obtain the same results with some complication.

Hamiltonian of this maximization is

$$H = \log c + \theta \log n + \epsilon v(a - \underline{a}) + \lambda\big[F(1-qn, a) - (1-\beta)ra - c - na\big], \quad (8)$$

where the co-state variable λ evolves according to

$$\dot{\lambda} = \rho\lambda - \frac{\partial H}{\partial a} = \big[\rho + n + (1-\beta)r - F_2(1-qn, a)\big]\lambda - \varepsilon v'(a - \underline{a}),$$
$$\lim_{t\to\infty} \lambda a e^{-\rho t} = 0. \quad (9)$$

The maximization of the Hamiltonian (8) by the control variables (c, n) for a given λ yields the first-order conditions

$$\frac{\partial H}{\partial c} = \frac{1}{c} - \lambda = 0, \qquad \frac{\partial H}{\partial n} = \frac{\theta}{n} - \big[qF_1(1-qn, a) + a\big]\lambda = 0.$$

Given these two equations, (3) and (7), we can replace λ by n as the co-state variable and define per capita consumption c as a function of productivity per capita, a, and the fertility rate n as follows:

$$c \doteq 1/\lambda = z(a,n)/\theta > 0, \qquad z(a,n) \doteq \big[qF_1(1-qn,a) + a\big]n > 0,$$
$$z_a \doteq \frac{\partial z}{\partial a} = (\underbrace{q}_{+}\ \underbrace{F_{12}}_{+} + 1)n > n > 0, \qquad z_n \doteq \frac{\partial z}{\partial n} = \underbrace{\frac{z}{n}}_{+} - \underbrace{q^2 n}_{+}\ \underbrace{F_{11}}_{-} > 0. \quad (10)$$

Inserting the function (10) into the differential equation (5), the change of per capita efficient land, $\dot{a}$, can be defined as a function of per capita efficient land a, the fertility rate n and the family's proportion of land, β, as follows:

$$\dot{a} = F(1-qn, a) - (1-\beta)ra - z(a,n)/\theta - na. \quad (11)$$

Noting (3) and (10), this function has the properties:

$$\frac{\partial \dot{a}}{\partial n} = -qF_1 - a - \frac{z_n}{\theta} < 0, \qquad \frac{\partial \dot{a}}{\partial a} = F_2 - n - \frac{z_a}{\theta}, \qquad \frac{\partial \dot{a}}{\partial \beta} = ra > 0. \quad (12)$$

3 The Dynamics of the Economy

In equilibrium, $\underline{a} = a$ holds true. Because the market for efficient land is competitive, the rent r is equal to the marginal product of efficient land for the given supply A of efficient land:[3]

$$r = F_2\big((1-qn)L, A\big) = F_2(1-qn, a). \quad (13)$$

[3]Because the production function F is homogeneous of degree one [cf. (3)], its partial derivative F_2 must be homogeneous of degree zero. This and $a = A/L$ yields (13).

Consider the evolution of the economy. Given $\underline{a} = a$, (7), (10) and (13), we can transform the differential equation (9) into the following form:

$$\begin{aligned}
&\rho + n - \beta F_2(1-qn,a) - \frac{\varepsilon}{\theta} z(a,n) \\
&= \rho + n + (1-\beta)r - F_2(1-qn,a) - \frac{\varepsilon}{\lambda} \\
&= \rho + n + (1-\beta)r - F_2(1-qn,a) - v'(0)\frac{\varepsilon}{\lambda} = \frac{\dot{\lambda}}{\lambda} = \frac{d\log\lambda}{dt} \\
&= -\frac{d}{dt}\log z(a,n) = -\frac{z_a}{z}\dot{a} - \frac{z_n}{z}\dot{n}.
\end{aligned} \tag{14}$$

Thus

$$\rho + n = \beta F_2(1-qn,a) + \varepsilon z(a,n)/\theta \quad \Leftrightarrow \quad \dot{a} = \dot{n} = 0. \tag{15}$$

Rearranging terms in (14) and noting $0 \le \beta < 1$, (3), (10), (12), (13) and (15), we obtain the change of the fertility rate, $\dot{n}$, as a function of per capita efficient land a, the fertility rate n and the family's proportion of land, β,

$$\dot{n} = \frac{z}{z_n}\left[\beta F_2(1-qn,a) + \frac{\varepsilon}{\theta} z(a,n) - n - \rho\right] - \frac{z_a}{z_n}\dot{a}, \tag{16}$$

with the following partial derivatives:

$$\begin{aligned}
\left.\frac{\partial \dot{n}}{\partial a}\right|_{\dot{a}=\dot{n}=0} &= \frac{z}{z_n}\left(\beta F_{22} + \frac{\varepsilon}{\theta} z_a\right) - \frac{z_a}{z_n}\frac{\partial \dot{a}}{\partial a} \\
&= \frac{z}{z_n}\left(\beta F_{22} + \frac{\varepsilon}{\theta} z_a\right) - \frac{z_a}{z_n}\left(F_2 - n - \frac{z_a}{\theta}\right) > 0 \\
&\Leftrightarrow \quad \frac{\varepsilon}{\theta} > \frac{1}{z}\left(F_2 - n - \frac{z_a}{\theta}\right) - \beta F_{22}/z_a,
\end{aligned} \tag{17}$$

$$\begin{aligned}
\left.\frac{\partial \dot{n}}{\partial n}\right|_{\dot{a}=\dot{n}=0} &= -\frac{z}{z_n}\beta F_{12} q + \frac{z}{z_n}\left(\frac{\varepsilon}{\theta} z_n - 1\right) - \frac{z_a}{z_n}\frac{\partial \dot{a}}{\partial n} \\
&= -\frac{z}{z_n}\beta F_{12} q + \frac{z}{z_n}\left(\frac{\varepsilon}{\theta} z_n - 1\right) + \frac{z_a}{z_n}\left(qF_1 + a + \frac{z_n}{\theta}\right) \\
&= -\frac{z}{z_n}\beta F_{12} q + \frac{\varepsilon}{\theta} z - \frac{z}{z_n} + \frac{z_a}{z_n}\underbrace{(qF_1 + a)}_{=z/n} + \frac{z_a}{\theta} \\
&= -\frac{z}{z_n}\beta F_{12} q + \frac{\varepsilon}{\theta} z - \frac{z}{z_n} + \frac{z_a}{z_n}\frac{z}{n} + \frac{z_a}{\theta} \\
&= \frac{\varepsilon}{\theta} z + \frac{z}{z_n}\Bigg(\underbrace{\frac{z_a}{n} - 1}_{qF_{12}} - \beta F_{12} q\Bigg) + \frac{z_a}{\theta}
\end{aligned}$$

$$= \underbrace{\frac{\varepsilon}{\theta}z}_{+} + \underbrace{(1-\beta)}_{+} \underbrace{\frac{zq}{z_n}}_{+} \underbrace{F_{12}}_{+} + \underbrace{\frac{z_a}{\theta}}_{+} > 0, \tag{18}$$

$$\frac{\partial \dot{n}}{\partial \beta} = \frac{z}{z_n}F_2 - \frac{z_a}{z_n}\frac{\partial \dot{a}}{\partial \beta} = \frac{z}{z_n}F_2 - \frac{z_a}{z_n}ra = \frac{z}{z_n}F_2 - \frac{z_a}{z_n}F_2 a = \left(1 - \frac{z_a}{z}a\right)\frac{z}{z_n}F_2. \tag{19}$$

The ambiguous sign of the partial derivative (17) can be explained as follows. If the status effect is strong enough (i.e. if $\frac{\varepsilon}{\theta}$ is high enough), then the family invests in wealth rather than in children. In that case, higher per capita wealth (= per capita efficient land a) generates higher income and a lower marginal utility of wealth, $\dot{\lambda} < 0$, and a lower opportunity cost of child rearing. This encourages to transfer labor from farming to child rearing, increasing the fertility rate, $\dot{n} > 0$. On the other hand, the sign of (18) is always clear. If the fertility rate n is over its short-run equilibrium level, then labor input in farming and the level of income are low, generating a lower marginal utility of wealth and a lower opportunity cost of child rearing, thus encouraging to transfer from farming to child rearing, i.e. $\dot{n} > 0$.

4 Long Run Effects of a Land Reform

The system (11) and (16) of per capita efficient land, a, and the fertility rate n can be linearized in the neighborhood of the steady state $\dot{a} = \dot{n} = 0$ as follows:

$$\begin{pmatrix} \partial \dot{a}/\partial a & \partial \dot{a}/\partial n \\ \partial \dot{n}/\partial a & \partial \dot{n}/\partial n \end{pmatrix} \begin{pmatrix} da \\ dn \end{pmatrix} + \begin{pmatrix} \partial \dot{a}/\partial \beta \\ \partial \dot{n}/\partial \beta \end{pmatrix} d\beta = 0.$$

If the Jacobian in this equation is negative,

$$\mathcal{J} \doteq \frac{\partial \dot{a}}{\partial a}\frac{\partial \dot{n}}{\partial n} - \frac{\partial \dot{a}}{\partial n}\frac{\partial \dot{n}}{\partial a} < 0, \tag{20}$$

then the system has a *saddle point*: there is only one initial value of the jump variable n that leads to the steady state. This is assumed to be the case in the following.

Consider now a land reform that increases the family's proportion of land, β. Noting (3), (10), (12), (13), and (20), the steady state values of per capita efficient land, a^*, and the fertility rate, n^*, are functions of preferences concerning status relative to children, ε/θ, and the family's proportion of land, β, with the properties

$$\frac{\partial a^*}{\partial \beta} = -\frac{1}{\mathcal{J}}\begin{vmatrix} \partial \dot{a}/\partial \beta & \partial \dot{a}/\partial n \\ \partial \dot{n}/\partial \beta & \partial \dot{n}/\partial n \end{vmatrix} = -\frac{1}{\mathcal{J}}\begin{vmatrix} \partial \dot{a}/\partial \beta & \partial \dot{a}/\partial n \\ \frac{z}{z_n}F_2 & -\frac{z}{z_n}\beta F_{12}q + \frac{z}{z_n}(\frac{\varepsilon}{\theta}z_n - 1) \end{vmatrix}$$

$$= -\frac{1}{\mathcal{J}}\frac{z}{z_n}\begin{vmatrix} ra & -qF_1 - a - \frac{z_n}{\theta} \\ F_2 & \frac{\varepsilon}{\theta}z_n - \beta F_{12}q - 1 \end{vmatrix} = -\frac{1}{\mathcal{J}}\frac{z}{z_n}F_2\begin{vmatrix} a & -qF_1 - a - \frac{z_n}{\theta} \\ 1 & \frac{\varepsilon}{\theta}z_n - \beta F_{12}q - 1 \end{vmatrix}$$

$$= -\underbrace{\frac{1}{\mathcal{J}}}_{-}\underbrace{\frac{z}{z_n}}_{+}\underbrace{F_2}_{+}\left[\frac{\varepsilon}{\theta}az_n + qF_1 + \frac{z_n}{\theta} - a\beta F_{12}q\right] > 0$$

$$\Leftrightarrow \quad \frac{\varepsilon}{\theta} > \underbrace{\beta q}_{+}\underbrace{F_{12}}_{+}/\underbrace{z_n}_{+} - \underbrace{\frac{1}{a}}_{+}(\underbrace{q}_{+}\underbrace{F_1}_{+}/\underbrace{z_n}_{+} + \underbrace{1/\theta}_{+}), \tag{21}$$

$$\frac{\partial n^*}{\partial \beta} = -\frac{1}{\mathcal{J}}\begin{vmatrix} \partial\dot{a}/\partial a & \partial\dot{a}/\partial\beta \\ \partial\dot{n}/\partial a & \partial\dot{n}/\partial\beta \end{vmatrix} = -\frac{1}{\mathcal{J}}\begin{vmatrix} \partial\dot{a}/\partial a & \partial\dot{a}/\partial\beta \\ \frac{z}{z_n}(\beta F_{22} + \frac{\varepsilon}{\theta}z_a) & \frac{z}{z_n}F_2 \end{vmatrix}$$

$$= -\frac{1}{\mathcal{J}}\frac{z}{z_n}\begin{vmatrix} -qF_1 - a - \frac{z_n}{\theta} & ra \\ \beta F_{22} + \frac{\varepsilon}{\theta}z_a & F_2 \end{vmatrix} = -\frac{1}{\mathcal{J}}\frac{z}{z_n}F_2\begin{vmatrix} -qF_1 - a - \frac{z_n}{\theta} & a \\ \beta F_{22} + \frac{\varepsilon}{\theta}z_a & 1 \end{vmatrix}$$

$$= -\underbrace{\frac{1}{\mathcal{J}}}_{-}\underbrace{\frac{z}{z_n}}_{+}\underbrace{F_2}_{+}\left[-\left(qF_1 + a + \frac{z_n}{\theta}\right) - a\left(\frac{\varepsilon}{\theta}z_a + \beta F_{22}\right)\right] < 0$$

$$\Leftrightarrow \quad \frac{\varepsilon}{\theta} > \underbrace{\frac{1}{z_a}}_{+}\left[-\left(\underbrace{\frac{q}{a}F_1}_{+} + 1 + \underbrace{\frac{z_n}{a\theta}}_{+}\right) - \underbrace{\beta}_{+}\underbrace{F_{22}}_{-}\right]. \tag{22}$$

From (21) and (22) it follows that

$$\lim_{\beta\to 0}\frac{\partial a^*}{\partial\beta} > 0, \qquad \lim_{\beta\to 0}\frac{\partial n^*}{\partial\beta} < 0, \qquad \lim_{(\varepsilon/\theta)\to\infty}\frac{\partial n^*}{\partial\beta}\Big/\frac{\partial a^*}{\partial\beta} = -\frac{z_a}{z_n}. \tag{23}$$

The results (21), (22) and (23) can be rephrased as follows:

Proposition 1 *In the long run, a land reform (i.e. an increase in the family's proportion of land, β) increases per capita efficient land a^*, but decreases the population growth rate n^* if and only if either of the following conditions holds true:*

(i) *The family's initial proportion of land is small enough, $\beta \to 0$.*
(ii) *The status-effect is strong enough (i.e. $\frac{\varepsilon}{\theta}$ is high enough) for the inequalities in* (21) *and* (22) *to hold.*

This result can be interpreted as follows. A land reform definitely increases the income of the peasant family as rent payments decrease. This increases the demand for children as these are normal goods. If the status-effect of efficient land is weak, then the income effect dominates and the number of children increases after the land reform. On the other hand, if the status-effect is strong, then the peasant family limits its size and invests the extra income to improve the efficiency of land.

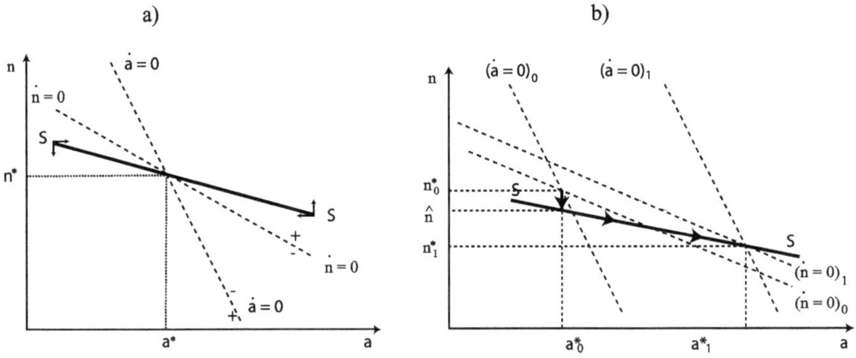

Fig. 1 The phase diagram: (**a**) the dynamics of the model and (**b**) the effects of a land reform

5 Short-Run Effects of a Land Reform

The saddle-point condition (20) is equivalent to

$$\frac{\partial \dot{a}}{\partial a}\frac{\partial \dot{n}}{\partial n} < \frac{\partial \dot{a}}{\partial n}\frac{\partial \dot{n}}{\partial a}. \tag{24}$$

Noting (12), (17) and (18), this implies

$$\frac{\partial \dot{a}}{\partial a} < \underbrace{\frac{\partial \dot{a}}{\partial n}}_{-}\underbrace{\frac{\partial \dot{n}}{\partial a}}_{+}\Big/\underbrace{\frac{\partial \dot{n}}{\partial n}}_{+} < 0.$$

Assume first that the system is initially in the steady state (a_0^*, n_0^*). Once β increases, the steady state moves to (a_1^*, n_1^*). Given (21), the status a rises but the fertility rate n falls, $a_0^* < a_1^*$ and $n_0^* > n_1^*$. Given (12), (18) and (21), both singular curves $(\dot{a} = 0)$ and $(\dot{n} = 0)$ are decreasing, but $(\dot{a} = 0)$ falls more steeply: in the (a, n) space:

$$\left.\frac{\partial n}{\partial a}\right|_{\dot{a}=0} = -\frac{\partial \dot{a}}{\partial a}\Big/\underbrace{\frac{\partial \dot{a}}{\partial n}}_{-} < -\underbrace{\frac{\partial \dot{n}}{\partial a}}_{+}\Big/\underbrace{\frac{\partial \dot{n}}{\partial n}}_{+} = \left.\frac{\partial n}{\partial a}\right|_{\dot{n}=0} < 0.$$

Since $\partial\dot{a}/\partial n < 0$ by (12), the variable a increases (decreases) below (above) the singular curve $(\dot{a} = 0)$. Since $\partial\dot{n}/\partial n > 0$ by (18), the variable n increases (decreases) above (below) the singular curve $(\dot{n} = 0)$. Hence, the stable saddle path SS is downward sloping (cf. Fig. 1a).

Noting (12) and (16), an increase in β shifts both singular curves $(\dot{a} = 0)$ and $(\dot{n} = 0)$ *upwards* in the (a, n) plane (cf. Fig. 1b):

$$\left.\frac{dn}{d\beta}\right|_{\dot{n}=0} = -\underbrace{\frac{\partial \dot{n}}{\partial \beta}}_{-}\Big/\underbrace{\frac{\partial \dot{n}}{\partial n}}_{+} > 0, \qquad \left.\frac{dn}{d\beta}\right|_{\dot{a}=0} = -\underbrace{\frac{\partial \dot{a}}{\partial \beta}}_{+}\Big/\underbrace{\frac{\partial \dot{a}}{\partial a}}_{-} > 0.$$

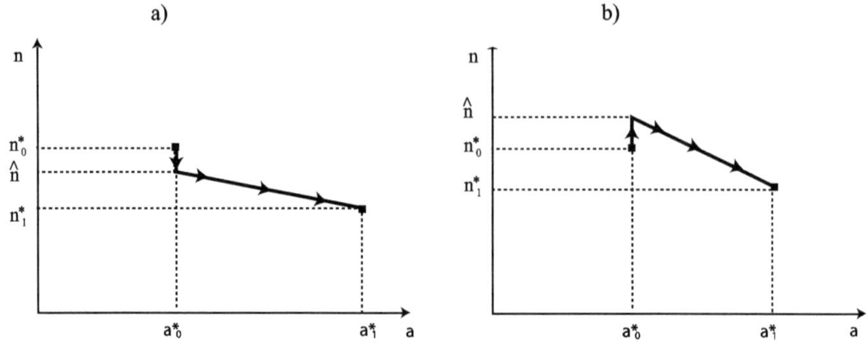

Fig. 2 The development of per capita productivity (a) and population growth (n) after a land reform

Figure 2 illustrates, that two types of developments are possible. In Fig. 2a, population growth undershoots.[4] In this case, population growth starts to decrease immediately after the land reform. Furthermore, the initial decrease may be considerable, i.e. population growth falls drastically. Nevertheless, population growth may also adopt a reverse course in the short run (cf. Fig. 2b): it overshoots, indicating that the income effect dominates over the status effect immediately after the land reform.

Given (11), (16), (21) and (22), the population growth rate n undershoots (cf. Fig. 2a), if and only if

$$\underbrace{\frac{\partial n^*}{\partial \beta}}_{-} \Big/ \underbrace{\frac{\partial a^*}{\partial \beta}}_{+} = \frac{dn^*}{da^*} < \underbrace{\frac{dn}{dt}}_{-} \Big/ \underbrace{\frac{da}{dt}}_{+} < 0, \tag{25}$$

where $\frac{dn^*}{da^*}$ is the slope of the line between points (n_0, a_0) and (n_1, a_1) and $\frac{dn}{dt}/\frac{dn}{dt}$ is the slope of the saddle path from $(a_0, \widehat{n})$ to (n_1, a_1). Furthermore, given (16) and (23), it holds true that

$$\lim_{(\varepsilon/\theta)\to\infty} \left(\frac{dn}{dt} \Big/ \frac{da}{dt} \right) = \underbrace{\frac{z}{z_n} \lim_{(\varepsilon/\theta)\to\infty} \left(F_2 + \frac{\varepsilon}{\theta} z - n - \rho \right)}_{+} \Big/ \underbrace{\frac{da}{dt}}_{+} - \frac{z_a}{z_n}$$

$$> -\frac{z_a}{z_n} = \lim_{(\varepsilon/\theta)\to\infty} \frac{\partial n^*}{\partial \beta} \Big/ \frac{\partial a^*}{\partial \beta}.$$

Thus, the inequality (25) corresponding to undershooting (cf. Fig. 2a) holds for high enough values of $\frac{\varepsilon}{\theta}$. This result can be rephrased as follows:

[4]This case is illustrated in Fig. 1b.

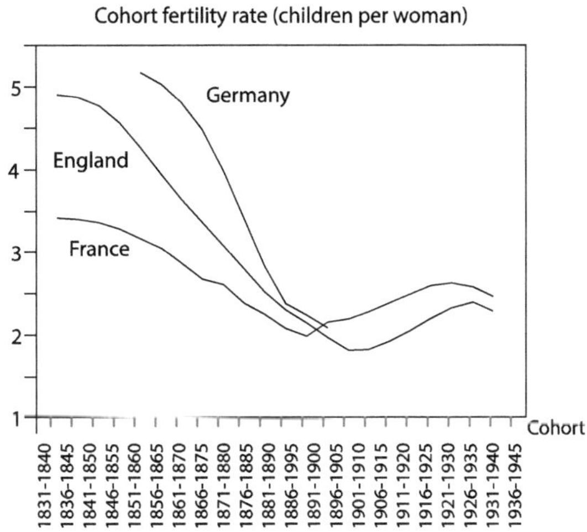

Fig. 3 The decline of fertility in France, England and Germany. Source: Festy (1979) (pages 266–267, 262 and 222)

Proposition 2 *If the status-effect is strong enough (i.e.* $\frac{\varepsilon}{\theta}$ *is high enough), then the land reform (i.e. an increase in* β*) decreases the population growth rate n immediately (cf. Fig.* 2a).

If the status effect is very strong, then the family generates status immediately by transferring resources from child rearing into investment in efficient land a.

6 Supportive Evidence

In this section, we provide suggestive evidence in favor of the landowning hypothesis from European history.

One of the greatest puzzles in demographic history is why fertility declined in rich and urbanized England much later than in poor and rural France.[5] Figure 3 illustrates the fertility trends in England, France, and Germany from 1831–1840 to 1936–1945, showing that even though fertility was declining everywhere, its level in 1831–1840 was much lower and its decrease much faster in France.[6] If economic factors were the driving forces of the fertility decline, this should have started first in England. Nevertheless, this was not the case. In 1831–1840, the fertility in England was more than 40 % higher than in France. Furthermore, it took over 30 years for England to reach the 1831–1840 numbers in France. On the other hand, England was ahead of Germany as one expected (cf. Fig. 3). Why was the fertility rate so low in France?

[5]In 1820, the GDP per capita in England was 1.4 times larger than that in France, and the advantage of England only increased towards the end of the century (Maddison 1995, 194–196, and Guinnane 2011).

[6]The cohort fertility rate in Fig. 3 gives the total number of births given by women born in the time period indicated in the figure.

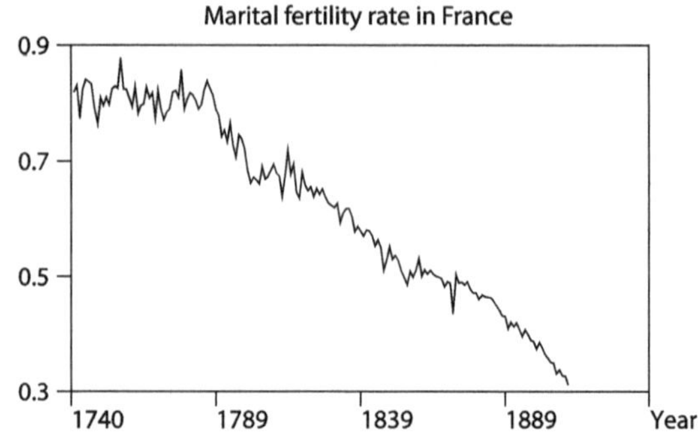

Fig. 4 The marital fertility rate expressed as the share of the maximum fertility rate (1.00) in France. Source: Weir (1994)

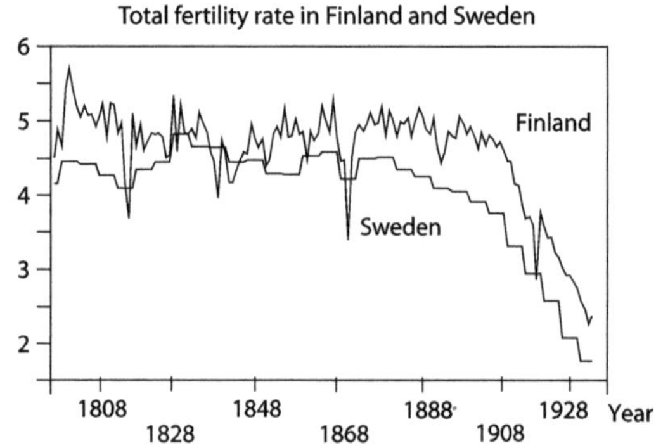

Fig. 5 The total fertility rate (children per woman) from 1776 to 1935 in Finland and Sweden. Sources: Statistics Finland (2013), Statistics Sweden (2013)

Figure 4 presents the (marital) fertility in France from 1740 to 1911. It shows that fertility declined sharply at the time of the land reform during the Great Revolution 1789–1799, while no land reform occurred in England or Germany: in 1830, 63 % of the population was landowning peasants in France, while in Britain the share of landowners was only 14 % (Chesnais 1992, p. 337). Actually, the widespread ownership of land was a unique feature of France (Cummins 2013). For the new rural bourgeoisie class, fertility control supplied a powerful method for social rise. Thus, it is likely that the fertility decline in France was due to the decline in the child demand among the peasants (Cummins 2013). Furthermore, by associating early wealth and fertility data, Gummings shows that those peasants who had the greatest land property also had the lowest fertility and their fertility decline was the fastest, indicating that status-seeking may have played an important role.

Another example comes from Finland and Sweden. Finland was part of Sweden from 1150 to 1809, thus sharing many social institutions and cultural features with the latter. Figure 5 shows that, once onset, the phase of the fertility decline was fast in both countries. In Sweden, fertility decreased steadily from 1880 to 1935, falling from 4.5 to 1.765 children per woman. In Finland, however, fertility remained high (4.72 children per woman) until 1908, but then started declining, reaching the number 2.37 in 1935. The decline of 2.35 children in only 27 years is one of the fastest in Western countries, and may be associated with the land reform which started in

1908 as the tenants were allowed to buy their farms.[7] Note that both France and Finland exhibit strong undershooting, i.e., a sudden downward jump in fertility after the land reform, indicating that the status effect may have been strong in both countries.

7 Conclusions

This paper examines the effects the landowning and land reforms by a family-optimization model with endogenous fertility and status-seeking. A land reform decreases the costs of farming by decreasing the rents, generating more income for peasants. The outcome of this depends on preferences. If the role of status is small, then the peasants rear more children which are normal goods for them. This leads to a persistent stagnation of income and productivity. But if the role of status is sufficient, then peasants limit their family size and invest in efficient land. If status-seeking is strong enough, then fertility decreases immediately after the land reform.

The demographic history in Europe provides some supportive evidence for this landowning hypothesis. Fertility declined in rich and urbanized England much later than in poor and rural France due to the land reform in the latter during Great Revolution 1789–1799. The fertility control, which supplied a powerful method for social rise for the new rural bourgeois class, led to an exceptional fertility decline in France. There is evidence that the peasants with the greatest land property had the lowest fertility indicating they were subject to strong status-seeking. In Finland as well, the land reform in 1908 generated one of the most drastic fertility declines in Western countries.

Acknowledgements The authors thank the reviewer for constructive comments.

References

Becker, G. S. (1991). *A treatise on the family*. Cambridge: Harvard University Press. Enlarged edition.
Chesnais, J.-C. (1992). *The demographic transition: stages, patterns and economic implications*. Oxford: Oxford University Press.
Corneo, G., & Jeanne, O. (2001). On relative-wealth effects and long-run growth. *Research in Economics*, *55*(4), 349–358.
Cummins, N. (2013). Marital fertility and wealth during the fertility transition: rural France, 1750–1850. *The Economic History Review*, *66*(2), 449–476. doi:10.1111/j.1468-0289.2012.00666.x.

[7]Unfortunately, the land reform experienced some drawbacks which, together with the general unrest of the time, led to an outburst of a civil war in 1918. One of the conditions for the later social cohesion in Finland was the famous Lex Kallio in 1922 which made larger and wealthier farms possible for the peasants.

Festy, P. (1979). *La féconfité des pays occidentaux de 1870 à 1970*. Paris: Presses Universitaires de France.

Fisher, W. H., & Hof, F. X. (2005). Status seeking in a small open economy. *Journal of Macroeconomics*, *27*(2), 209–232.

Galor, O., Moav, O., & Vollrath, D. (2009). Inequality in land ownership, the emergence of human capital promoting institutions and the great divergence. *Review of Economic Studies*, *76*(1), 143–179.

Guinnane, T. W. (2011). The historical fertility transition: a guide for economists. *Journal of Economic Literature*, *49*(3), 589–614.

Kurz, M. (1968). Optimal economic growth and wealth effects. *International Economic Review*, *9*(3), 348–357.

Lehmijoki, U., & Palokangas, T. (2009). Population growth overshooting and trade in developing countries. *Journal of Population Economics*, *22*(1), 43–56.

Lehmijoki, U., & Palokangas, T. (2010). Trade, population growth, and the environment in developing countries. *Journal of Population Economics*, *23*(4), 1351–1373.

Lucas, R. E. J. (2002). *Lectures on economic growth*. Cambridge: Harvard University Press.

Maddison, A. (1995). *Monitoring the world economy, 1820–1992*. Paris: OECD.

Malthus, Th. R. (1798). *An essay on the principle of population, as it affects the future improvement of society with remarks on the speculations of Mr. Godwin, M. Condorcet, and other writers*. London: St. Paul's Church-Yard. Printed for J. Johnson.

Razin, A., & Ben-Zion, U. (1975). An intergenerational model of population growth. *The American Economic Review*, *65*(5), 923–933.

Smith, A. (1776). *An inquiry into the nature and causes of the wealth of nations* (1st ed., Vol. 1). London: W. Strahan.

Statistics Finland (2013).

Statistics Sweden (2013).

Weir, D. R. (1994). New estimates of nuptiality and marital fertility in France, 1740–1911. *Population Studies*, *48*(2), 307–331.

Optimal Harvesting of Size-Structured Biological Populations

Olli Tahvonen

Abstract The question of harvesting size-structured biological resources is generic in resource economics but purely understood. This study is based on a well known density-dependent size-structured population model that includes an age-structured model as a special case. Harvest from each size class can be chosen independently. Mathematically the model is an any number of state and control variables discrete-time optimization problem. While earlier studies have analysed the Maximum Sustainable Yield (MSY) steady states using problem-specific optimization procedures, this study applies non-linear programming and analyses the dynamic economic problem. It is shown that with two size classes, there may exist six steady state regimes. The optimal steady state is shown to be either unique or a continuum implying that earlier MSY-theorems are not entirely correct. Given a unique steady state the optimal solution converges toward a saddle point steady state or a stationary cycle. Optimal harvest of single individuals deviates from Faustmann-type timing, and a higher interest rate may cause a shift to harvesting older age classes. For the general specification with any number of size classes, equations for optimal steady states and a stability result are obtained.

1 Introduction

Economic analysis on biologically regenerating natural resources is heavily based on two distinct approaches. These are the optimal rotation model (Faustmann 1849; Samuelson 1976) and the dynamic biomass harvesting model (Gordon 1954; Clark 1990). The optimal rotation approach solves the question when to clear-cut a group of even-aged trees and it leads to harvest every 50 years, for example. The dynamic biomass approach describes how to optimize the rate of harvest of a biomass (e.g. fish population). Both approaches have produced extensive literature including theoretical extensions and practical applications. In spite of this success the limitation of the rotation approach is the *a priory* commitment to point-input, point-output

O. Tahvonen (✉)
Department of Forest Sciences, University of Helsinki, 00014 Helsinki, Finland
e-mail: olli.tahvonen@helsinki.fi

E. Moser et al. (eds.), *Dynamic Optimization in Environmental Economics*,
Dynamic Modeling and Econometrics in Economics and Finance 15,
DOI 10.1007/978-3-642-54086-8_15,

structure and silence on whether there could be other possibilities to cope with biological density dependence, to organize the tree age class structure and the resulting timing of harvest and regeneration. The limitation of the biomass model is that it cannot answer the question on whether one should direct the harvesting activity to some age or size group and save the others.

Overcoming these limitations requires extending the description of the biologically regenerating natural resource under harvest. This can be done by specifying the population as an age- or size-structured system. One route toward such models can be based on the well-known studies by Leslie (1945) and Usher (1966), who pioneered in developing discrete-time structured population models that are currently considered basic workhorses of population ecology (Cushing 1998; Caswell 2001). In biological sciences these models are used e.g. for fish, plant, mammal and insect populations.

For clarity it is important to distinguish at least two different classes of discrete time structured optimal harvesting models. The first class is based on Leslie (1945) or Usher (1966) age- or size structured population models and on various density dependence assumptions. In these models harvest may be age class specific (perfect selectivity assumption) or models may include some specific harvesting technology and "effort" as a single control variable (Getz and Haight 1989, pp. 45, 143). The second class of models has a centuries long historical background in forest sciences and forest planning (Reed 1986; Getz and Haight 1989, p. 308). These models consist of a number of biologically separate "even-aged stands" in a larger region (e.g. a country). The economic problem is to allocate the total land area over time and over the stands (with varying ages) in order to find optimal harvest and an equilibrium market price of timber. Clearly, the land allocation model is developed to analyse a very different problem compared to the single population model and not recognizing this difference has caused a rather serious confusion (Goetz and Xabadia 2011; Xabadia and Goetz 2010; Goetz et al. 2011).

The first steps in understanding the problem of harvesting a single population were taken by Baranov (1918), Beverton and Holt (1957) and Walters (1969). Beddington and Taylor (1973) and Rorres and Fair (1975) developed a model where harvesting activity is age-class specific (the prefect selectivity assumption). While these studies apply the linear Leslie matrix model (and *ad hoc* restrictions), Reed (1980), Getz (1980) and Getz and Haight (1989) base their models on Walters (1969) and Allen and Basasibwaki (1974), and include density dependence in recruitment. In contrast Clark (1990) studies the problem following Beverton and Holt (1957) and assumes constant exogenous recruitment. In spite of the various extensions of this model (see Getz and Haight 1989) and its use in many empirical studies (Haight 1985; Tahvonen 2011a), the main theoretical results are still offered by the works of Reed (1980) and Getz (1980).

Both of these studies solve the Maximum Sustainable Yield (MSY) steady state and apply a problem-specific solution method that separates the model into linear and non-linear programming problems. Their main theorem states that optimal MSY harvesting is bimodal, i.e. it involves a partial or total harvest of only one age class, or a partial harvest of one age class and a total harvest of another, older age class.

Getz and Haight (1989) generalized this MSY approach to size-structured models, and proceeded to a general discussion of how the analysis could be extended by applying the discrete-time Maximum Principle. However, they did not present any further results.

In spite of its merits, the solution procedure in Reed (1980) and Getz (1980) is designed for steady-state analysis and cannot be applied for obtaining dynamic solutions. Its specific nature has perhaps additionally discouraged further developments in the analytical understanding of this problem.

This study analyses the general dynamic economic problem by applying nonlinear programming and the Karush-Kuhn-Tucker (KKT) theorem. A study by Wan (1994) and further studies by Salo and Tahvonen (2002, 2003) present an example in which most theoretical properties of a structured model (land allocation forestry model) can be found by including only two classes. Applying this method for the population level model, this study shows the existence of six different steady-state regimes. These steady states are unique, except for a case of a steady state continuum. This continuum is not recognized in the main theorem by Reed (1980) or Getz (1980).

The model specifies a problem in optimal harvest timing, but the solution is not within the realm of the Faustmann rotation model: it is optimal to harvest the cohorts after they have reached their maximum discounted value, and an increase in interest rate may result in harvesting of older age classes rather than younger ones. Comparing solutions with the biomass approach shows that neglecting information on population interior structure leads to suboptimal steady states and misleading "optimal extinction" results (cf. Clark 1973).

Given zero interest rate the interior steady states are shown to be local saddle points. Under some additional restrictions, the saddle point property is shown to also hold with positive interest rates. However, given "high" discounting, the optimal solution may converge toward a stationary cycle. In contrast, a boundary-type regime is independent of (small) changes in interest rate and the steady state is locally stable. Earlier literature does not present any stability results.

The analysis of two solution regimes is extended to steady-state equations with any number of age or size classes. Given the age-structured specification, it is possible to show a local stability result independently of the number of age classes. A numerical example demonstrates that the optimal harvest level within the structured approach may equal zero, even if the initial population biomass exceeds its optimal steady state level.

2 The Size- and Age-Structured Optimization Problem

Let x_{st}, $s = 1, \ldots, n$, $t = 0, 1, \ldots$ denote the number of individuals in size class s in the beginning of period t. A fraction $0 < \alpha_s < 1$, $s = 1, \ldots, n$ moves to class $s + 1$ by the end of period t and a fraction $0 \leq \beta_s < 1$, $s = 1, \ldots, n$ remains in class s. The remaining fraction, i.e. $0 \leq 1 - \alpha_s - \beta_s < 1$, equals natural mortality. The number of offspring x_{0t} is given as

$$x_{0t} = \sum_{s=1}^{n} \gamma_s x_{st}, \tag{1}$$

where $\gamma_s \geq 0$, $s = 1, \ldots, n$ denotes fecundity. Offspring are vulnerable to density, and let the twice continuously differentiable function φ denote the number of offspring that survive over their first period. This function satisfies:

$$\begin{aligned} &(\text{A1}): \quad \varphi(0) = 0, \\ &(\text{A2}): \quad 0 < \varphi'(0) \leq 1, \\ &(\text{A3}): \quad \varphi'' < 0, \\ &(\text{A4}): \quad \lim_{x_0 \to \infty} \varphi'(x_0) = 0. \end{aligned}$$

Let $h_{st} \geq 0$, $s = 1, \ldots, n$, $t = 0, 1, \ldots$. Denote the number of individuals harvested at the end of any period t. The size-structured population model is now given by (1) and

$$x_{1,t+1} = \varphi(x_{0t}) + \beta_1 x_{1t} - h_{1t}, \tag{2}$$

$$x_{s+1,t+1} = \alpha_s x_{st} + \beta_{s+1} x_{s+1,t} - h_{s+1,t} \quad s = 1, \ldots, n-1. \tag{3}$$

The (valuable) size of individuals in units of weight or volume is $f_s \geq 0$, $s = 1, \ldots, n$. The total harvested (valuable) biomass is $H_t \geq 0$ and

$$H_t = \sum_{s=1}^{n} f_s h_{st}. \tag{4}$$

The utility function U is twice continuously differentiable, increasing and concave in H. Let $b = (1 + r)^{-1}$ denote the discount factor and r the interest rate. The problem is to

$$\max_{\{h_{st}, s=1,\ldots,n, t=0,1,\ldots\}} \sum_{t=0}^{\infty} U(H_t) b^t,$$

subject to (1)–(4), the nonnegativity conditions $h_{st} \geq 0$, $x_{st} \geq 0$, $s = 1, \ldots, n$, $t = 0, 1, \ldots$ and the given initial size distribution x_{s0}, $s = 1, \ldots, n$.

If the recruitment function φ was linear and $\varphi' = 1$, the population model (1)–(3) would be the standard size-classified model in population ecology (Caswell 2001, p. 59). Given assumptions (A1)–(A4), the population model is a size-classified specification with density dependence in recruitment (Caswell 2001, p. 504). A concave increasing recruitment function is common in fishery models (Beverton and Holt 1957), but also possible for shade-tolerant trees. Assuming $\beta_s = 0$, $s = 1, \ldots, n-1$, the model yields the standard density-dependent age-structured model as a special case. Note that the harvest levels h_{st}, $s = 1, \ldots, n$ can be chosen independently.

This is natural e.g. with trees, but sometimes it is also possible in fisheries, if specific harvesting gear types exist or different cohorts can be found from different locations.

When the utility function is bounded and $b < 1$, the existence of optimal solutions follows from Theorem 4.6 in Stokey and Lucas (1989, p. 79). The Lagrangian and the Karush-Kuhn-Tucker (KKT) conditions are written as

$$L = \sum_{t=0}^{\infty} b^t \Bigg[U\left(\sum_{s=1}^{n} f_s h_{st}\right) + \lambda_{1t}\big(\varphi(x_{0t}) + \beta_1 x_{1t} - h_{1t} - x_{1,t+1}\big)$$
$$+ \sum_{s=1}^{n-1} \lambda_{s+1,t}(\alpha_s x_{st} + \beta_{s+1} x_{s+1,t} - h_{s+1,t} - x_{s+1,t+1}) \Bigg],$$

$$\frac{\partial L}{\partial h_{st}} b^{-t} = U' f_s - \lambda_{st} \le 0, \tag{5a}$$

$$h_{st} \ge 0, \tag{5b}$$

$$h_{st} \frac{\partial L}{\partial h_{st}} b^{-t} = 0, \quad s = 1, \ldots, n, \tag{5c}$$

$$\frac{\partial L}{\partial x_{s+1,t+1}} b^{-t} = b\lambda_{1,t+1}\varphi' \gamma_{s+1} + b\lambda_{s+1,t+1}\beta_{s+1}$$
$$+ b\lambda_{s+2,t+1}\alpha_{s+1} - \lambda_{s+1,t} \le 0, \quad s = 0, \ldots, n-1, \tag{6a}$$

$$x_{s+1,t+1} \ge 0, \tag{6b}$$

$$\frac{\partial L}{\partial x_{s+1,t+1}} b^{-t} x_{s+1,t+1} = 0, \quad s = 0, \ldots, n-1, \tag{6c}$$

where $t = 0, 1, \ldots$ and $\alpha_n \equiv 0$. Because the utility and recruitment functions are concave and other functions are linear, these conditions are sufficient for optimality given that the discrete analogue of the transversality condition in the Mangasarian (1966) sufficiency theorem is satisfied (Sydsaeter et al. 2008, p. 447). This holds for solutions converging to a steady state or stationary cycle. Numerical solutions will be computed by an interior point algorithm described and tested by Byrd et al. (1999) and Wächter and Biegler (2006).

3 Two Size Classes

3.1 Steady State Regimes

When $n = 2$, the optimality conditions are

$$U'(H_t) f_s - \lambda_{st} \le 0, \tag{7a}$$

$$h_{st} \geq 0, \tag{7b}$$

$$h_{st} \frac{b^{-t} \partial L}{\partial h_{st}} = 0, \quad s = 1, 2, \tag{7c}$$

$$b\lambda_{1,t+1}\varphi'(x_{0,t+1})\gamma_1 + b\lambda_{1,t+1}\beta_1 + b\lambda_{2,t+1}\alpha_1 - \lambda_{1t} \leq 0, \tag{8a}$$

$$x_{1,t+1} \geq 0, \tag{8b}$$

$$x_{1,t+1} \frac{b^{-t} \partial L}{\partial x_{1,t+1}} = 0, \tag{8c}$$

$$b\lambda_{1,t+1}\varphi'(x_{0,t+1})\gamma_2 + b\lambda_{2,t+1}\beta_2 - \lambda_{2t} \leq 0, \tag{9a}$$

$$x_{2,t+1} \geq 0, \tag{9b}$$

$$x_{2,t+1} \frac{b^{-t} \partial L}{\partial x_{2,t+1}} = 0, \tag{9c}$$

$$x_{1,t+1} = \varphi(\gamma_1 x_{1t} + \gamma_2 x_{2t}) + \beta_1 x_{1t} - h_{1t}, \tag{10}$$

$$x_{2,t+1} = \alpha_1 x_{1t} + \beta_2 x_{2t} - h_{2t}. \tag{11}$$

Given a steady state the time subscripts can be canceled. Equations (10) and (11) determine the admissible steady state levels of x_1 and x_2 and the possible solution regimes. By (10) any steady state satisfies $h_1 = \varphi - x_1(1 - \beta_1)$ implying the boundary: $h_1 = \varphi - x_1(1 - \beta_1) = 0$. At the (x_1, x_2)-plane $h_1 = 0$ defines x_2 as a convex function of x_1 (Fig. 1). From (11) any steady state satisfies $h_2 = \alpha_1 x_1 - x_2(1 - \beta_2)$. Denote the boundary by $h_2 = \alpha_1 x_1 - x_2(1 - \beta_2) = 0$. A carrying capacity equilibrium with $x_1 > 0$, $x_2 > 0$ exists, if the slope of $h_1 = 0$ is below the slope of $h_2 = 0$ at the origin of the (x_1, x_2)-plane, i.e.

$$\text{(A5)}: \quad \frac{1 - \beta_1 - \varphi'(0)\gamma_1}{\varphi'(0)\gamma_2} < \frac{\alpha_1}{1 - \beta_2}.$$

Denote this equilibrium by x_1^c, x_2^c. If $\partial h_1/\partial x_1 = \varphi'(0)\gamma_1 - 1 + \beta_1 > 0$, the slope of $h_1 = 0$ is negative at the origin and the function intersects the x_1-axis with some $\hat{x}_1 > 0$ (Fig. 1).

At admissible steady states, $h_1 \geq 0$ and $h_2 \geq 0$, i.e. the steady states exist between or on the lines $h_1 = 0$, $h_2 = 0$ but above or on the x_1-axis. When $\hat{x}_1 > 0$ the population can survive even if all individuals entering size class 2 are harvested (at the end of every period) implying $x_2 = 0$ and $\alpha_1 x_1 = h_2$. Note that the states $0 < x_1 < \hat{x}_1$, $x_2 = 0$ require $h_1 > 0$, $h_2 > 0$. In addition to the carrying capacity level (with no harvest) Fig. 1 shows the six different steady state solution regimes:

$$A: \quad x_1 > 0, \qquad x_2 > 0, \qquad h_1 > 0, \qquad h_2 = 0,$$

$$B: \quad x_1 > 0, \qquad x_2 > 0, \qquad h_1 = 0, \qquad h_2 > 0,$$

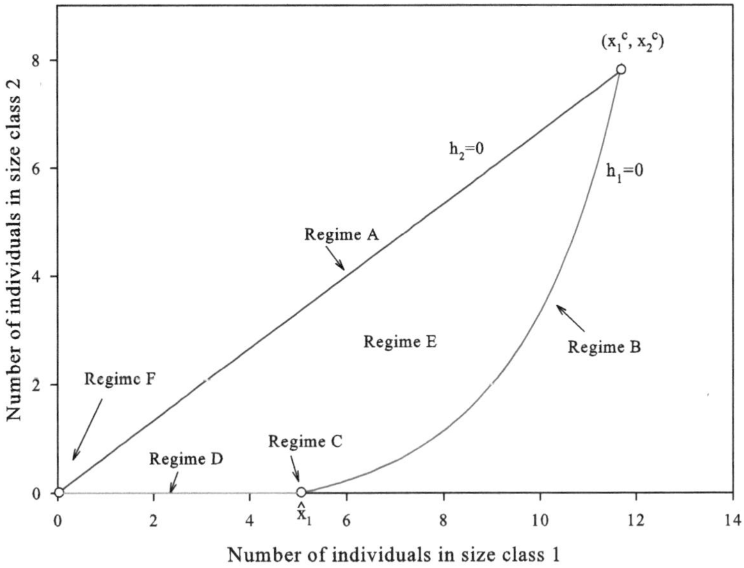

Fig. 1 Steady state regimes

$$
\begin{array}{lllll}
C: & x_1 > 0, & x_2 = 0, & h_1 = 0, & h_2 > 0, \\
D: & x_1 > 0, & x_2 = 0, & h_1 > 0, & h_2 > 0, \\
E: & x_1 > 0, & x_2 > 0, & h_1 > 0, & h_2 > 0, \\
F: & x_1 = 0, & x_2 = 0, & h_1 = 0, & h_2 = 0.
\end{array}
$$

Excluding regime F, the steady states satisfy $x_1 > 0$ and (8a) holds as an equality. In addition, either $h_1 = 0$ or $h_2 = 0$ or both $h_1 > 0$ and $h_2 > 0$. Define

$$\mu_1(x_1, x_2) \equiv f_1\{1 - b[\varphi'(\gamma_1 x_1 + \gamma_2 x_2)\gamma_1 + \beta_1]\} - b\alpha_1 f_2.$$

Applying (7a)–(7c) together with (8a) as an equality it follows that

$$\mu_1 > 0 \Rightarrow h_1 > 0,\ h_2 = 0, \tag{12a}$$

$$\mu_1 < 0 \Rightarrow h_1 = 0,\ h_2 > 0, \tag{12b}$$

$$\mu_1 = 0 \Rightarrow h_1 \geq 0,\ h_2 \geq 0, \tag{12c}$$

where in (12c) the case $h_1 = h_2 = 0$ is excluded. It is possible that μ_1 is positive or negative for all $x_1 \geq 0$, $x_2 \geq 0$ or that $\mu_1 = 0$ defines a decreasing straight line in the (x_1, x_2)-plane with slope $-\gamma_1/\gamma_2$ as shown in Fig. 2a. Above the $\mu_1 = 0$ curve it holds that $\mu_1 > 0$ and $h_1 > 0$ and below the curve that $\mu_1 < 0$ and $h_2 > 0$.

Given $x_1 > 0$ and $x_2 > 0$ both (8a) and (9a) hold as equalities and imply

$$\mu_2(x_1, x_2) = (1 - b\beta_1) - b\varphi'(\gamma_1 x_1 + \gamma_2 x_2)\left(\gamma_1 + \frac{b\gamma_2\alpha_1}{1 - b\beta_2}\right) = 0, \tag{13}$$

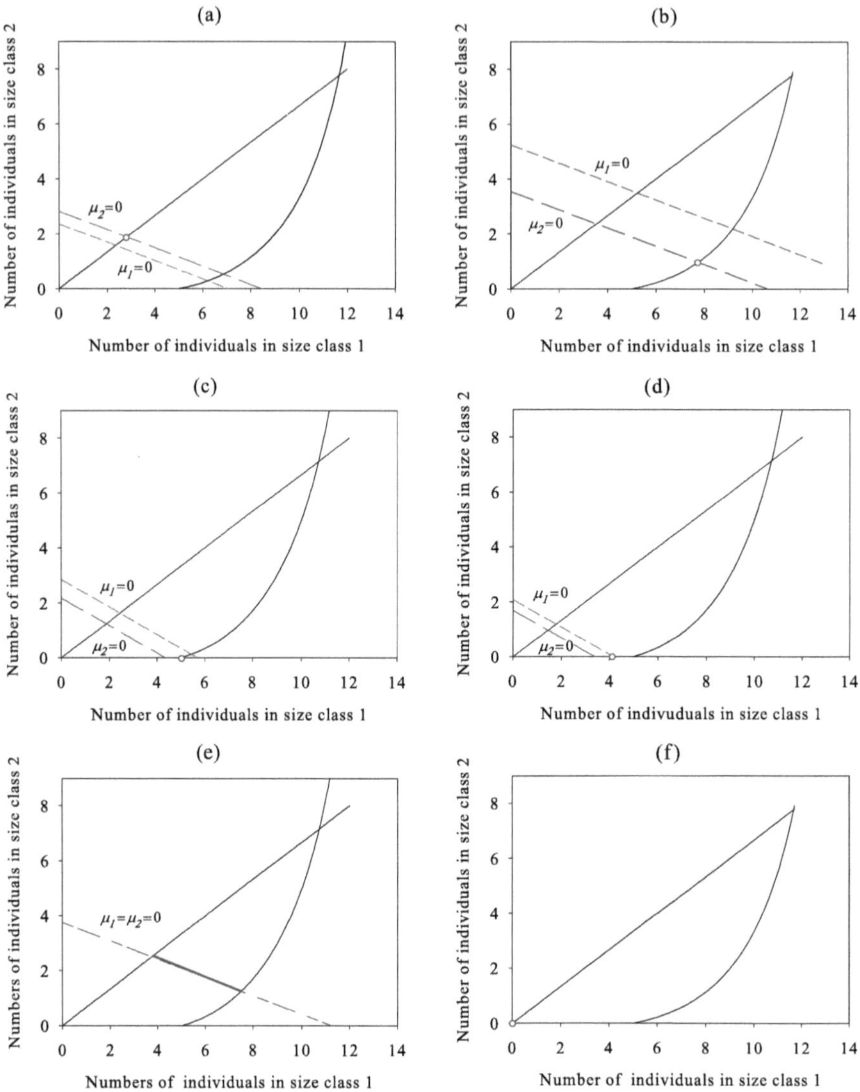

Fig. 2 Optimality of steady state regimes

while for steady states with $x_1 > 0$, $x_2 = 0$ it must hold that

$$\mu_2 > 0. \tag{14}$$

Differentiation shows that $\mu_2(x_1, x_2) = 0$ is a straight line in the (x_1, x_2)-plane with slope $-\gamma_1/\gamma_2$. Above (below) the line $\mu_2 > 0$ ($\mu_2 < 0$). Assumptions (A3)–(A5) imply that the locus of $\mu_2(x_1, x_2) = 0$ always exists below the population's carrying capacity level.

Any steady state with $x_1 > 0$, $x_2 > 0$ must satisfy (13), i.e. $\mu_2 = 0$. Given the locus of $\mu_1 = 0$ below the locus of $\mu_2 = 0$, (12a) implies that $h_1 > 0$ and $h_2 = 0$, i.e. the optimal steady state exists in regime *A* and on the line $h_2 = 0$ (Fig. 2a). Assume that the locus of $\mu_1 = 0$ exists above the locus of $\mu_2 = 0$ and that the latter exists above point $(\hat{x}_1, 0)$. By (12b) the optimal steady state satisfies $h_1 = 0$ and $h_2 > 0$, i.e. the optimal steady state exists in regime *B* and on line $h_1 = 0$ (Fig. 2b).

If $\hat{x}_1 > 0$, the optimal steady state may also exist in regimes *C* or *D*. Given $\mu_1(\hat{x}_1, 0) < 0$ and $\mu_2(\hat{x}_1, 0) > 0$, conditions (12b) and (14) imply that the optimal steady state satisfies $h_1 = 0$, $h_2 > 0$, $x_1 = \hat{x}_1 > 0$ and $x_2 = 0$. This is regime *C* in Fig. 2c. The other possibility is that $\mu_1(\hat{x}_1, 0) > 0$ and $\mu_2(\hat{x}_1, 0) > 0$, i.e. regime *D* (Fig. 2d) and $h_1 > 0$, $h_2 > 0$. If the locus of $\mu_1 = 0$ coincides with the locus of $\mu_2 = 0$, the steady states exist in regime *E*, which is a continuum between curves $h_2 = 0$ and $h_1 = 0$ (Fig. 2e). The findings so far can be summarized as

Proposition 1 *Given* (A1)–(A5) *and*

1. $\mu_2(0,0) < 0$, $\mu_1(x_1, x_2)|_{\mu_2(x_1,x_2)=0} > 0$, *A is optimal,*
2. $\mu_2(\hat{x}_1, 0) < 0$, $\mu_1(x_1, x_2)|_{\mu_2(x_1,x_2)=0} < 0$, *B is optimal,*
3. $\hat{x}_1 > 0$, $\mu_2(\hat{x}_1, 0) > 0$, $\mu_1(\hat{x}_1, 0) \le 0$, *C is optimal,*
4. $\hat{x}_1 > 0$, $\mu_1(\hat{x}_1, 0) > 0$, $\mu_1(0,0) < 0$, $\mu_2(x_1, x_2)|_{\mu_1(x_1,x_2)=0} > 0$, *D is optimal,*
5. $\mu_2(0,0) < 0$, $\mu_1(x_1, x_2)|_{\mu_2(x_1,x_2)=0} = 0$, *E is optimal.*

3.2 *Interpretations*

Regime *A* ($x_1 > 0, x_2 > 0, h_1 > 0, h_2 = 0$) satisfies (from (8a)–(8c)–(11))

$$\frac{f_1[\varphi'(x_0)\gamma_1 - (1-\beta_1)] + \alpha_1 f_2}{f_1} < r, \tag{15a}$$

$$\varphi'(x_0)\left(\gamma_1 + \frac{b\alpha_1\gamma_2}{1 - b\beta_2}\right) - (1-\beta_1) = r, \tag{15b}$$

$$h_1 = \varphi(x_0) - x_1(1-\beta_1), \tag{15c}$$

where $x_0 = \gamma_1 x_1 + \gamma_2\alpha_1 x_1/(1-\beta_2)$. The derivative of the LHS of (15b) w.r.t. x_1 is negative, i.e. the steady state is unique. Given the level of x_1, Eq. (15c) determines h_1. The LHS of (15a) is the marginal rate of return when decreasing h_1 in order to harvest one unit of x_2. It is lower than the interest rate, i.e. it is optimal to harvest class 1 instead of class 2. In addition to the return in terms of $h_2(\alpha_1 f_2)$, decreasing h_1 increases h_1 of the next period by increasing the recruitment net of the fraction that dies or moves to class 2. Equation (15c) defines a sustainable harvest and is concave in x_1 and zero with $x_1 = 0$ or x_1 sufficiently large. Maximizing h_1 using (15c) results in the LHS of (15b) if $b = 1$. Increasing x_1 marginally increases the net output of x_1 by increasing the recruitment net of the fraction not remaining in class 1, i.e. by $\varphi'\gamma_1 - (1-\beta_1)$. In addition, an increase in x_1 causes an increase in x_2, which after a delay, increases the level of x_1 and h_1. Delays in term $\frac{b\alpha_1\gamma_2}{1-b\beta_2}$ in the LHS of (15b) are discounted (over infinite periods). Thus by (15b) the rate of interest equals the present value of the marginal surplus production of the size-structured population. Differentiation of (15b) shows that the levels of x_0, x_1, x_2 and h_1 decrease with the discount rate.

The inequality sign in (15a) is reversed in regime *B*, and it is thus worth saving all individuals in class 1 and only harvesting class 2. The unique x_1 and x_2 levels are given by $x_1 = \varphi(x_0) + \beta_1 x_1$ and (15b), and the harvest by $h_2 = \alpha_1 x_1 - x_2(1-\beta_2)$. When regime *C* is optimal, the LHS of (15a) is higher and the LHS of (15b) is lower than the interest rate and $x_0 = \gamma_1 \hat{x}_1$. Although it is optimal to save all individuals in class 1 and harvest them from class 2, the discounted marginal surplus production is too low for retaining class 2 for reproduction. Finally, in regime *D* the latter condition still holds, but the marginal rate of return of harvesting all individuals from class 2 is too low, and it is therefore optimal to harvest a fraction of class 1. Optimal x_1 is given by the equality of the LHS of (15a) and the discount rate and the harvests by $x_1 = \varphi(x_0) + \beta_1 x_1 - h_1$ and $h_2 = \alpha_1 x_1$.

When the optimal steady state is in regime *C*, decreasing the discount factor decreases the LHS of (15a) and (15b), implying that an increase in the interest rate may cause a switch from regime *C* to regime *D*. Decreasing the discount factor in regime *D* decreases the steady state level of x_1, implying lower x_1, h_1 and h_2. These comparative static results are intuitive: if an increase in the interest rate has an effect, it causes a lower steady-state population size and the harvesting of individuals from (size or age) class 1 rather than class 2.

Equation (15b) holds for regimes *A* and *B*. It yields x_0 as an increasing function of b, i.e. $x_0 = x_0(b)$, $x_0'(b) > 0$. Thus, the lower the interest rate, the lower is the steady state number of recruits in these regimes. Equation (15a) and its derivative w.r.t. b can be written as

$$\begin{aligned} y &\equiv f_1\big(b\varphi'(x_0)\gamma_1 + b\beta_1 - 1\big) + bf_2\alpha_1, \\ \partial y/\partial b &= f_1\big(\varphi'\gamma_1 + b\varphi'' x_0'\gamma_1 + \beta_1\big) + f_2\alpha_1 \\ &= \frac{f_1\gamma_1[\alpha_1\gamma_2(b^2\beta_1\beta_2 - 1) - \beta_1\gamma_1(b\beta_2 - 1)^2]}{[\alpha_1 b\gamma_2 - \gamma_1(b\beta_2 - 1)]^2} + \alpha_1 f_2 + \beta_1 f_1, \end{aligned}$$

where the last line is obtained by (15b). If $\gamma_1 = 0$ or low, $\partial y/\partial b > 0$ implies that increasing the interest rate (decreasing the discount factor) may result in a switch from regime *B* to regime *A*, but not *vice versa*. Thus, in these cases a higher interest rate implies harvesting from class 1, where individuals are younger and/or smaller. However, when $\gamma_1 > 0$, the sign of $\partial y/\partial b$ is indeterminate. If it is negative, a decrease in the discount factor may cause a switch from regime *A* to regime *B*, i.e. a switch to harvesting older or larger individuals under a higher interest rate. An example is shown in Fig. 3. When the interest rate is below $r = 0.111$, the LHS of (15a) is negative, regime *A* and harvesting only from class 1 are optimal. Increasing the interest rate above this level implies that (15a) becomes positive, i.e. regime *B* is optimal and harvesting is only optimal from class 2.

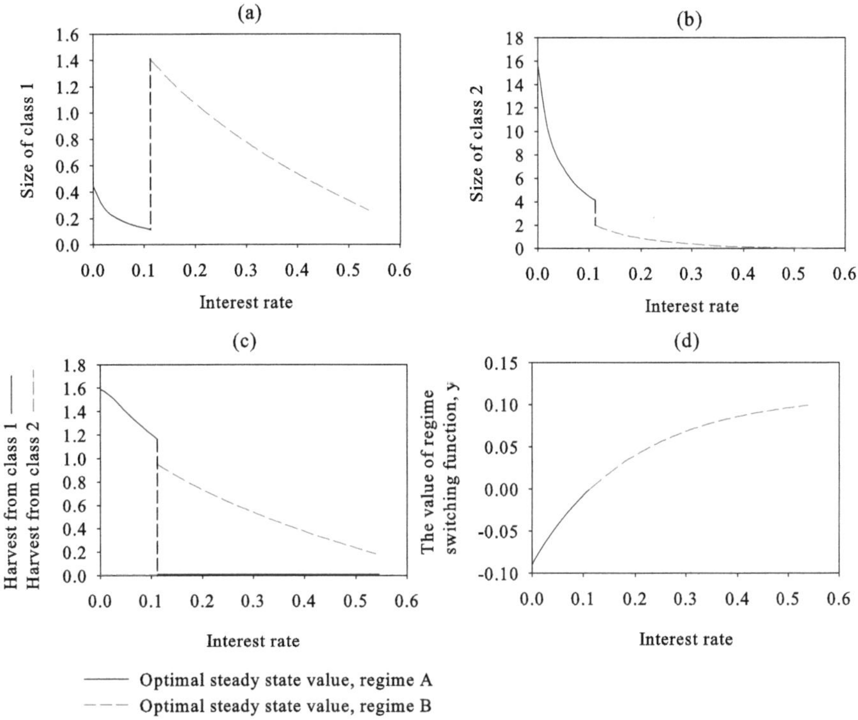

Fig. 3 Optimal steady state and the case where higher interest rate implies a switch of harvest from class 1 to class 2. Note: $\varphi = x_0/(1+0.4x_0)$, $\gamma_1 = 1$, $\gamma_2 = 0.6$, $\alpha = 0.7$, $\beta_1 = 0.1$, $\beta_2 = 0.98$, $f_1 = 1$, $f_2 = 1.1$

The intuition is that when a higher interest rate decreases the number of offspring ($x_0'(b) > 0$), the marginal rate of return from saving class 1 individuals to class 2 increases and this effect may dominate the "normal" effects of a higher discount rate.

The next question concerns the possible non-existence of steady states with strictly positive population size and harvesting:

Proposition 2 *Regime F* ($x_i = h_i = 0, i = 1, 2$) *is the optimal steady state if and only if* $\mu_2(0,0) \geq 0$ *and* (i) $1 - \beta_1 - \varphi'(0)\gamma_1 > 0$ *or* (ii) $\mu_1(0,0) \geq 0$.

Proof Appendix 1. □

Condition $\mu_2(0,0) \geq 0$ is equivalent to the LHS of (15b) being lower than the interest rate, even when both x_1 and x_2 are zero. Thus, $x_2 > 0$ cannot be optimal. Yet this is not sufficient for $x_1 = 0$ and regime F to be optimal. It must either hold that regimes $x_1 > 0$, $x_2 = 0$ are not admissible ($\hat{x}_1 = 0$), or if admissible, they must not be optimal. The former case follows if $1 - \beta_1 - \varphi'(0)\gamma_1 > 0$, and the latter if $\mu_1(0,0) \geq 0$, i.e. if the LHS of (15a) is always lower than the interest rate. Note that

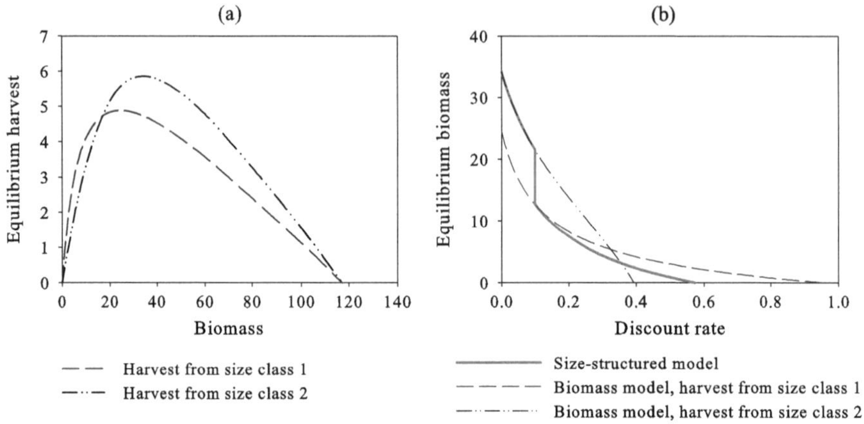

Fig. 4 Comparison of steady states between the size-structured and the biomass model. Note: $\varphi = 0.9\gamma_2 x_2/(1+0.1\gamma_2 x_2)$, $\gamma_1 = 0$, $\gamma_2 = 2$, $\alpha_1 = 0.4$, $\beta_1 = 0.5$, $\beta_2 = 0.4$, $f_1 = 1$, $f_2 = 1.5$

in the case where regime D is always non-optimal (e.g. if $f_1 = 0$), the solution will remain in regime C even if $r \to \infty$.

The size-structured population model can be used to develop an equilibrium biomass model. Assume that the population biomass is given as $X = f_1 x_1 + f_2 x_2$. It is not possible to optimize harvesting between age classes within the biomass framework. Firstly assume that only size class 2 is harvested, implying $h_2 = \alpha_1 x_1 - x_2(1-\beta_2)$ and $\varphi(\gamma_1 x_1 + \gamma_2 x_2) - x_1(1-\beta_1) = 0$. The last equation and the biomass equation can be used to obtain both x_1 and x_2 as functions of population biomass X. The remaining function for h_2 yields harvest as a function of population biomass. Similarly, if only size class 1 is harvested, $h_1 = \varphi(\gamma_1 x_1 + \gamma_2 x_2) - x_1(1-\beta_1)$ and $\alpha_1 x_1 - x_2(1-\beta_2) = 0$, it is once again possible to obtain the equilibrium harvest h_1 as a function of population biomass X. Denote these equilibrium harvests as $h_s = F_s(X)$, $s = 1, 2$. The optimal steady state condition within the biomass model is $F_s'(X) = r$, $s = 1, 2$. Fig. 4 shows numerical examples of these steady states and the optimal steady states obtained by the size-structured model. The maximum sustainable harvest is obtained by harvesting size class 2 only (Fig. 4a). Thus, the steady states of the size-class and biomass models are equal when $r = 0$ and $h_1 = 0$, $h_2 > 0$ in the biomass model (Fig. 4b). When $0 < r \leq 0.1$, the equilibrium biomass in the size-class model is slightly lower than in the biomass model. When $r = 0.1$, the equilibrium continuum is optimal in the size-class model (regime E) and when $0.1 < r < 0.577$, it is optimal to harvest only from size class 1 (regime A). As shown, the biomass model based on $h_1 = 0$, $h_2 > 0$ yields "optimal extinction" when $r > 0.39$ while the critical discount rate in the size-structured model is 0.578. The comparison of the two models becomes different if $h_1 > 0$, $h_2 = 0$ in the biomass model. In this case the biomass model yields a lower MSY steady state, but the biomass becomes higher when $r > 0.1$. In addition, the "optimal extinction discount rate" in the biomass model is now higher compared to the size-structured model.

3.3 Discussion

In their main theorems Reed (1980) and Getz (1980) state that assuming the maximization of sustainable yield, either a single age class or two age classes are harvested. In the latter case, the harvest is partial for the younger class and total for the older class. Reed's theorem (1980) explicitly states that in the former case where only one age class is harvested, the harvest may be partial or total. Here only a single age class is harvested in regimes *A*, *B* and *C*, with the harvest being partial in regimes *A* and *B* and total in regime *C*. Regime *D* is a case where two classes are harvested, the younger one partially and the older one totally. However, in addition to these regimes, regime *E* has two partially-harvested classes. This is not covered in the theorems by Getz (1980) and Reed (1980). They study MSY steady states and the age-structured model, i.e. the special case $b = 1$, $\beta_1 = \beta_2 = 0$. However, this difference is not essential, and it is possible to obtain regime *E* as an optimal solution in an age-structured model with zero interest rate. For example, if

$$\varphi = Ax_0/(1 + Bx_0), \quad A = 1, \ B = 1/5,$$

$$\gamma_1 = 1, \ \gamma_2 = 2, \ \alpha_1 = 7/10, \ f_2 = 5f_1/4, \ b = 1,$$

both $\mu_1 = 0$ and $\mu_2 = 0$ imply $x_2 = \sqrt{10} - 5/2 - x_1/10$, i.e. regime *E* is optimal.

Clark (1990) analyses the famous dynamic pool or cohort model by Beverton and Holt (1957). Their model coincides with the age-structured special case studied here assuming interest rate is zero, recruitment is an exogenous constant and the analysis is restricted to steady states. Clark presents the result that given non-selective harvesting gear, pulse fishing is optimal, instead of continuous harvesting as suggested by Beverton and Holt (1957). The period length between pulses is solved by an optimal rotation solution for the Faustmann (1849) formula. In his review, Wilen (1985) suggests this to be incorrect, because nothing similar to the land area constraint exists in the given problem and thus the problem must be viewed as a Fisherian "single-shot" model. However, the argument by Wilen (1985) appears to be incorrect: the assumption of non-selective fishing gear has exactly the same effect as the land area constraint in the rotation model (Tahvonen 2011b).

A similar question related to the connection with the Faustmann model can be asked here. Assuming exogenous and constant recruitment (and the age class structure), Eq. (15a) can be written as $f_1 - bf_2\alpha_1 \geq 0$. This is simply the optimality condition for harvesting age class 1 in a discrete-time two-period Fisherian "single shot" optimal timing problem. Given endogenous recruitment, (15a) can be written as $f_1(1 - b\varphi'\gamma_1) - bf_2\alpha_1 \geq 0$. As adding endogenous recruitment decreases the LHS of this equation (increases the LHS of (15a)), it implies that cohorts should be harvested after they reach their maximum (discounted) biomass. This result is a consequence of the fact that saving individuals to be harvested from older age classes increases recruitment. Theorem 2 in Reed (1980) states the same result with zero interest rate. The land area constraint in the Faustmann model implies that cohorts should be harvested sooner than at the age of maximum (discounted) biomass. Together this shows that the model studied here does not include any Faustmann-type rotation structure.

Reed (1980) expects that introducing discounting would decrease the age of the harvested cohort. Our result shows that due to endogenous recruitment this need not be the case (Fig. 3).

4 Stability of Steady States

The steady state stability analysis calls for an analysis of the four non-linear difference equations (conditions (7a)–(7c)–(11)). However, some special cases where the optimal solution can be found in closed form exist. Denote the steady state harvests in regimes A and B by h_1 and h_2, respectively.

Proposition 3 *Given* (A1)–(A5), $\mu_2(\hat{x}_1, 0) < 0, \gamma_1 = \gamma_2, \alpha_1 + \beta_1 = \beta_2$ *and* $\mu_2(x_{10}, x_{20}) = 0$, *the optimal solution is* (a) $h_{1t} = h_1, h_2 = 0, t = 0, 1, \ldots,$ *if* $\mu_1(x_{10}, x_{20}) > 0$, *and* (b) $h_{2t} = h_2, h_{1t} = 0, t = 0, 1, \ldots,$ *if* $\mu_1(x_{10}, x_{20}) < 0$.

Proof Case (a): Recall that the slope of $\mu_2(x_1, x_2) = 0$ in the (x_1, x_2)-plane is $-\gamma_2/\gamma_1$. Denote this linear function by $x_2 = \eta - x_1$, where η is a constant and positive by $\mu_2(\hat{x}_1, 0) < 0$. When $\mu_1(x_{10}, x_{20}) > 0$, the steady state satisfies $x_1 = \varphi + \beta_1 x_1 - h_1, x_2 = \alpha_1 x_1 + \beta_2 x_2$ and $x_2 = \eta - x_1$. This yields $h_1 = \varphi - \eta(1 - \beta_2)$. Given $\mu_2(x_{10}, x_{20}) = 0$, it follows that $x_{1,t+1} = \varphi(\eta) + \beta_1 x_{1t} - [\varphi(\eta) - \eta(1 - \beta_2)]$, and $x_{2,t+1} = \alpha_1 x_{1t} + \beta_2 x_{2t}$ for $t = 0$. Thus $x_{11} + x_{21} = \eta$ and then by induction $x_{1,t+1} + x_{2,t+1} = \eta$ for $t = 1, 2, \ldots$. In addition, $x_{1t} \to \varphi(\eta)(1 - \beta_2)/(1 - \beta_1)$ and $x_{2t} \to \alpha\eta(1 - \beta_1)$, as $t \to \infty$ since $0 < \beta_1 < 1$. The constants $\lambda_1 = U' f_1$ and $\lambda_2 = f_1 U'(1 - b\varphi'(\eta) - b\beta_1)/(b\alpha)$ solve (7a)–(7c)–(9a)–(9c) by $\mu_2(x_{10}, x_{20}) = 0$ and $\mu_1(x_{10}, x_{20}) > 0$, implying that the solution is optimal. Case (b) is analogous to case (a). □

In the case of Proposition 3, the size classes are symmetric in the sense that fecundities and natural mortalities coincide, although the usable size of harvested individuals may differ. The symmetry property, together with specific initial states, implies that harvest and recruitment levels are constant over time. Numerical examples of these solutions are given in Fig. 5 by the solid straight lines.

When applying conditions (7a)–(7c)–(11) more generally for interior solutions, solutions in regime A are determined by the system:

$$x_{1,t+1} = \varphi(\gamma_1 x_{1t} + \gamma_2 x_{2t}) + \beta_1 x_{1t} - h_1(\lambda_{1t}), \tag{16}$$

$$x_{2,t+1} = \alpha_1 x_{1t} + \beta_2 x_{2t}, \tag{17}$$

$$\lambda_{1,t+1} = \frac{\lambda_{2t}\alpha_1 - \lambda_{1t}\beta_2}{b\varphi'(x_{1,t+1}, x_{2,t+1}, \lambda_{1,t+1})(\alpha_1\gamma_2 - \beta_2\gamma_1) - b\beta_1\beta_2}, \tag{18}$$

$$\lambda_{2,t+1} = \frac{\varphi'(x_{1,t+1}, x_{2,t+1}, \lambda_{1,t+1})(\lambda_{1t}\gamma_2 - \lambda_{2t}\gamma_1) - \lambda_{2t}\beta_1}{b\varphi'(x_{1,t+1}, x_{2,t+1}, \lambda_{1,t+1})(\alpha_1\gamma_2 - \beta_2\gamma_1) - b\beta_1\beta_2}, \tag{19}$$

$$\begin{aligned}&\varphi'(x_{1,t+1}, x_{2,t+1}, \lambda_{1,t+1})\\ &\quad = \varphi'\big(\gamma_1\big[\varphi(\gamma_1 x_{1t} + \gamma_2 x_{2t}) + \beta_1 x_{1t} - h_1(\lambda_{1t})\big] + \gamma_2(\alpha_1 x_{1t} + \beta_2 x_{2t})\big),\end{aligned} \tag{20}$$

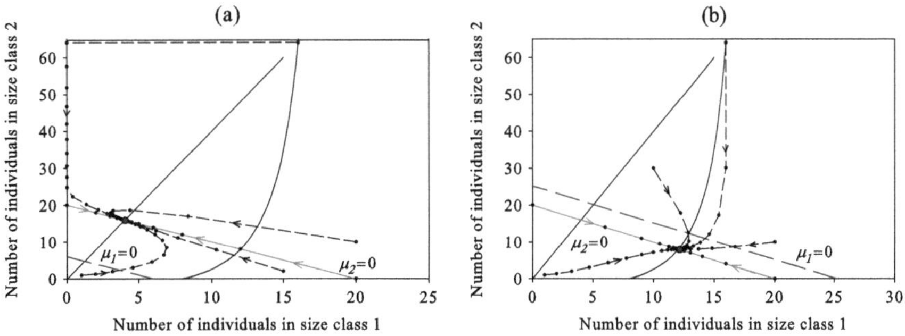

Fig. 5 Optimal solutions under the assumptions in Propositions 3 and 4. Note: $\varphi = \frac{0.9x_{0t}}{1+0.1x_{0t}}$, $x_{0t} = x_{1t} + x_{2t}$, $b = 1$, $U = H_t^{0.9}$, $\alpha_1 = 0.4$, $\beta_1 = 0.5$, $\beta_2 = 0.9$, $f_1 = 1$, $f_2 = 0.5$ in (**a**), $f_2 = 2$ in (**b**). Equilibrium in (**a**): $x_1 = 4$, $x_2 = 16$, $h_1 = 4$, $h_2 = 0$, and in (**b**): $x_1 = 12$, $x_2 = 8$, $h_2 = 4$. Roots of the characteristic equation in (**a**): $r_1 = \frac{1}{2}$, $r_2 = 2$, $r_3 = \frac{3}{5}$, $r_4 = \frac{5}{3}$

$$h_1'(\lambda_{1t}) = 1/\big(U'' f_1\big), \tag{21}$$

and in regime B by (18), (19) and

$$x_{1,t+1} = \varphi(\gamma_1 x_{1t} + \gamma_2 x_{2t}) + \beta_1 x_{1t}, \tag{22}$$

$$x_{2,t+1} = \alpha_1 x_{1t} + \beta_2 x_{2t} - h_{2t}(\lambda_{2t}), \tag{23}$$

$$\begin{aligned}&\varphi'(x_{1,t+1}, x_{2,t+1}, \lambda_{2,t+1})\\ &\quad = \varphi'\big(\gamma_1\big[\varphi(\gamma_1 x_{1t} + \gamma_2 x_{2t}) + \beta_1 x_{1t}\big] + \gamma_2\big(\alpha_1 x_{1t} + \beta_2 x_{2t} - h_{2t}(\lambda_{2t})\big)\big),\end{aligned} \tag{24}$$

$$h_2'(\lambda_{2t}) = 1/\big(U'' f_2\big). \tag{25}$$

Proposition 4 *Given the conditions of Proposition* 3 *and* $U'' < 0$, *but excluding the initial state restrictions, the optimal equilibria in regimes A and B are local saddle points, i.e. two characteristic roots have absolute values below* 1 *and two above* 1.

Proof For regimes *A* and *B*, computing the Jacobian matrix and the associated characteristic equation yields the fourth order polynomial

$$\Omega_s(r) = \frac{(\beta_1 - r)[Q_s r + (1-r)(br-1)](b\beta_1 r - 1)}{b^2\beta_1},$$

where $Q_s = \lambda_s b^2 \varphi'' h'$, $s = 1, 2$. The characteristic roots are: $r_1 = \beta_1$, $r_2 = 1/(b\beta_1)$,

$$r_{2,3} = \Big[\sqrt{Q_s^2 + 2Q_s(b+1) + b^2 - 2b + 1} \pm (Q_s + b + 1)\Big]/(2b), \quad s = 1, 2.$$

Obviously $0 < r_1 < 1$ and $1 < r_2$. When $Q_s = 0$, $s = 1, 2$, the value of r_{s3} $(= \sqrt{\bullet} + Q_s + b + 1)$ equals b^{-1} and the value of r_{s4} $(= \sqrt{\bullet} - Q_s - b - 1)$ equals 1. Since $\partial r_{s3}/\partial Q_s > 0$ and $\partial r_{s4}/\partial Q_s < 0$ and $\lim_{Q_s \to \infty} r_{s4} = 0$ for both $s = 1, 2$ the steady

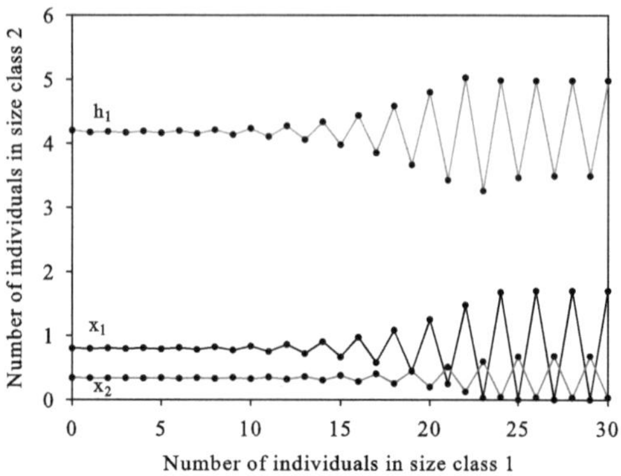

Fig. 6 Optimal stationary cycle. Note: $\varphi = \frac{0.9x_{0t}}{1+0.1x_{0t}}$, $x_{0t} = 5x_{1t} + 20x_{2t}$, $b = 0.45$, $U = H_t^{0.99}$, $\alpha_1 = 0.4$, $\beta_1 = 0.4$, $\beta_2 = 0.05$, $f_1 = 5$, $f_2 = 1$. Equilibrium: $x_1 = 0.80$, $x_2 = 0.34$, $h_1 = 4.2$, $h_2 = 0$. Roots: $r_1 = -1.70$, $r_2 = -1.31$, $r_3 = 0.01$, $r_4 = 206.55$

states are saddle points where two roots have absolute values above 1 and two roots have absolute values below 1. □

Examples under the assumptions in Proposition 4 are shown by black dashed lines in Fig. 5. The next result does not require symmetry, but is restricted to zero discounting and cases where the Jacobian matrix for system (16)–(21) can be evaluated at the steady state.

Proposition 5 *Given* (A1)–(A5), $b = 1$, $U'' < 0$ *and* $\tau_2 \equiv (\beta_2 - 1)(\alpha_1\gamma_2 - \beta_2\gamma_1) + \alpha_1\gamma_2\beta_1 \neq 0$, *the steady state in regime A is a (local) saddle point, i.e. two characteristic roots have absolute values below* 1 *and two above* 1.

Proof Appendix 2. □

The zero discount rate assumption in Proposition 5 is far from necessary for the steady state properties stated in the proposition. However, when the discount rate is high enough, the stability properties may change and three of the roots may become unstable.

The cyclical solution is depicted in Fig. 6. The initial state is close to the steady state, but because of a small deviation the solution diverges and approaches a stationary cycle.

When $\tau_2 = 0$, the denominators of (18)–(19) approach zero when the solution approaches the steady states. Since the Jacobian matrix becomes indeterminate at the steady state, the stability properties cannot be analysed by the linearization method. However, numerical computation suggests that optimal solutions are perfectly continuous around $\tau_2 = 0$. With the age-structured model (i.e. $\beta_1 = 0$), this special case implies that the difference equation system collapses into three equations, and the steady-state stability properties can be analysed using the ordinary linearization method:

Proposition 6 *Assume* (A1)–(A5), $b = 1$, $U'' < 0$ *and* $\beta_1 = 0$. *If* $\alpha_1\gamma_2 - \beta_2\gamma_1 \neq 0$, *the steady state in regime A is a local saddle point and two of the characteristic* *roots have absolute values above* 1 *and two have values below* 1. *If* $\alpha_1\gamma_2 - \beta_2\gamma_1 = 0$, *optimal solutions are determined by three difference equations with a steady state, where the absolute values of two characteristic roots are below* 1 *and one above* 1.

Proof When $\alpha_1\gamma_2 - \beta_2\gamma_1 \neq 0$, the structure of characteristic roots follows from Proposition 5. When $\alpha_1\gamma_2 - \beta_2\gamma_1 = 0$, optimal solutions are defined by (20)–(21) and

$$x_{1,t+1} = \varphi(\gamma_1 x_{1t} + \gamma_2 x_{2t}) - h_1(\lambda_{1t}), \qquad x_{2,t+1} = \alpha_1 x_{1t} + \beta_2 x_{2t},$$

$$\lambda_{1,t+1} = \frac{\lambda_{1t}}{\beta_2 + \varphi'(x_{1t}, x_{2t}, \lambda_{1t})\gamma_1}, \qquad \lambda_{2t} = \frac{\lambda_{1t}\gamma_2}{\gamma_1}.$$

The Jacobian matrix of this system evaluated at the steady state has two characteristic roots with absolute values below 1 and one above 1. □

Propositions 5 and 6 assume zero discounting. However, ordinary saddle point properties can be shown to hold over a wide range of positive discount rates:

Proposition 7 *Given* (A1)–(A5), $U'' < 0$, $\gamma_1 = 0$ *and* $\frac{1}{\beta_1+\beta_2+1} \leq b < \frac{1}{\beta_1+\beta_2}$, *the steady state in regime A is a* (*local*) *saddle point, i.e. two characteristic roots have absolute values below* 1 *and two above* 1.

Proof Appendix 3. □

Propositions 5, 6 and 7 show stability properties for the steady states of regime *A*. Identical propositions can also be given for the stability properties of regime *B* steady states, with only some minor changes in the proofs. The stability properties of regime *C* and *D* are given as:

Proposition 8 *Given* (A1)–(A5) *and* $U'' < 0$, *the steady state in regime C is locally stable and the steady state in regime D is a local saddle point.*

Proof Appendix 4. □

Note that given the steady state solution exists in these regimes, the qualitative stability properties are independent of the discount rate. In regime *C* the optimal harvest is simply $h_{1t} = \alpha_1 x_{1t}$, implying that the solution is unique despite the steady state being locally stable. Given the solution is already in this regime at the initial state, it is independent of changes in the discount rate and utility function.

Earlier studies of this model have not analysed the steady-state stability properties. Reed (1980) writes that such analysis appears to be difficult, but if initial numbers of individuals exceed the steady state levels in all age classes, a constant escapement policy where all age classes are harvested down to their steady state values "would certainly be an acceptable policy". However, even with a linear objective function such a solution is, in general, non-optimal. The reason is simply that if it is optimal to only harvest one age class at a steady state, a similar regime is optimal also in the vicinity of the steady state. Another simple example of the non-optimality of constant escapement is regime C, where the optimal solution is $h_1 = 0$, $h_2 = \alpha_1 x_{1t}$ in the vicinity of the steady state, implying that the steady state is approached asymptotically.

5 Any Number of Size Classes

A natural extension of the two classes specification is to assume that the optimal steady state harvest is targeted to some size class m such that $1 < m$ and $m + 1 < n$, i.e. the harvest targets interior size classes. Thus, suppose: $x_s > 0$, $s = 1, \ldots, m$, $x_s = 0$, $s = m + 1, \ldots, n$, and either $h_s = 0$, $s = 1, \ldots, m$, $h_{m+1} = \alpha_m x_m > 0$, $h_s = 0, s = m+1, \ldots, n$ or $h_m = \alpha_{m-1} x_{m-1} - (1 - \beta_m) x_m > 0$, $h_{m+1} = \alpha_m x_m > 0$. The former case is an extension of regime C and the latter an extension of regime D. The steady state satisfies

$$x_s = \mu_s x_{s-1}, \quad \mu_s = \alpha_{s-1}/(1-\beta_s),\ s = 2, \ldots, m-1,$$

$$x_m = \alpha_{m-1} x_{m-1}/(1-\beta_m) - h_m/(1-\beta_m) = \mu_m x_{m-1} - h_m/(1-\beta_m),$$

$$x_s = l_s x_1, \quad s = 1, \ldots, m-1, \qquad x_m = l_m x_1 - h_m/(1-\beta_m),$$

$$x_0 = x_1 \sum_{s=1}^{m} \gamma_s l_s - \gamma_m h_m/(1-\beta_m) \equiv R_0^m x_1 - \gamma_m h_m/(1-\beta_m),$$

where

$$l_s = \prod_{i=2}^{s} \mu_i, \quad s = 1, \ldots, m, \qquad l_1 = \prod_{i=2}^{1} \mu_i \equiv 1 \quad \text{and} \quad R_0^m = \sum_{s=1}^{m} \gamma_s l_s,$$

implying that

$$x_1 = \varphi\big(R_0^m x_1 - \gamma_m h_m/(1-\beta_m)\big) + \beta_1 x_1,$$

which directly yields equilibrium x_1 if $h_m = 0$. Note that R_0^m equals the expected steady state number of offspring individuals produce over their lifetime, and is called the net reproductive value (Fisher 1930; Samuelson 1977).

At the steady state $x_s > 0$, $s = 1, \ldots, m$ and the KKT conditions (6a)–(6c) can be written as

$$\lambda_{s+1} + \lambda_s (b\beta_s - 1)/(b\alpha_s) = -\lambda_1 b\varphi' \gamma_s/(b\alpha_s), \quad s = 1, \ldots, m,$$

where $\lambda_{m+1} = U' f_{m+1}$ and, if $h_m > 0$, in addition $\lambda_m \geq U' f_m$. Equation (19) is a linear non-autonomous difference equation for λ_s with $\lambda_{m+1} = U' f_{m+1}$ as the boundary condition. It can be solved iteratively starting from $s = m$ and proceeding toward λ_2. This yields

$$\lambda_s = \lambda_1 \Psi_{1s} + \Psi_{2s} U' f_{m+1}, \quad s = 2, \dots, m+1,$$

where

$$\Psi_{1s} = \sum_{j=s}^{m} \sigma_j \prod_{k=s}^{j-1} \eta_k, \qquad \Psi_{2s} = \prod_{i=s}^{m} \eta_i, \quad s = 2, \dots, m+1,$$

$$\sigma_j = \frac{b\varphi' \gamma_j}{1 - b\beta_j}, \ \eta_i = \frac{b\alpha_i}{1 - b\beta_i}, \ j, i = s, \dots, m+1.$$

Finally, solution (20) for λ_2 and (6a)–(6c) for $x_1 > 0$ yields

$$\lambda_1 = \frac{b\alpha_1 \Psi_{22} U' f_{m+1}}{1 - b\varphi' \gamma_1 - b\beta_1 - b\alpha_1 \Psi_{12}}.$$

In addition, the optimal solution must satisfy conditions (5a), (5b), i.e. $U' f_s \leq \lambda_s$, $s = 1, \dots, m$. Applying solutions (21) and (20) for λ_s, $s = 1, \dots, m$ yields:

$$\left(\frac{\alpha_1 \Psi_{22} \Psi_{1s}}{1 - b\varphi' \gamma_1 - b\beta_1 - b\alpha_1 \Psi_{12}} + b^{-1} \Psi_{2s} \right) \frac{f_{m+1}}{f_s} - 1 \geq r, \quad s = 1, \dots, m, \tag{26}$$

where $\Psi_{21} = 0$. When $h_m > 0$ Eq. (22) for $s = m$ holds as an equality, and it with (18) determines h_m and x_m. For size classes $s = m+1, \dots, n$ it must hold that $U' f_s - \lambda_s \leq 0$ and $\partial L / \partial x_{s,t+1} \leq 0$. Setting $U' f_s = \lambda_s$, $s = m+1, \dots, n$ yields conditions (6a)–(6c) in the form

$$\frac{\varphi' \gamma_s b\alpha_1 \Psi_{22} f_{m+1}}{(1 - b\varphi' \gamma_1 - b\beta_1 - b\alpha_1 \Psi_{12}) f_s} + \beta_s + \frac{\alpha_s f_{s+1}}{f_s} - 1 \leq r, \quad s = m+1, \dots, n, \tag{27}$$

where $\alpha_n \equiv 0$.

6 Interpretations

Conditions (26) state that saving any individual from classes $s = 1, \dots, m$ to be harvested as a size class $m+1$ individual is an investment with a marginal rate of return higher or at least equal to the discount rate. This equation is a generalization of (15a). Assuming constant recruitment ($\varphi' = 0$) implies that the first LHS quotient of (26) is zero, and in the case of the age-structured specification ($\beta_s = 0, s = 1, \dots, n$) the condition collapses to

$$b^s f_s \prod_{s=1}^{m-1} \alpha_s \leq b^{m-1} f_{m+1} \prod_{s=1}^{m} \alpha_s, \quad s = 1, \dots, m, \tag{28}$$

i.e. to the discrete-time Fisherian “single shot” optimality condition. Given $\varphi' > 0$, the first LHS quotient of (26) is positive (by (21) and (5a)–(5c)), implying that it is never optimal to harvest a cohort (at the steady state) before it reaches the maximum discounted biomass level. The size-structured case generalizes this set-up and the outcome is similar. Conditions (27) in turn state that given size classes $s = m+1, \ldots, n-1$, saving an individual from size class s to be harvested as a size class $s+1$ individual is an investment with a marginal rate of return lower than the discount rate.

In his analysis on multiple cohorts/selective gear and exogenous recruitment Clark (1990, p. 301) obtains a continuous time analogue of Eq. (28) and writes that including recruitment that depends on population level would lead to severe theoretical difficulties. Equations (26) and (27) represent solutions to such extension in discrete time.

7 On Steady-State Stability

In the vicinity of the steady state in regime C, the solution is defined as

$$x_{1,t+1} = \varphi(x_0) + \beta_1 x_1, \qquad x_{s,t+1} = \alpha_{s-1} x_{s-1,t} + \beta_s x_s,$$
$$s = 1, \ldots, m, \; h_{m+1,t} = \alpha_m x_{mt}. \tag{29}$$

This solution is optimal within the region of the state space where conditions (5a)–(5c) and (6a)–(6c) are satisfied, and it is not optimal to harvest other size classes besides $m+1$. Such a region can be guaranteed to exist, by assuming that f_s, $s = 1, \ldots, n$, $s \neq m+1$ are low enough. Given this solution regime the system of state variables (29) and conditions (5a)–(5c) and (6a)–(6c) are separate in the sense that state variable development determines the development of the Lagrange multipliers, but no feedback from the Lagrange multipliers to optimal harvesting, i.e. to the level of $h_{m+1,t}$ exist. Thus, it is possible to study the local stability of the steady state by simply analysing system (29), and to extend the regime C part of Proposition 8. For simplicity the analysis is restricted to the special case of the age-structured model.

Proposition 9 *Given* (A1)–(A7), $n \geq 2$, $\beta_s = 0$, $s = 1, \ldots, n$ *and that the optimal steady state is in regime C, the steady state is locally stable.*

Proof The Jacobian matrix of system (29) is

$$J = \begin{bmatrix} \varphi'\gamma_1 & \cdots & \varphi'\gamma_{m-1} & \varphi'\gamma_m \\ \alpha_1 & \cdots & 0 & 0 \\ \vdots & \ddots & \vdots & \vdots \\ 0 & \cdots & 0 & 0 \\ 0 & \cdots & \alpha_{m-1} & 0 \end{bmatrix}.$$

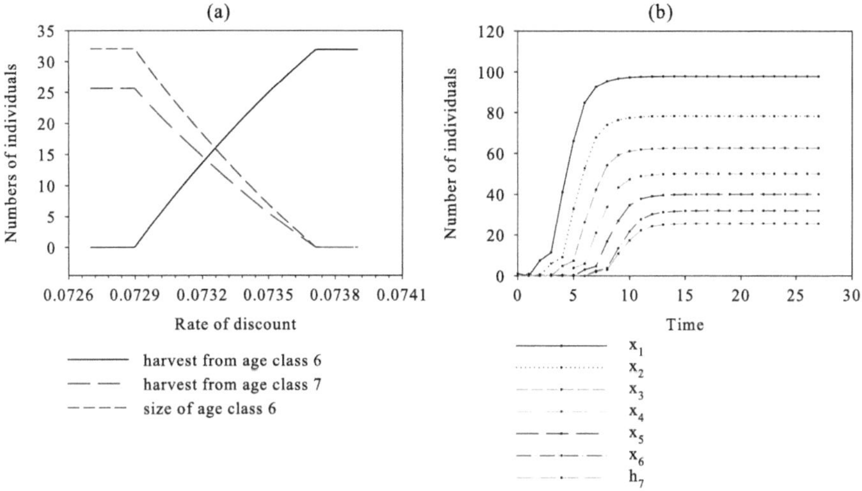

Fig. 7 A numerical example of solutions approaching a steady state in regime C. Note: $\alpha_s = 0.8$, $s = 1, \ldots, 10$, $f = 1, 2.7, 6.9, 16.7, 35.5, 60, 80.3, 91.7, 98.8, 100$, $\gamma = 0, 1, 2, 2, 2, 2, 2, 2, 2, 2$, $U = H_t$, $r = 0$, $\varphi = 10x_0/(1 - 0.1x_0)$

The characteristic equation can be given as

$$\sum_{s=1}^{m} \frac{\varphi' \gamma_s \mu_s}{r^s} = 1, \tag{30}$$

assuming that $r \neq 0$. This implies by the Perron-Frobenius theorem that the absolute values of all characteristic roots remain below 1 if

$$\varphi' \sum_{s=1}^{m} \gamma_s \mu_s = \varphi' \Gamma_m < 1.$$

The steady state must satisfy $\varphi(x_1 \Gamma_m) - x_1 = 0$. Since φ is a concave function, the steady state must satisfy $\varphi' \Gamma_m - 1 < 0$, implying that the Perron root is positive and lower than 1 and that the steady state is locally stable. □

The dynamic properties of this system coincide with those of an unharvested age-structured population when $n = m - 1$ and $\beta_{m-1} = 0$. This was studied by Getz and Haight (1989, p. 40), who derived the characteristic equation (30).

Figure 7a shows how increasing the discount rate transforms the optimal steady state solution from the $m = 6$ steady state to the $m = 5$ steady state via regime D, where it is optimal to simultaneously harvest from age classes 6 and 7. An example of the solution described in Proposition 9 is shown in Fig. 7b. Initially individuals exist only in age class 1. With no discounting it is optimal to apply a steady state harvest only in age class 7, i.e. $x_s > 0$, $s = 1, \ldots, 6$, $x_s = 0$, $s = 7, \ldots, 10$. Given the

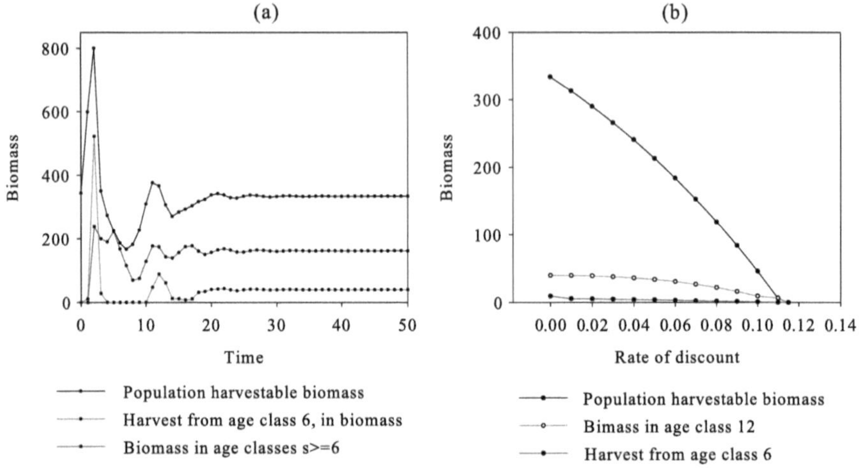

Fig. 8 Optimal dynamic solution (**a**) and the dependence of the steady state on discounting (**b**) using the data in Reed (1980)

linear utility function, this is the only age class that is harvested during the whole transition period.

Finally, Fig. 8 shows the optimal solutions for the data in Reed (1980). Given no discounting, the optimal steady state (i.e. MSY) harvest is to remove about 44 % from age class 6 (as obtained in Reed 1980). Thus, the number of individuals in all age classes remains positive and the solution is an example of regime *B*. Figure 8a shows the optimal dynamic solution (not computed by Reed 1980). Note that although the initial biomass slightly exceeds the steady state biomass, the initial harvest level is zero because the biomass is located in age classes that are too young for harvesting. Almost half of the biomass at the steady state is in age classes 6, ..., 12. Such a solution is clearly impossible to obtain within the unstructured biomass approach. Figure 8b shows that the optimal steady state exists only if the discount rate is below 11.5 %.

8 Summary

The objective of this study has been to extend the understanding of a generic discrete-time age- and size-structured optimization model where the harvest is perfectly selective. Given an increasing and concave function for recruitment, this model can be viewed as a theoretical description of harvesting shade-tolerant uneven-aged tree populations. Selective harvesting may be possible in fisheries if e.g. different age classes can be found in different locations. The model was first studied assuming two age or size classes, and it was shown that six different steady-state regimes exist. One of the steady states is a continuum. It was shown that the structured model can be viewed as a generalization of the biomass model. However, neglecting information on population structure will not yield correct optimal steady states.

The classic Faustmann (1849) model answers the question "at what age it is optimal to fell a tree". According to the answer (and assuming no regeneration cost), it is optimal to harvest the tree before it reaches its discounted maximum value, because waiting postpones all future harvests. The model studied here can be interpreted to answer the same question. However, the model does not include the Faustmann-type (1849) rotation structure, and it is optimal to harvest the tree after it reaches the discounted maximum value because waiting increases regeneration. In the Faustmann (1849) model, increasing the discount rate decreases the age of the harvested cohort, but a higher interest rate in the study at hand may imply a switch to harvesting an older age class.

Analysing steady-state stability reveals that with a zero or "low" interest rate the optimal steady states are saddle point stable. However, it was possible to give a numerical example showing that with a "high" interest rate the optimal solution may converge toward a limit cycle. Given a linear objective, constant escapement is optimal in the generic biomass model (Reed 1979) but it is not, in general, optimal in structured models.

Finally, the optimal steady-state equations were solved without limiting the number of age or size classes. These equations (not presented in earlier studies) further reveal the structure of the optimal solution: when the harvest is moved forward in the age or size classes, the yield *per capita* may increase but natural mortality has a negative effect on the number of individuals obtained. This decrease is compensated by increased recruitment due to an increased number of reproductive individuals. Finally, these trade-offs are influenced by discounting, but differently than in optimal rotation models.

Given the age-class specification and the optimal steady-state regime where only one age class is harvested and the harvest is total, it was possible to show that the steady state is locally stable, i.e. harvesting all individuals at the end of the period when they enter the (optimal harvestable) age class yields a convergence to a stable age structure. This suggests that instead of the constant escapement solution often proposed, a simple practically applicable harvesting rule may be just to harvest the individuals when they enter the age class that is the oldest at the optimal steady state.

Acknowledgements I thank Wayne Getz for discussions during the Fourth Meeting on Computational and Mathematical Population Dynamics in Taiyuan, China, 2013. In addition, I thank Tapio Palokangas for generous help. This study was initiated during my 2004–2005 stay in Brisbane, Australia. Financial support from Yrjö Jahnsson and Maj and Tor Nessing Foundations is gratefully acknowledged.

Appendix 1

Proof of Proposition 2 Case (i): Because $1 - \beta_1 - \varphi'(0)\gamma_1 > 0$ the condition (7a), i.e. $\lambda_1(b\varphi'(0)\gamma_1 + b\beta_1 - 1) + b\lambda_2\alpha_1 \leq 0$ will always be satisfied as an equality by some choice $\lambda_1 \geq U'(0)f_1$, $\lambda_2 \geq U'(0)f_2$. Eliminating λ_1 from (8a) shows that this condition is satisfied as an inequality when $\mu_2(0,0) \geq 0$. Thus, regime F satisfies the necessary optimality conditions and must be optimal. Case (ii): Setting $\lambda_1 = U'(0)f_1$ and $\lambda_2 = U'(0)f_1[1 - b(\varphi'\gamma_1 + \beta_1)]/(b\alpha_1)$ implies by $\mu_2(0,0) \leq 0$ that (8a) is satisfied with $x_1 = x_2 = 0$. The solutions for λ_1 and λ_2 imply that (6a) is satisfied for $i = 1$ as an equality and $h_1 = 0$ and by $\mu_1(0,0) \leq 0$ for $i = 2$ as an inequality with $h_2 = 0$ and (7a) as an equality, i.e. regime F is optimal. Thus, conditions $\mu_2(0,0) \geq 0$ and (i) or (ii) are sufficient for regime F to be optimal. If condition $\mu_2(0,0) \geq 0$ holds but neither conditions (i) or (ii) hold, the optimal steady state is in regime D or E. If condition $\mu_2(0,0) \geq 0$ does not hold, the optimal solution is in regime A, B, C, D or E. Thus conditions are necessary for the optimality of regime F. □

Appendix 2

Proof of Proposition 5 Given (16)–(21) and $b = 1$, the characteristic equation evaluated at the steady state is

$$\Omega_1(r) = r^4 + y_1 r^3 + y_2 r^2 + y_1 r + 1,$$

where

$$y_1 = \left[\lambda_1\varphi'' h'\gamma_1(\alpha_1\gamma_2 - \beta_2\gamma_1)\tau_1^2 - \alpha_1^2\gamma_2^2\beta^2 + \gamma_1\left(\beta_2^2 - 1\right)\left(2\alpha_1\gamma_2\beta - \gamma_1\beta_2^2 + \gamma_1\right)\right]/\tau_2,$$

$$y_2 = \left\{\begin{array}{l} \lambda_1\varphi'' h'\tau_1^2(\gamma_2^2\alpha_1^2 - 2\beta_2\gamma_1\gamma_2\alpha_1 + \gamma_1^2\beta_2^2 + \gamma_1^2) + 2[\alpha_1^2\gamma_2^2(\beta^2 - \beta + 1)] \\ \quad + \alpha_1\gamma_1\gamma_2(1 - \beta_2)(2\beta_1\beta_2 + \beta_1 + 2\beta_2^2 + 1) + \gamma_1^2(\beta_2 - 1)^2(\beta_2^2 + \beta_2 + 1) \end{array}\right\}/\tau_2,$$

where $\tau_1 = \alpha_1\gamma_2 - \gamma_1\beta_2 + \gamma_1$, $\tau_2 = \tau_1\tau_3$, $\tau_3 = (\beta_2 - 1)(\alpha_1\gamma_2 - \beta_2\gamma_1) + \alpha_1\gamma_2\beta_1 \neq 0$, $\beta = \beta_1 + \beta_2$. Computing yields:

$$\Omega_1(1) = \frac{\lambda_1\varphi'' h'\tau_1^3}{\tau_2},$$

$$\Omega_1(-1) = \frac{\lambda_1\varphi'' h'\tau_1^2(\gamma_2\alpha_1 - \gamma_1\beta_2 - \gamma_1)^2 + 4[\gamma_2\alpha_1\beta + \gamma_1(1 - \beta_2)(1 + \beta_2)]^2}{\tau_2}.$$

Thus, if u is a root then $1/u$ is a root. The numerators of both $\Omega(1)$ and $\Omega(-1)$ are strictly positive. Thus, when $\tau_3 < 0$, it holds that $\Omega(1) < 0$, $\Omega(-1) < 0$ and by $\Omega(0) = 1$ there exist four real roots, two with absolute values above one and two below one. Write $\Omega_1(r)r^{-2} = r^2 + w_1 r + w_2 + w_1 r^{-1} + r^{-2}$ and $|r| = 1$, $\Rightarrow r = e^{ia} = \cos(a) + i\sin(a)$ implying

$$\Omega(r)r^{-2} = \frac{\lambda_1\varphi'' h'}{\tau_2}\Theta_1 + 4\left(\cos^2(a) - 1\right) + \Theta_2\left(\cos(a) - 1\right),$$

$$\Theta_1 = \{[2\gamma_1\tau_1^2(\alpha_1\gamma_2 - \beta_1\gamma_1)]\cos(a) + \tau_1^2(\gamma_2^2\alpha_1^2 - 2\beta_2\gamma_1\gamma_2\alpha_1 + \gamma_1^2\beta_2^2 + \gamma_1^2)\},$$
$$\Theta_2 = -\frac{2}{\tau_2}[\gamma_2\alpha_1\beta + \gamma_1(1-\beta_2^2)]^2.$$

The minimum of term Θ_1 is zero and is attained when $a = \pi$ and $\gamma_2 = \gamma_1(\beta_2+1)/\alpha_1$. Given these values for a and γ_1, it follows that $\Omega(r)r^{-2} > 0$ by $\tau_3 > 0$. The minimum of $4(\cos^2(a) - 1) + \Theta_2(\cos(a) - 1)$ is attained when $a = 0$ and is zero because $\Theta_2 \leq 8$. When $a = 0$ it holds that $\frac{\lambda_1\varphi'' h'}{\tau_2}\Theta_1 > 0$. Thus, no root lying on the unit circle can solve the polynomial implying that the steady state must be a saddle point. □

Appendix 3

Proof of Proposition 7 The characteristic equation is $\Omega(r) = r^4 + r^3 y_1 + r^2 y_2 + r y_1/b + 1/b^2 = 0$, where

$$y_1 = \frac{\beta(\beta b + b - 1)}{1 - b\beta},$$
$$y_2 = \frac{\lambda_1\alpha_1^2 b^4 \varphi'' \gamma_2^2 h'^3 \beta^2 + b^2(\beta_1^2 + 2\beta_1\beta_2 + \beta_2^2 + 1) - 2b\beta + 1}{b^2(b\beta - 1)}.$$

This yields

$$\Omega(1) = \frac{\lambda_1\alpha_1^2 b^2 \varphi'' \gamma_2^2 h'}{b\beta - 1},$$
$$\Omega(-1) = \frac{\lambda_1\alpha_1^2 b^3 \varphi'' \gamma_2^2 h' + 2\beta[b^2(\beta+1)] + b\beta - 1}{b^2\beta - b}.$$

Recall that $\beta = \beta_1 + \beta_2$. Given $b < \frac{1}{\beta_1+\beta_2}$ the denominators of both $\Omega(1)$ and $\Omega(-1)$ are negative. The numerator of $\Omega(1)$ is positive and $\Omega(1) < 0$. The numerator of $\Omega(-1)$ is positive if $\frac{1}{\beta_1+\beta_2+1} \leq b$, implying together with $\Omega(0) = 1/b^2$ that under these conditions roots are real and the absolute value of two roots are below 1 and the absolute values of two roots are above 1, i.e. the steady state is a saddle point. □

Appendix 4

Proof of Proposition 8 Regime C is defined by $x_{1,t+1} = \varphi(\gamma_1 x_{1t}) + \beta_1 x_{1t}$, $h_{2t} = \alpha_1 x_{1t}$. This regime exists only when $\partial\sigma_2(0,0)/\partial x_1 = 1 - \beta_1 - \varphi'(0)\gamma_1 < 0$ and at the steady state $x_1 = \hat{x}_1$ it must hold that $\partial\sigma_2(\hat{x}_1, 0)/\partial x_1 = 1 - \beta_1 - \varphi'(\hat{x}_1)\gamma_1 > 0$ implying that $\varphi'(\hat{x}_1)\gamma_1 + \beta_1 < 1$, i.e. that the steady state is locally stable.

In regime D $\lambda_{1t} = U'f_1$, $\lambda_{2t} = \lambda_{1t} f_2/f_1$, $\lambda_{1,t+1}(b\varphi'\gamma_1 + b\beta_1 + b\alpha_1 f_2/f_1) - \lambda_{1t} = 0$, $x_{1,t+1} = \varphi(\gamma_1 x_{1t}) + \beta_1 x_{1t} - h_{1t}$ and $h_{2t} = \alpha_1 x_{1t}$. This leads to the system:

$$U'(H_{1,t+1}) f_1 \big(b\varphi'(\gamma_1 x_{1,t+1})\gamma_1 + b\beta_1 + b\alpha_1 f_2/f_1\big) - U'(H_t) = 0,$$

$$x_{1,t+1} = \varphi(\gamma_1 x_{1t}) + \beta_1 x_{1t} - h_{1t}.$$

The characteristic equation is given as:

$$\Omega(r) = r^2 - r\big(\delta_1 + \varphi'\gamma_1 + \beta_1\big) + \delta_1\big(\varphi'\gamma_1 + \beta_1\big) + \delta_2,$$

where

$$\delta_1 = 1 + \frac{f_2\alpha_1}{f_1} + \frac{U'b\varphi''\gamma_1^2}{U''f_1}$$

and

$$\delta_2 = -\frac{f_2\alpha_1(\varphi'\gamma_1 + \beta_1 - 1)}{f_1} - \frac{U'b\varphi''\gamma_1^2(\varphi'\gamma_1 + \beta_1)}{U''f_1}.$$

Since $\Omega(0) = \varphi'\gamma_1 + \beta_1 + \frac{f_2\alpha_1}{f_1} > 0$, and $\Omega(1) = -\frac{U'b\varphi''\gamma_1^2}{U''f_1} < 0$, both roots are positive and the value of one root is below 1 and the value of the other is above 1, i.e. the steady state is a saddle point. □

References

Allen, R. L., & Basasibwaki, P. (1974). Properties of age structure models for fish populations. *Journal of the Fisheries Research Board of Canada*, *31*, 1119–1125.

Baranov, T. I. (1918). On the question of the biological basis of fisheries. *Issledovatel'skie Ikhtiologicheskii Intitut Izvestiya*, *1*, 81–128.

Beddington, J. R., & Taylor, D. B. (1973). Optimum age specific harvesting of a population. *Biometrics*, *29*, 801–809.

Beverton, J. H. R., & Holt, S. J. (1957). On the dynamics of exploited fish populations. Ministry of Agriculture, Fish and Food, London.

Byrd, R. H., Hribar, M. E., & Nocedal, J. (1999). An interior point algorithm for large scale nonlinear programming. *SIAM Journal on Optimization*, *9*, 877–900.

Caswell, H. (2001). *Matrix population models*. Massachusetts: Sinauer

Clark, C. W. (1973). Profit maximization and the extinction of animal species. *Journal of Political Economy*, *81*, 950–961.

Clark, C. W. (1990). *Mathematical bioeconomics: the optimal management of renewable resources*. New York: Wiley.

Cushing, J. M. (1998). *An introduction to structured population dynamics*. Philadelphia: SIAM.

Faustmann, M. (1849). Berechnung des Werthes, welchen Waldboden, sowie noch nicht haubare Holzbestände für die Waldwirtschaft besitzen. *Allgemeine Forst- und Jagd-Zeitung*, *25*, 441–455.

Fisher, R. A. (1930). *The general theory of natural selection*. Oxford: Clarendon Press.

Getz, W. M. (1980). The ultimate sustainable yield problem in nonlinear age-structured populations. *Mathematical Biosciences*, *48*, 279–292.

Getz, W. M., & Haight, R. G. (1989). *Population harvesting: demographic models for fish, forest and animal resources*. Princeton: Princeton University Press.

Goetz, R., & Xabadia, A. (2011). Optimal management of a size-structured forest with respect to timber and non-timber values. In R. Boucekkine, N. Hritonenko, & Y. Yatsenko (Eds.), *Optimal control of age-structured populations in economy, demography, and the environment* (pp. 207–228). Oxon: Routledge.

Goetz, R., Xabadia, A., & Calvo, E. (2011). Optimal forest management in the presence of intraspecific competition. *Mathematical Population Studies*, *18*, 151–171.

Gordon, H. S. (1954). The economic theory of a common property resource: the fishery. *Journal of Political Economy*, *62*, 124–142.

Haight, R. G. (1985). A comparison of dynamic and static economic models of uneven aged stand management. *Forest Science*, *31*, 957–974.

Leslie, P. H. (1945). On the use of matrices in certain population mathematics. *Biometrical Journal*, *33*, 183–212.

Mangasarian, O. L. (1966). Sufficient conditions for the optimal control of nonlinear systems. *SIAM Journal on Control*, *4*, 139–152.

Reed, W. E. (1979). Optimal escapement levels in stochastic and deterministic harvesting models. *Journal of Environmental Economics and Management*, *6*, 350–363.

Reed, W. E. (1980). Optimum age-specific harvesting in a nonlinear population model. *Biometrics*, *36*, 579–593.

Reed, W. E. (1986). Optimal harvesting models in forest management. *Natural Resource Modelling*, *1*, 55–79.

Rorres, C., & Fair, W. (1975). Optimal harvesting for an age-specific population. *Mathematical Biosciences*, *24*, 31–47.

Salo, S., & Tahvonen, O. (2002). On equilibrium cycles and normal forests in optimal harvesting of tree vintages. *Journal of Environmental Economics and Management*, *44*, 1–22.

Salo, S., & Tahvonen, O. (2003). On the economics of forest vintages. *Journal of Economic Dynamics & Control*, *27*, 1411–1435.

Samuelson, P. (1976). Economics of forestry in an evolving society. *Economic Inquiry*, *14*, 466–492.

Samuelson, P. (1977). Generalizing Fisher's "reproduction value": linear differential and difference equations of "dilute" biological systems. *Proceedings of the National Academy of Sciences of the United States of America*, *74*, 5189–5192.

Stokey, N., & Lucas, R. E. (1989). *Recursive methods in economic dynamics*. Cambridge: Harvard University Press.

Sydsaeter, K., Hammond, P., Seierstad, A., & Strom, A. (2008). *Further mathematics for economic analysis*. Gosport: Prentice Hall.

Tahvonen, O. (2011a). Optimal structure and development of uneven-aged Norway spruce forests. *Canadian Journal of Forest Research*, *41*, 2389–2402.

Tahvonen, O. (2011b). Age-structured optimization models in fisheries bioeconomics: a survey. In R. Boucekkine, N. Hritonenko, & Y. Yatsenko (Eds.), *Optimal control of age-structured populations in economy, demography, and the environment* (pp. 140–173). Oxon: Routledge.

Usher, M. B. (1966). A matrix approach to the management of renewable resources, with special reference to selection forests-two extensions. *Journal of Applied Ecology*, *6*, 346–347.

Wächter, A., & Biegler, L. T. (2006). On the implementation of an interior-point filter line-search algorithm for large scale nonlinear programming. *Mathematical Programming*, *106*, 25–57.

Walters, C. J. (1969). A generalized computer simulation model for fish population studies. *Transactions of the American Fishery Society*, *98*, 505–512.

Wan, Y. H. (1994). Revisiting the Mitra-Wan tree farm. *International Economic Review*, *35*, 193–198.

Wilen, J. E. (1985). Bioeconomics of renewable resource use. In A. V. Kneese & J. L. Sweeney (Eds.), *Handbook of natural resource and energy economics* (Vol. I, pp. 61–124). Amsterdam: Elsevier.

Xabadia, A., & Goetz, R. (2010). The optimal selective logging regime and the Faustmann formula. *Journal of Forest Economics*, *16*, 63–82.

MIX
Papier aus verantwortungsvollen Quellen
Paper from responsible sources
FSC® C105338

If you have any concerns about our products,
you can contact us on
ProductSafety@springernature.com

In case Publisher is established outside the EU,
the EU authorized representative is:
Springer Nature Customer Service Center GmbH
Europaplatz 3, 69115 Heidelberg, Germany

Printed by Libri Plureos GmbH
in Hamburg, Germany